Achieving sustainable cultivation of hemp

It is widely recognised that agriculture is a significant contributor to global warming and climate change. Agriculture needs to reduce its environmental impact and adapt to current climate change whilst still feeding a growing population, i.e. become more 'climate-smart'. Burleigh Dodds Science Publishing is playing its part in achieving this by bringing together key research on making the production of the world's most important crops and livestock products more sustainable.

Based on extensive research, our publications specifically target the challenge of climate-smart agriculture. In this way we are using 'smart publishing' to help achieve climate-smart agriculture.

Burleigh Dodds Science Publishing is an independent and innovative publisher delivering high quality customer-focused agricultural science content in both print and online formats for the academic and research communities. Our aim is to build a foundation of knowledge on which researchers can build to meet the challenge of climate-smart agriculture.

For more information about Burleigh Dodds Science Publishing simply call us on +44 (0) 1223 839365, email info@bdspublishing.com or alternatively please visit our website at www.bdspublishing.com.

Related titles:

Achieving sustainable greenhouse cultivation
Print (ISBN 978-1-78676-280-1); Online (ISBN 978-1-78676-282-5, 978-1-78676-283-2)

Advances in plant factories
Print (ISBN 978-1-80146-316-4); Online (ISBN 978-1-80146-317-1, 978-1-80146-318-8)

Chapters are available individually from our online bookshop:
https://www.bdspublishing.com/webshop/

climate•SMART•publishing
IN AGRICULTURAL SCIENCE

BURLEIGH DODDS SERIES IN AGRICULTURAL SCIENCE
NUMBER 175

Achieving sustainable cultivation of hemp

Edited by Professor Emerit Jay Stratton Noller, Oregon State University, USA

burleigh dodds
SCIENCE PUBLISHING

Published by Burleigh Dodds Science Publishing Limited
82 High Street, Sawston, Cambridge CB22 3HJ, UK
www.bdspublishing.com

Burleigh Dodds Science Publishing, 1518 Walnut Street, Suite 900, Philadelphia, PA 19102-3406, USA

First published 2026 by Burleigh Dodds Science Publishing Limited

Library of Congress Control Number: 2026935170

British Library Cataloguing in Publication Data
A catalogue record for this book is available from the British Library

ISBN 978-1-83545-083-3 (print)
ISBN 978-1-83545-084-0 (ePub)
ISBN 978-1-83545-085-7 (PDF)
ISSN 2059-6936 (print)
ISSN 2059-6944 (online)

DOI: 10.19103/AS.2025.0161

Typeset by Deanta Global Publishing Services, Dublin, Ireland

Contents

Part 2 Applications, sustainability and nutraceutical properties of hemp

Series list

Title	Series number
Achieving sustainable cultivation of maize - Vol 1 From improved varieties to local applications Edited by: Dr Dave Watson, CGIAR Maize Research Program Manager, CIMMYT, Mexico	001
Achieving sustainable cultivation of maize - Vol 2 Cultivation techniques, pest and disease control Edited by: Dr Dave Watson, CGIAR Maize Research Program Manager, CIMMYT, Mexico	002
Achieving sustainable cultivation of rice - Vol 1 Breeding for higher yield and quality Edited by: Prof. Takuji Sasaki, Tokyo University of Agriculture, Japan	003
Achieving sustainable cultivation of rice - Vol 2 Cultivation, pest and disease management Edited by: Prof. Takuji Sasaki, Tokyo University of Agriculture, Japan	004
Achieving sustainable cultivation of wheat - Vol 1 Breeding, quality traits, pests and diseases Edited by: Prof. Peter Langridge, The University of Adelaide, Australia	005
Achieving sustainable cultivation of wheat - Vol 2 Cultivation techniques Edited by: Prof. Peter Langridge, The University of Adelaide, Australia	006
Achieving sustainable cultivation of tomatoes Edited by: Dr Autar Mattoo, USDA-ARS, USA and Prof. Avtar Handa, Purdue University, USA	007
Achieving sustainable production of milk - Vol 1 Milk composition, genetics and breeding Edited by: Dr Nico van Belzen, International Dairy Federation (IDF), Belgium	008
Achieving sustainable production of milk - Vol 2 Safety, quality and sustainability Edited by: Dr Nico van Belzen, International Dairy Federation (IDF), Belgium	009
Achieving sustainable production of milk - Vol 3 Dairy herd management and welfare Edited by: Prof. John Webster, University of Bristol, UK	010
Ensuring safety and quality in the production of beef - Vol 1 Safety Edited by: Prof. Gary Acuff, Texas A&M University, USA and Prof. James Dickson, Iowa State University, USA	011
Ensuring safety and quality in the production of beef - Vol 2 Quality Edited by: Prof. Michael Dikeman, Kansas State University, USA	012
Achieving sustainable production of poultry meat - Vol 1 Safety, quality and sustainability Edited by: Prof. Steven C. Ricke, University of Arkansas, USA	013
Achieving sustainable production of poultry meat - Vol 2 Breeding and nutrition Edited by: Prof. Todd Applegate, University of Georgia, USA	014

Introduction

Despite deriving from the same plant species (*Cannabis sativa*), hemp and recreational cannabis are recognised as two entirely different products. Plants with a tetrahydrocannabinol (THC) content of 0.3% or less are classified as hemp, whereas plants with a THC content over 0.3% are classified as cannabis. The global industrial hemp market is estimated to reach $18.6 billion by 2027, with over 40 countries now actively involved in growing the crop which is utilised by an array of industries, including the construction, livestock feed and textiles sectors. Hemp has been shown to be a particularly sustainable crop, with recent research highlighting its ability to absorb and store up to twice as much carbon when compared to trees. However, despite hemp's versatility, there remain significant issues surrounding regulation, improving cultivation practices and perhaps most importantly consumer acceptance of hemp-derived products.

This collection reviews the wealth of research on production of industrial hemp to allow it to optimise its role as a versatile and sustainable product with a wide range of applications. Part 1 covers physiology, breeding and cultivation, with chapters on the taxonomy and classification of hemp, its nutrient and water requirements, breeding techniques and outdoor hemp cultivation techniques. Part 2 chapters examine advances in the use of hemp fibre in construction and building, the use of hemp grain livestock and fish feed, and the environmental sustainability of hemp. Chapters in Part 3 include discussions of integrated pest management in hemp, integrated disease and weed management. The final part of the book examines social, business, economic and political issues. Beginning with a chapter on Canada's regulatory stance on industrial hemp production, Part 4 then moves on to a review of the consumer perceptions of hemp-based products. Finally, there is an exploration of building sustainable biobased manufacturing supply chains, in order to help hemp reach its full global economic potential.

Part 1 Physiology, breeding and cultivation

The opening chapter of the book focuses on the taxonomy and classification of the genus *Cannabis*. Chapter 1 reviews key methods, including biological, biochemical and genetic approaches, and examines how the expansion of the Cannabaceae family has redefined *Cannabis* within a broader phylogenetic context. The chapter also looks at the cladistic classification of *C. Sativa* L., with a focus on resolving issues relating to cannabis taxonomy through genomic and computational advances.

Chapter 2 discusses the nutrient and water requirements of hemp. The chapter discusses primary and secondary macronutrient requirements and current understanding of micronutrient demands for optimal yields. The chapter explores the interactions between nutrients and water supply, as well as their impacts on the environment. The chapter then examines the various methods used to assess and manage crop nutrient and water status, including soil testing, as well as fertilisation and irrigation strategies. By providing a comprehensive overview, the chapter aims to contribute to the sustainable production and utilisation of this valuable crop.

The third chapter highlights the fact that, despite the resurgence of interest in the cultivation of hemp, there has been a lag in advances in breeding techniques. Chapter 3 focuses initially on the challenging but promising potential use of marijuana lines to develop new hemp varieties. It then looks at key issues and developments in the range of available breeding techniques, from basic use of clones through to advances in pedigree and hybrid breeding. The chapter also discusses issues in seed and grain production before concluding by looking at ways advances in breeding can improve resistance to biotic and other stresses.

The final chapter in Part 1 provides a comprehensive overview of outdoor hemp cultivation, focusing on agronomic principles, physiological traits, and regulatory landscapes that define successful production systems for fiber, grain and cannabinoid/essential oil end-uses. Chapter 4 begins with a definition and classification of hemp, and then moves on to the different types of hemp suitable for outdoor cultivation. Next, the chapter explores tailored agronomic practices for particular end uses. It then moves on to discuss the regenerative approaches to more sustainable hemp cultivation.

Part 2 Applications, sustainability and nutraceutical properties of hemp

The first chapter of Part 2 focuses on the use of hemp fibre in the construction and building industry. Chapter 5 starts off with a synopsis of current research trends, followed by an overview of the different types of hemp-based building materials, and the evolution and advances of hemp construction. It then explores the regulatory framework surrounding the use of hemp in construction in Europe, North America and other international countries. The chapter also provides an overview of current drivers and barriers to the use of hemp-based construction materials.

Chapter 6 examines the advances in hemp use in livestock feed. To begin with, it outlines common terms, and the legal status of hemp in livestock diets. It then reviews hemp's composition, nutritional value and cannabinoid content. This chapter also explores hemp byproducts as alternative feed ingredients in ruminant diets and their effects on ruminant health, performance, meat and milk quality. Finally, it discusses hemp in monogastric diets and how hemp byproducts can have positive effects on health, performance and animal byproduct quality as well.

The next chapter explores the advances of hemp usage as feed for fish. Chapter 7 starts off with an analysis of alternative protein sources used in aquaculture, and then moves on to an overview of hempseed components and nutrient compositions. Next, it outlines the cannabinoids and antinutrient factors in hemp seed, as well as hempseed meal and oil as an alternative ingredient for aquaculture species. Chapter 7 goes on to display some exploratory case studies, closing with a discussion regarding the sustainability and the economic potential of industrial hempseed meal for the aquafeed industry.

The final chapter of Part 2 discusses the environmental sustainability of hemp, focusing on common metrics and a consideration of cultivation geography. Chapter 8 covers environmental sustainability metrics, such as hemp's water use and nutrient requirements, whilst also reviewing its need for pesticides and herbicides in pest and weed management. The chapter also discusses the environmental benefits of hemp in agriculture. A case study on cultivation geography is also provided, focusing specifically on California, USA.

Part 3 Hemp diseases, pests and weeds

Part 3 begins with a chapter that discusses integrated pest management in hemp. Chapter 9 first provides an overview of some of the key approaches that are needed to improve integrated pest management of hemp. The chapter then moves on to provide a synopsis of pest identification, focusing specifically on chewing and sucking pests. A section on integrated pest management is also included, drawing attention to some of the key techniques involved. It then discusses chemical control methods for current pests and looks to future genetic and breeding efforts to improve hemp resistance to pests. The chapter concludes with a case study on the corn earworm, a pest that is widespread in North American hemp production.

Chapter 10 examines integrated disease management in hemp cultivation. It first provides an overview of basic concepts in plant pathology, and then goes on to highlight key diseases of hemp, breaking them down into how their various life-cycle stages can affect hemp. The chapter also provides examples of disease management strategies in field and greenhouse hemp cultivation, looking specifically at the 'disease triangle' and how this can be manipulated to manage diseases effectively. The chapter concludes with an overview of case studies in hemp pathology using leaf spot disease, Fusarium head blight and powdery mildew as discussion points.

The final chapter of Part 3 discusses weed management in hemp cultivation. Chapter 11 begins with general recommendations for optimal crop establishment, and then moves on to chemical and mechanical weed control. Next it details competitive cultivars for weed suppression, followed by increased plant populations and narrow row spacing for weed suppression. Chapter 11 also provides an overview of the weed management strategies of crop rotation, cover crops, fertilisation and irrigation management.

Part 4 Social, business, economic and political issues

Part 4 begins with a chapter on Canada's regulatory stance on industrial hemp production. Chapter 12 introduces the history of the regulation of cannabis, marijuana and hemp. Following this is an exploration of current licensing and law related to hemp in Canada, focusing on aspects such as current hemp statistics and hemp and hemp-related disputes. The chapter ends with an analysis of Canada's regulatory framework for medical cannabis, drawing specific attention to medical marijuana before and after the *Cannabis Act* in 2018 was put in place. It also highlights current concerns associated with medical marijuana and its efficacy.

Chapter 13 reviews current understanding of consumer demand, trust and sustainability for hemp-based products. The chapter begins by providing an overview of the process that involved the legalisation of hemp products which then led to the need to understand potential consumer demand. This chapter also includes a case study to support the need to meet these demands whilst also ensuring that consumers can trust the products whilst also highlighting the hemp foods and cannabidiol (CBD)'s safety and sustainability.

The final chapter of the book focuses on building sustainable biobased manufacturing supply chains to help hemp reach its global economic potential. Chapter 14 begins with a brief history of hemp supply chains, and recent policy changes in the US. This is followed by a synopsis of building sustainable biobased manufacturing supply chains. The chapter then goes on to explain the value-chain analysis and technology road-mapping approaches to understanding value creation in supply chains and production network development. It describes how value-chain analysis helps to identify gaps and opportunities for companies and policy makers in designing policies and investment incentives to nurture the scaling of hemp-based industries. Chapter 14 then discusses the promotion of biobased manufacturing and creating supply chain resilience. Finally, it touches on paths forward with diversified product sectors to create a bright future for hemp.

Part 1

Physiology, breeding and cultivation

Chapter 1

Taxonomy and classification of hemp

Mehdi Babaei, Adrian Monthony and Davoud Torkamaneh, Laval University, Canada

1 Introduction

Cannabis (*Cannabis sativa* L.) is among the earliest domesticated and cultivated plants in human history (Babaei et al., 2022; Crocq, 2020). Its extensive history is deeply intertwined with humanity's evolving social, cultural, and economic developments, reflecting significant shifts over time (Barcaccia et al., 2020). Records of the use of cannabis date back as early as 3000–6000 years ago in China where it was used for seed production, fiber, and oil (Anwar et al., 2006; Russo et al., 2008; Sun, 2023). Cannabis has been known by various local and regional names throughout history. In Iran and Afghanistan, for instance, the plant is referred to as 'shahdaneh' (شاهدانه) or 'shahdanaj' (شهدانج) for industrial and non-drug purposes, while the term 'hashish' (حشیش) is used to describe its resinous product (Block, 2023; Clarke, 1998). Similar historical names exist in other ancient civilizations, as reviewed by Block (2023), including 'A.ZAL.LA' in Sumerian, 'azallû' in Akkadian, '麻' (ma) in Chinese, 'भङ्ग' (bhang) in Sanskrit, and 'κάνναβις' (cannabis) in Greek. Documented common names for cannabis in different languages are detailed in Figure 1. In addition to these geographical naming conventions, cannabis is also classified and named based on its applications, commonly referred to as hemp, marijuana, marihuana, Indian hemp, industrial hemp, fiber type, and drug type (Babaei et al., 2022; de Candolle, 1867; Small, 2015). This diverse nomenclature reflects

http://dx.doi.org/10.19103/AS.2025.0161.01

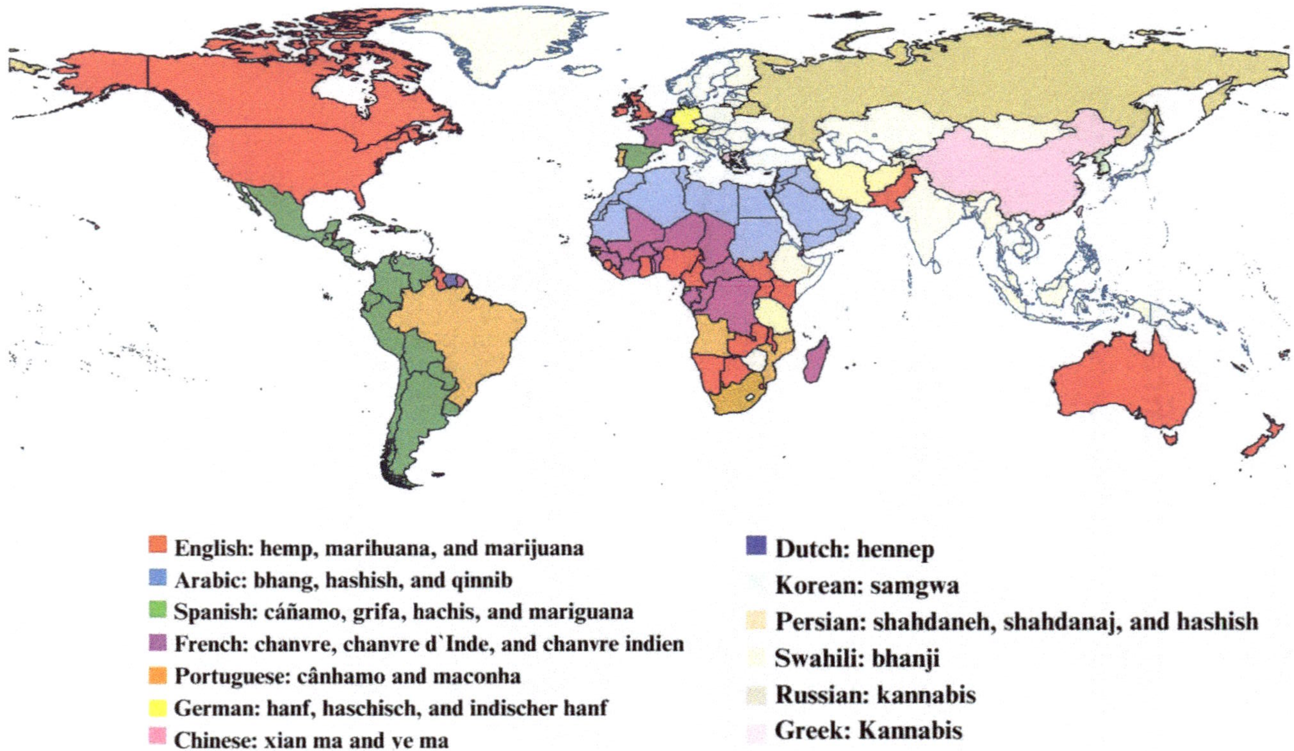

Figure 1 Global distribution of common names for *Cannabis* in various languages. The map illustrates the diverse terminology used to refer to *Cannabis* across different regions and cultures, reflecting its widespread use and cultural significance. The names are color-coded to correspond with the countries or regions where these terms are predominantly used. Based on Fordjour et al. (2023) and Lata et al. (2023), with updates; created using QGIS.

the plant's significant historical and cultural roles across different civilizations, setting the stage for a deeper exploration of the even more varied and complex approaches to its taxonomy and classification. In this chapter, we will refer to the genus specifically by its scientific name, *Cannabis*, while using 'cannabis' when referring to the plants in a more general context.

Taxonomy and classification involve defining and assigning entities to groups, and living organisms are particularly complex due to historical evolutionary relationships (Luisa, 2012; Small, 2017). Taxonomic classifications are more than just terminological distinctions; they hold significant implications for biological comprehension and future research (Van der Gulik et al., 2017). Modern taxonomy was defined by Linnæus; in his book *Philosophia botanica*, he developed an ordered method for classification and nomenclature in which each species is assigned a binomial system that includes the generic name (first) and the specific name (second). The Linnaean approach has evolved to incorporate evolutionary and genealogical studies, with some aspects being abandoned or modified to allow for specialized analysis using concepts from neo-systematics and cladistics (Lapierre et al., 2023b; Von Linné, 2005). An accurate taxonomy should reflect our best scientific knowledge at every hierarchical level. Importantly, it also influences our scientific interpretations and experimental focus (Van der Gulik et al., 2023).

Plant species can be classified using artificial, natural, or phylogenetic systems of classification, with the latter emphasizing the evolutionary relationship among different taxa. The classification system follows the *International Code of Nomenclature for algae, fungi, and plants* (ICN) hierarchy, which includes Kingdom, Division, Subdivision, Class, Order, Family, Genus, and Species (Alamgir, 2017). Artificial or mechanical systems categorize taxa using one or a few characters, while natural systems use several characters. In addition to considering the morphology, cytology, biochemistry, and physiology of the relevant plant groupings, phylogenetic systems also consider the evolutionary connections among taxa (Alamgir, 2017; Doyle, 2018; Rouhan and Gaudeul, 2021). Biological taxonomists favor natural classification, which suggests ideal classifications should reflect nature's fundamental structure (Small, 2017). The value of a classification depends on its utility for specific purposes, which applies to biological classifications.

Taxonomy has changed dramatically during the last few decades and has moved beyond the conventional Linnaean morphological model, adopting a more systematic and interdisciplinary approach (Cuperus, 2022). Changes in biological systematics have followed a Darwinian rivalry among three ideological and methodological approaches: numerical taxonomy (phenetics), cladistics, and evolutionary systematics. These shifts in systematic approaches have been necessary and inevitable, aiming to achieve maximum operationality and objectivity in results. Despite the dominance of the phylogenetic approach in system construction, modern practical taxonomy is essentially pseudo-cladistics, as taxonomic decisions are based on statistical methods for reconstructing phylogenies and delineating species rather than on the analysis of homologies (synapomorphies) (McKelvey, 2022; Vinarski, 2022). Two main schools of thought dominate plant systematics: cladistics (phylogenetics) and traditional taxonomy. These schools employ different principles, criteria, and types of evidence, which often lead to conflicting conclusions (McKelvey, 2022; Wiley and Lieberman, 2011). Despite this, some phylogeneticists interpret their results by incorporating taxonomic insights based on morphology, and conversely, some taxonomists integrate molecular cladograms into their analyses (Grant, 2003; McKelvey, 2022). This suggests that these two perspectives, though distinct, can be used in a complementary manner. In botanical classification, organisms are divided into various taxa (the basic units in taxonomy) based on similarities and differences. In contrast, the cladistics system divides organisms into different groups (clades), which share a common ancestor, following a hierarchical pattern of evolutionary branching. This approach emphasizes a different aspect of natural diversity compared to traditional taxonomy (Grant, 2003). The conventional taxonomic method, which integrates new evidence with existing knowledge and relies on informed judgment, often produces more comprehensive and reliable outcomes. However, the effectiveness of any classification system is fundamentally dependent on the use of internationally accepted and unambiguous scientific names. Without such consensus, the vast body of information on global biodiversity gathered from past and present research becomes irretrievable (McNeill, 1996). Linnæus (1938), in his work *Critica botanica*, translated by Sir Arthur Hort (see also Cook (2010)), emphasized the critical need for agreement among botanists on Latin names: 'I do not object to any nation retaining its own vernacular names for plants; what I do earnestly desire is that all learned botanists should agree over the Latin names.'

As a species with many vernacular names, the importance of Linnaeus's words becomes even more evident when discussing cannabis – a plant that, despite its long history and widespread use, has yet to achieve a taxonomic consensus. The classification of the cannabis plant has always been challenging for researchers due to its unique position in human society throughout history. Cannabis differs from typical crops, which are usually classified based on species, cultivars, subspecies, and hybrids (Babaei and Ajdanian, 2020; Dach et al., 2015; Mudge et al., 2018; Peng and Shahidi, 2021; Small, 2017; Upton, 2023). Due to its significance in producing medicinal compounds and its unique molecular properties, cannabis is also subject to two additional classification systems: a legal system imposed arbitrarily by governments (distinguishing between hemp and drug types (Lapierre et al., 2023b; Pant et al., 2024)) and a cladistic system based on biological characteristics, including phenetic (e.g.

chemotypes) and molecular traits (Dawidowicz et al., 2024; Fonseca et al., 2018).

Given these complexities, current taxonomic classifications for cannabis, in particular, can greatly benefit from advanced technologies such as genome sequencing, electron microscopy, chromatography, and mathematical and statistical models. This chapter delves into the historical development of hemp classification, current taxonomic frameworks, and their implications for research, breeding, and regulatory frameworks in the growing cannabis industry, emphasizing the importance of understanding its diverse genetic makeup and identifying distinct phenotypes.

- **Taxonomic classification**: a systematic arrangement of organisms into hierarchical categories based on shared characteristics and evolutionary relationships. These categories are organized in a hierarchy that includes Kingdom, Division, Subdivision, Class, Order, Family, Genus, and Species.
- **Species**: a separately evolving metapopulation identified by analyzing the genetic history of individual allele populations. Species are independent of phylogenetic lineages.
- **Cladistics**: a method of classification that organizes organisms based on common ancestry and evolutionary branching, emphasizing shared derived traits (synapomorphies).
- **Cannabis**: a genus composed of a single species, *Cannabis sativa* L., known for its economic and cultural importance.
- **Cannabis varieties**: the primary phenotypic ancestral groupings of cannabis, traditionally classified as sativa, indica, and ruderalis. These groupings are based on morphological and geographical traits but are often debated due to extensive hybridization.
- **Cannabis forma**: sub-varieties or hybrids within cannabis varieties, distinguished by unique morphological, biochemical, or phenological features.
- **Dioecious**: plants that produce either exclusively male (XY) or exclusively female (XX) flowers on separate individuals.
- **Monoecious**: plants that produce both male and female flowers on the same individual.
- **Hermaphroditism**: a condition in which dioecious plants, under specific environmental or genetic conditions, revert to a monoecious-like state, producing both male and female flowers on the same plant. In hermaphroditic flowers, stamens (male, containing pollen sacs) and pistils (female, including stigma, style, and ovary) coexist within the same flower.
- **Chemotypes**: chemical phenotypes of cannabis categorized based on the relative proportions of cannabinoids like THC and CBD. Common

chemotypes include THC-dominant, CBD-dominant, and balanced varieties.

2 Legal classification and impact on taxonomy of cannabis

Cannabis played an important role in human history since its domestication, although studies on it substantially declined during the mid-twentieth century, as it became illegal in most Western countries (Warf, 2014). Criminalization of cannabis, largely driven by negative perceptions surrounding its psychoactive properties, had shifted focus away from its many agricultural applications for many decades (Mudge et al., 2018; Torkamaneh and Jones, 2021). As a consequence, cannabis has not benefited from modern scientific advances to the degree that other agronomically important crops have, resulting in a massive knowledge gap surrounding the basic biology of cannabis. Governments primarily regulate cannabis based on its Δ^9-tetrahydrocannabinol (THC) levels, despite the plant containing at least 150 diverse organic compounds known as cannabinoids (Dawidowicz et al., 2024; Fonseca et al., 2018). For instance, the US Food and Drug Administration (FDA) regulates THC in specific products but has not approved its use in foods or dietary supplements due to its Schedule I status as a controlled substance (Cascini et al., 2012; FDA, 2020; Geweda et al., 2024). Today, the most commonly recognized method of classifying cannabis is by THC concentration, which distinguishes industrial-type or hemp from drug-type cannabis (Lapierre et al., 2023b; Sawler et al., 2015). Drug-type plants typically have higher THC levels than cannabidiol (CBD) levels (THC: CBD ratio > 1) by dry weight of the flower buds. Whereas hemp is distinguished by its low THC content (<0.3%), despite often having higher levels of other cannabinoids, such as CBD, cannabigerol (CBG), cannabichromene (CBC), and cannabinol (CBN), none of which are currently regulated in any jurisdictions (Canada, 2015; Cherney and Small, 2016; Dawidowicz et al., 2024; De Meijer and Hammond, 2005; Hajizadeh, 2016; Mark et al., 2020; Monthony et al., 2021; Torkamaneh and Jones, 2021). To further differentiate these types genetically, early research has focused on the THC or CBD synthase loci, which are key genetic markers used to distinguish hemp from drug-type cannabis (Sawler et al., 2015).

The prolific classification of cannabis into *sativa* and *indica* has historically been a source of confusion (McPartland and Guy, 2017; Small, 2015). Individuals involved in illicit cultivation have associated specific characteristics with the names used in the taxonomy of cannabis species and subspecies. These associations were based on traits intended to convey product qualities to consumers, rather than being grounded in scientific principles (Sawler et al., 2015). For example, they claim that *indica* is calming and antipsychotic due to its higher CBD content, while *sativa* is more stimulating and psychoactive because of its higher THC content (Small, 2017). Historically, however, taxonomists used *sativa* for non-narcotic plants like hemp and *indica* for narcotic species like marijuana (McPartland and Guy, 2017). Furthermore, the extensive hybridization between perceived *indica* and *sativa* types over a prolonged period of restrictions on cannabis cultivation has made it impossible to differentiate between them (Murovec et al., 2022; Small, 2017; Upton, 2023). As a result, it is misleading to claim that *sativa* always has high THC content or that *indica* consistently has high CBD levels. This confusion surrounding cannabis nomenclature has even reached high-ranking academic journals, further complicating understanding for both consumers and scientists (Gould, 2015; McPartland, 2018).

One indirect effect of cannabis illegality on its taxonomy is the significant reduction in genetic diversity over the years. Restrictions on cultivation and breeding, along with limited access to natural germplasm resources, have created a genetic bottleneck, leading to genetic erosion in cannabis (Halpin-McCormick et al., 2024). According to a review by McPartland (2017), in the 1970s and 1980s, a limited number of plants of low genetic diversity were brought from Afghanistan and India to the US and the Netherlands. These limited gene pools formed the backbone of breeding programs in these regions, where Indian and Afghan landraces were valued for traits such as rapid maturation and cold tolerance. By the late 1980s, most drug-type cannabis in the USA, Canada, and Europe were hybrids of Indian and Afghan landraces, making pure Indian and Afghan lines difficult to find (Small, 2017). These genetic restrictions have resulted in the exploitation of a narrow genetic potential. Repeated hybridization among these limited genetic varieties has resulted in highly heterozygous genomes, aimed at producing new phenotypic variations. This has made it challenging to establish clear classifications in cannabis. Furthermore, repeated selection pressures have shifted diversity toward specific traits, such as high THC content (de Ronne et al., 2023; Lapierre et al., 2023a; Naim-Feil et al., 2022, 2023). As a result, today's commercial cultivars are derived from a narrow genetic pool, maintained through continuous propagation.

3 Botanical taxonomy of *Cannabis*

Cannabis sativa L. is an annual flowering member of the order Urticales and the family Cannabaceae (Ahmed et al., 2008; Schilling et al., 2020). *Cannabis* is predominantly dioecious, meaning an individual plant will produce either exclusively male (XY) or exclusively female flowers (XX). Some plants are monoecious, mostly hemp, producing both male and female flowers on the same plant, and under specific conditions dioecious plants can revert to a monoecious-like state called hermaphroditism or hermies (Carpentier et al., 2012). The comprehensive taxonomic framework of *Cannabis*, from the Kingdom to the forma level, is illustrated in Figure 2. This hierarchical diagram traces the botanical lineage of *Cannabis*, highlighting its position within the Cannabaceae family and showcasing the species' (single or multiple) diversity, including subspecies, varieties, and forms (Catalogue of Life (COL), 2024; Erdtman, 2023; Fordjour et al., 2023; Fu et al., 2023; Ii, 2003; Koren et al., 2020; Kovalchuk et al., 2020; Lapierre et al., 2023b; McPartland, 2018; Sambamurty, 2013; Verma, 2010).

3.1 Expansion and revision of the family Cannabaceae

Cannabaceae family, which now comprises ten genera including *Cannabis* and *Humulus*, has expanded to include eight genera previously classified under the Celtidaceae family: *Celtis*, *Pteroceltis*, *Aphananthe*, *Chaetachme*, *Gironniera*, *Lozanella*, *Trema*, and *Parasponia*. This reclassification, based on genetic (DNA) evidence, was implemented by the Angiosperm Phylogeny Group in 1998 (Ii, 2003; Kovalchuk et al., 2020). As a result, the Cannabaceae family now includes approximately 170 accepted species, with *Humulus lupulus* L. (common hop) being a notable member (Kovalchuk et al., 2020; McPartland, 2018).

Building on this reclassification, one of the most extensive phylogenetic studies to date on the Cannabaceae family utilized a hybridization-based sequencing (Hyb-Seq) technique that captures specific genomic regions (Fu et al., 2023). This study analyzed 90 nuclear genes and 82 chloroplast loci from samples representing 83 species across all genera within Cannabaceae. The study revealed that all genera within the Cannabaceae family were monophyletic, except for *Celtis*, which formed two distinct clades: 'Celtis I' and 'Celtis II.' The researchers proposed recognizing 'Celtis II' as a separate genus named *Sparrea* within this family. Additionally, they supported merging *Parasponia* into *Trema*. These changes led to a revised generic classification of Cannabaceae comprising ten monophyletic genera: *Aphananthe*, *Cannabis*, *Celtis*, *Chaetachme*, *Gironniera*, *Humulus*, *Lozanella*, *Pteroceltis*, *Sparrea*, and *Trema* as illustrated in Figure 2 (Fu et al., 2023). These revisions reflect the ongoing refinement of botanical classification driven by advances in genetic research, providing a clearer understanding of the evolutionary relationships within the Cannabaceae family.

3.2 Cannabis (Cannabis) and hops (Humulus): evolutionary genus siblings

Cannabis and its close relative *Humulus* (hops) share several important characteristics within the Cannabaceae family. *Humulus* includes three species: *H. lupulus*, *H. japonicus*, and *H. yunnanensis* (Shephard et al., 2004). Like

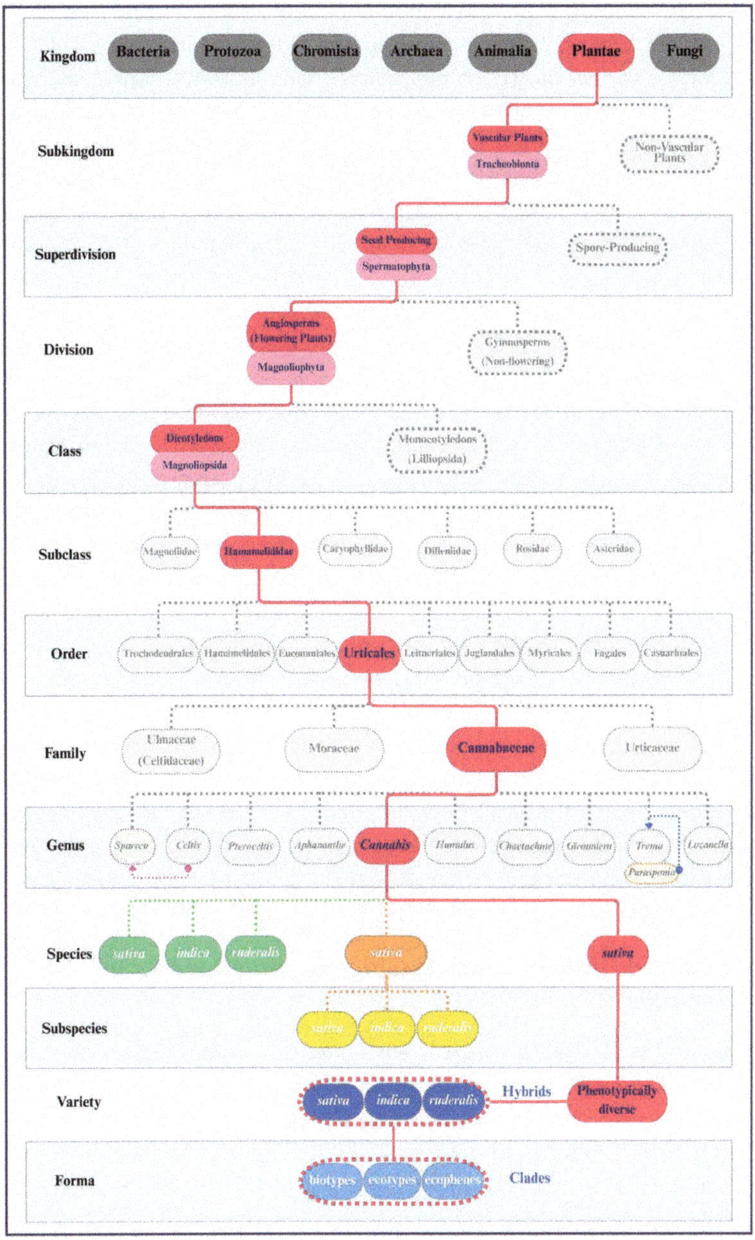

Figure 2 Comprehensive taxonomic framework of *Cannabis*, illustrating its hierarchical classification from kingdom to forma level. The pink path highlights the most likely classification of the plant at the species, subspecies, variety, and forma levels based on the most up-to-date review of the literature. Dashed gray lines represent taxonomic classes unrelated to *Cannabis*. The dashed blue and purple lines indicate recent taxonomic revisions in the Cannabaceae family, while the dashed orange and green lines represent different species-level hypotheses within *Cannabis*.

Cannabis, *Humulus* is dioecious and features both alternate and opposite leaf arrangements, with both genera producing achenes (dry fruits). They share several key traits, including triporate pollen grains, imbricate flower buds, and a persistent perianth (though domesticated *Cannabis* has a deciduous perianth). However, significant differences exist as well, such as in the stipule arrangement and seed coat characteristics. These shared and distinct traits within the Cannabaceae family highlight the evolutionary ties and unique adaptations of *Humulus* and *Cannabis* (Kovalchuk et al., 2020; McPartland, 2018; Yang et al., 2013).

Recent studies have further underscored the close relationship between *Cannabis* and *Humulus,* particularly in their sex chromosome system. *H. lupulus* shares a similar XX/XY system with *Cannabis* (Prentout et al., 2021). Both *Cannabis* and *Humulus* possess heteromorphic sex chromosomes, with *H. lupulus* sharing a similar XX/XY system with *Cannabis*, while *H. japonicus* has XX/XY_1Y_2 chromosomes (Razumova et al., 2023). Prentout et al. (2021) identified 265 XY-linked genes in *H. lupulus* using the SEX-DETector tool (Muyle et al., 2016), including 112 genes that are also sex-related in *Cannabis*. Additionally, the oldest documented plant sex chromosome system, dating back 12–28 million years, has been described in *Cannabis*, and it is predicted that this system is also shared by the genus *Humulus* (Prentout et al., 2021). Genomic in situ hybridization (GISH) painting of the Y chromosomes in *H. lupulus*, *H. japonicus*, and *C. sativa* revealed extensive male-specific regions (MSY) with an accumulation of repetitive DNA on these chromosomes (Razumova et al., 2023). These similarities, along with the preserved nature of their sex chromosomes since the last common ancestor, represent a rare occurrence in the plant kingdom that opens new avenues for investigating phylogenetic classification hypotheses.

3.3 Taxonomic debates and classification frameworks of Cannabis

The taxonomy of the *Cannabis* genus has been the subject of ongoing debate, with questions remaining about whether it is a monospecific genus (*C. sativa* L.) or poly-specific, comprising *C. sativa*, *C. indica*, and *C. ruderalis,* as can be seen in Figure 2 (Anwar et al., 2006; Flores-Sanchez and Verpoorte, 2008; Lapierre et al., 2023b; McPartland, 2018; McPartland and Guy, 2017; Schwabe and McGlaughlin, 2019; Small and Beckstead, 1973). In addition to these views, some researchers argue that traditional species and subspecies classifications are insufficient, and that classification should instead focus on the genetic diversity reflected in phenotypic traits (Burbrink et al., 2022; Patten, 2015; Vinarski, 2020). This approach would require categorizing *Cannabis* into different groups (clades) based on observed characteristics, which have been influenced by extensive crossbreeding within *C. sativa*.

In response to these debates, several taxonomies have emerged, primarily based on morphological, environmental, and chemical traits. These have broadly been grouped into three frameworks (Lapierre et al., 2023b).

1. The first framework consolidates all *Cannabis* plants into a single species with three subspecies;
2. The second framework proposes multiple distinct species within the genus, with ongoing discussion about whether there are two or three species.
3. The third framework considers *C. sativa* as the sole species, without subspecies, acknowledging a wide range of phenotypic and genetic variability.

The third framework will be further discussed in Section 4 'Cladistic classification of *C. sativa* L.'

3.3.1 Monospecific

The taxonomy of *Cannabis* as *Cannabis sativa* L. began with Linnæus's systemic approach in *Species Plantarum* in 1753. The protologue designation of *C. sativa* L. as a single species holds priority under the ICN, being the first *Cannabis* taxon established after the commencement of botanical nomenclature. Although *C. sativa* was first coined by Barbaro (1516), this designation was not recognized until 23 years after his death. Linnæus' classification set a precedent for the monotypic view, which recognized *C. sativa* as the sole species within the genus focusing on its flower parts and seeds (Lapierre et al., 2023b; McPartland, 2018; McPartland and Guy, 2017). Later, in 1807, botanist Christiaan Hendrik Persoon contributed to the monotypic classification of *Cannabis* by categorizing *C. indica* as a subspecies of *C. sativa* (Erkelens and Hazekamp, 2014). Subsequently, in 1838, Lindley supported Linnæus' monotypic classification by identifying *C. sativa* as the only species in the genus *Cannabis* (Lindley, 2011). In 1843, William O'Shaughnessy, in his essay on the medical applications of *Cannabis*, used the term 'Indian hemp' (*C. indica*) in the title. Despite using the *C. indica* designation, O'Shaughnessy argued that there was no significant distinction between *C. sativa* and *C. indica*, thereby supporting the monotypic classification, under the name *C. indica*. His use of the term indica might have been influenced by the term 'Indian hemp,' which referred to the plant's historical medical use in India and neighboring regions (Lapierre et al., 2023b). In 1867, De Candolle affirmed the monotypic classification of *C. sativa* as a single species. Nonetheless, he suggested that this species could be subdivided into four categories: (α) Kif, (β) Vulgaris, (γ) Pedemontana, and (δ) Chinensis. According to his system, southern hemp with significant psychoactive effects was classified as group α (Kif). Varieties of northern hemp used primarily for fiber were categorized into groups γ (Pedemontana) and δ (Chinensis), while intermediate varieties that exhibited characteristics of all the groups were placed in group β (Vulgaris). Despite these distinctions, these categories fell under the single species *C. sativa* (de Candolle, 1867; Lapierre et al., 2023b).

3.3.2 Poly-specific

The debate surrounding the poly-specific nature of *Cannabis* includes discussions on whether it comprises two or three distinct species. As reviewed by Lapierre et al. (2023b), the term indica was introduced by J. B. Lamarck in 1785, and the concept of a ruderal species was later discussed by D. E. Janischewsky in 1924. Both contributions were based on additional morphological features such as trichome structure, leaf shape, and geographic distribution. These observations led to the development of a new perspective that considers *Cannabis* as polyspecific. Lamarck proposed a taxonomy that recognized *C. sativa* and *C. indica* as distinct species. He distinguished *C. indica* from *C. sativa* based on several morphological traits such as its shorter height, woodier stems, alternate branch ramifications, narrower leaflets, and a villous calyx on the female flowers (de Lamarck, 1786; McPartland and Guy, 2017). In 1940, Serebriakova proposed a poly-species taxonomy for *Cannabis*, distinguishing *C. sativa* and *C. indica* and further categorizing them into subspecies and 'varieties' (Koren et al., 2020). Similarly, in 1950, Zhukovskii proposed a two-species system, including *C. sativa* and *C. ruderalis*, which had been introduced by Janischewsky in 1924 (Zhukovskii, 1971).

4 Cladistic classification of *C. sativa* L.

Understanding the differences between *Cannabis* plants involves examining both genetic and physical traits, which reflect variations across isolated groups, within populations, and through evolutionary changes (Orgogozo et al., 2015). A cladistic approach to *Cannabis* classification should consider its genetic diversity, phenotypic variations, and environmental adaptations, which manifest as distinct morphological, phenological, biochemical, and ecological characteristics. This approach provides a framework for categorizing hybrid forms resulting from crossbreeding, thus encapsulating the wide array of traits observed in *C. sativa* plants. By examining cladistic classification, researchers and cultivators can better understand the functional and regulatory implications of cannabis, aiding in its legal classification, practical applications, and breeding strategies.

4.1 Phenetic characters: morphological and phenological approach

In *Cannabis*, a wide range of morphological traits can be examined at both the macro level (such as plant height, internode length, and leaf area) and the micro level (such as trichome structure). These play a role in distinguishing individuals across different trait association panels, which can be used to enhance classification and identification (Babaei et al., 2024). However, relying on any single trait for classification can result in groups that differ significantly from those based on another trait. Much like how selective breeding in dogs (*Canis lupus familiaris*) has produced an extraordinary range of breeds with diverse physical traits (Freedman and Wayne, 2017), selective breeding in *Cannabis sativa* has similarly led to a broad spectrum of phenotypic diversity (Figure 3). However, just as we do not classify each dog breed as a separate species despite their incredible diversity, it is not enough to simply describe groups of *Cannabis* phenotypes as a distinct species. When considering which traits to use in *Cannabis* taxonomy, it is crucial to carefully evaluate whether these traits genuinely distinguish species from one another or merely reflect variations within a single species. For example, while one grouping might focus on plant habit, leaf size, and arrangement, another might prioritize flower number, flower timing, and arrangement, leading to different classifications (Babaei et al., 2024; De Meijer and Keizer, 1996; Raman et al., 2017). Which traits to consider, and their relative importance in cannabis, remains an open question and the subject of ongoing research.

Recent research has extensively characterized cannabis varieties, from medicinal to industrial cultivars, as well as landrace populations, by analyzing agronomic, morphological, and phenological traits. Using methods like principal component analysis (PCA) and clustering with heatmaps, researchers have identified key traits that form distinct clusters and assessed their importance in differentiating these groups (Babaei et al., 2024; Carlson et al., 2021; Lapierre et al., 2023a; Naim-Feil et al., 2021; Petit et al., 2020). Commercially available drug-type *Cannabis* genotypes are primarily short-day plants, with flowering typically induced by 12 to 14-hour photoperiods (Moher et al., 2021). However, this trait varies when looking at hemp genotypes and landraces, with many hemp and some drug-type cultivars exhibiting photoperiod insensitivity and autoflowering, a characteristic often attributed to wild *Cannabis* designated as *ruderalis*, that likely inhabited higher-latitude regions with longer day lengths during summer months (Dowling et al., 2024; McPartland, 2018). Babaei et al. (2024) identified two indigenous populations of cannabis in Iran as

Figure 3 Phenetic diversity within a species: above, diverse dog breeds (*Canis lupus familiaris*) illustrate the extensive morphological and phenotypic variation that can arise from selective breeding. Similarly, below, distinct phenotypes of *Cannabis* demonstrate the wide-ranging morphological differences cultivated through targeted breeding practices over time, suggesting the potential for significant variation within a single species. Dog images sourced from Pexels (pexels.com). *Cannabis* images are original, provided by Éliana Lapierre and Mehdi Babaei.

early-flowering that reach reproductive maturity before the photoperiod switch is triggered by shortening days. The diversity in flowering times is influenced by the cultivar's intended use and environmental conditions, such as latitude and weather. For instance, early-flowering cultivars suit high-latitude regions with short growing seasons, while late-flowering cultivars are preferred in low-latitude areas for fiber production (Amaducci et al., 2005; Salentijn et al., 2015; Struik et al., 2000). Similarly, the trait of ramification, which refers to the extent and pattern of branching, also produces different groupings. Hemp (fiber) cultivars, for example, have longer internodes and fewer branches, maintaining their stature even in open cultivation environments with wider spacing. In contrast, seed and medicinal cultivars exhibit shorter internodes and more branches across all planting conditions (Small and Marcus, 2002).

These examples illustrate how the use of different traits – whether it is flowering time, branching patterns, or internode length – can lead to distinctly different classification outcomes in *Cannabis*.

The debate over which traits to use for classification highlights the complexity of cannabis taxonomy. Relying on a single trait may lead to artificial groupings that do not fully capture the plant's diversity. On the other hand, combining multiple traits using morphometric or biometric methods can provide a more comprehensive classification but can also complicate the process by including less significant traits (Christodoulou et al., 2020). As Grant (2003) suggests, while a comprehensive approach has its advantages, it is often more effective to focus on the most valuable and differentiating traits to avoid misleading classifications (Small and Marcus, 2002).

4.2 Phenetic characters: biochemical approach

A valuable differentiating trait that is often used in cannabis classification is cannabinoid profile. Cannabinoids are organic compounds with bioactive properties that interact with various receptors throughout the body and are responsible for its psychoactive and medicinal properties (Dawidowicz et al., 2024; Ramzy and Priefer, 2021; Silver, 2019). Today, over 150 cannabinoids have been identified in cannabis (Dawidowicz et al., 2024). However, the classification has traditionally focused on the levels and ratios of two key cannabinoids, CBD and THC (Lapierre et al., 2023b).

In 1971, Fetterman et al. developed a two-way chemical classification of *Cannabis* based on physical and chemical traits to differentiate between drug-type and fiber-type *C. sativa* (Fetterman et al., 1971). In 1976, Small and Cronquist introduced a taxonomic system incorporating physical characteristics, geographical distribution, and THC content (chemotype) (Small and Cronquist, 1976). Their two-stage system first divides *C. sativa* into hemp and drug types (i.e. two subspecies: *C. sativa* subsp. *sativa* and *C. sativa* subsp. *indica*) based on THC and CBD contents, and then further separates them into wild and domesticated varieties in each subspecies (McPartland, 2018; McPartland et al., 2018, 2019; Small and Cronquist, 1976). The model further categorizes *Cannabis* into four main groups based on cannabinoid levels: (i) non-narcotic plants from Europe and Western Asia with low THC and high CBD, domesticated for fiber and oilseeds; (ii) non-narcotic plants from East Asia with low to moderate THC and significant CBD, classified as *C. sativa* subsp. *sativa* var. *sativa* and *C. sativa* subsp. *sativa* var. *spontanea*; (iii) narcotic plants from South-Central Asia with high THC and distinct domestication traits, classified as *C. sativa* subsp. *indica* var. *indica*; and (iv) narcotic plants from South Asia and the Middle East with similar THC and CBD levels, identified as *C. sativa* subsp.

indica var. *karifiristanica* (Lapierre et al., 2023b). Recent updates suggest that hybrids among these categories might form fifth and sixth groups (Small, 2015).

This elaborate chemotype system, which classifies plants into chemovars based on cannabinoid levels and phenotype (Turner et al., 1979), has faced criticism even from those who have used it (De Meijer et al., 2003), suggesting that relying on chemotypes alone as a taxonomic criterion remains 'very disputable.' Similarly, Hazekamp and Fischedick (2012) analyzed 28 chemical compounds in cannabis samples using PCA and proposed a chemotaxonomic classification model, but they observed significant chemical variability among samples that could impact classification accuracy.

Despite skepticism, classification based on cannabinoid profiles remains a focal point in cannabis research. For instance, Jin et al. (2021) combined leaf traits with cannabinoid content to differentiate CBD-dominant and intermediate cultivars, finding that CBD-dominant cultivars had narrower leaflets and lighter green coloration, while intermediate cultivars exhibited hybrid characteristics. They argued that this approach of combining morphological and chemical data provides a more robust framework for classifying cannabis varieties and understanding their genetic background. However, the proliferation of chemotype classification systems, each with its own proposed categories and subcategories, with no apparent convergence on a standardized system, has led to overly complicated and arguably artificial distinctions within a highly variable, widely cultivated, and domesticated plant. This complexity is particularly problematic given that cannabinoid levels, the basis for many of these divisions, are highly influenced by environmental factors, developmental stages, and selective breeding pressures.

Given the complexities and artificial distinctions introduced by these chemotype systems, others have suggested considering how certain cannabinoid ratios, rather than absolute levels, might offer a more stable basis for classification. Although absolute levels of cannabinoids fluctuate due to environmental factors, it has been shown that the ratio of CBD to THC remains consistent throughout the plant's life cycle and independent of environmental influences (Baker et al., 1982; Pacifico et al., 2008). Furthermore, since CBN is a degradation product of THC that forms over time during the storage of dried flowers, it is included in the ratio as THC + CBN/CBD (Bachir et al., 2022). However, this approach has also struggled to differentiate between fiber and drug types, arising from the existence of variable intermediate chemotypes and outlier plants (Vergara et al., 2020).

4.3 Molecular characters

In *Cannabis*, diploids consist of nine pairs of autosomes and one pair of polymorphic sex chromosomes, with karyotype genomes of 2n = 18 + XX for females and 2n = 18 + XY for males. The haploid genome size is around 818 Mb for females and 843 Mb for males (Braich et al., 2020). With the increasing availability and decreasing costs of molecular tools in the 2000s, many researchers began focusing on classifying *Cannabis* by examining cultivars from various geographic regions, using genomic DNA and molecular markers to better understand its genetic diversity. In 2004, Mandolino and Carboni aimed to identify chemotype-specific markers by crossing inbred lines with THC and CBD chemotypes. Their study revealed a Mendelian segregation pattern in the F2 generation, indicating a single gene with two codominant alleles, BT and BD. They identified RAPD markers, which led to the development of a SCAR marker (B190/B200), useful for identifying chemotypes within the studied population, though it lacked applicability across other populations (Mandolino and Carboni, 2004). Following this, Hillig (2005) introduced a molecular marker-based classification system, rejecting the idea of *Cannabis* as a monotypic genus. They identified two species, *C. sativa* and *C. indica*, and suggested a potential third, *C. ruderalis*, based on allozyme variations at 17 genomic loci. The PCA of 52 alleles in 157 samples from diverse geographical locations supported a polytypic theory, aligning with earlier taxonomic theories but proposing three species and seven putative taxa: *C. sativa* subsp. *sativa* var. *sativa*, *C. sativa* subsp. *sativa* var. *spontanea*, *C. sativa* subsp. *indica* var. *kafiristanica*, *C. indica*, *C. indica* sensu, *C. chinensis*, and *C. ruderalis*.

Gilmore et al. (2007) investigated the phylogeny of 188 cannabis plants derived from 76 populations, encompassing various plant-use groups and geographical regions, to distinguish drug types from fiber types. Using chloroplast and mitochondrial markers, they identified six minor haplotypes and three major haplotypes despite the uniparental (maternal) inheritance and low mutation rate in organelles. These haplotypes were highly indicative of both crop use and geographic origin. In their findings, two organelle haplotypes (I and II) were prevalent in European populations, primarily fiber types. In contrast, other haplotypes were more common in *C. indica* drug types from southern Europe and Asia, with haplotypes V and VI being typical for drug types from Africa, India, Southeast Asia, and Mexico.

In 2013, Clarke and Merlin proposed a taxonomic model integrating morphological characteristics, molecular markers, and chemotypes. Building on Hillig's (2005) nomenclature, they discovered mutations leading to THC and CBD production in *Cannabis* evolution. Their classification model, known as 'dope versus rope,' distinguishes between narrow leaf drug (NLD), broad leaf drug (BLD), and narrow leaf hemp (NLH) cultivars. Although hemp produces negligible THC, both hemp and drug-type *Cannabis* possess alleles for THC (BT) and CBD (BD) synthesis. They found that BT was weakly expressed in *C. sativa* but strongly expressed in *C. indica*, whereas BD was more pronounced in *C. sativa*. They proposed reclassifying current drug-type *Cannabis* under the single species *C. indica* due to the higher presence of the BT allele in *C. indica* compared to *C. sativa*, and suggested *C. ruderalis* as the wild ancestor of both species. However, their conclusions were based on limited samples and genomic analyses (Clarke and Merlin, 2013; Lapierre et al., 2023b).

Further supporting the significance of genetics in cannabis classification, variations in single nucleotides (SNPs) within the single-exon *THCAS* and *CBDAS* genes help distinguish between industrial (hemp-type) and medicinal (drug-type) cannabis, based on chemotype activity (Fulvio et al., 2021). Studies show genotype significantly influences a cannabis chemotype and cannabinoid profile, with allelic status of two loci B and O determining cannabis chemotype. The allelic status and cannabinoid profile of five common *Cannabis sativa* chemotypes are as follows: Chemotype I primarily produces THC(A) with a B locus of (B_T/B_T or B_T/B_0) and an O locus of (O/O or O/o). Chemotype II has a CBD(A) to THC(A) ratio of approximately 1.5:1 with a B locus of (B_T/B_D) and an O locus of (O/O or O/o). Chemotype III predominantly produces CBD(A) with a B locus of (B_D/B_D or B_D/B_0) and an O locus of (O/O or O/o). Chemotype IV mainly produces CBG(A) with a B locus of (B_0/B_0) and an O locus of (O/O or O/o). Chemotype V is cannabinoid-free with any B locus and an O locus of (o/o) (De Meijer and Hammond, 2005; Mandolino and Carboni, 2004; Stack et al., 2021).

With the advent of advanced sequencing techniques and genomic analysis, along with the use of indigenous germplasm, more details about the domestication process of cannabis and the genetic mechanisms controlling cannabinoids have been uncovered. In this context, recent pangenome research has revealed that different genomic regions of cannabis exhibit high diversity (Michael et al., 2024). Still, the cannabinoid synthase genes are located in a variable region with low genetic diversity. In another study, the domestication history of the plant was examined using 110 accessions from various parts of the world (Ren et al., 2021). This analysis revealed that all current hemp- and drug-type cultivars have diverged from a common gene pool, now represented by wild plants and landraces found in China. The study also showed that the complete loss of the *THCAS* or *CBDAS* genes occurred during intensive selection for increased fiber production or psychoactive properties (Ren et al., 2021). The new findings indicate that the current binary classification, which divides plants into hemp and marijuana, may need reconsideration, as cannabinoid synthases are inherited independently of other traits selected for fiber and psychoactive properties. These results also highlight the importance of selecting and preserving Asian wild species in future research.

5 Resolving cannabis taxonomy: genomic advances and computational insights

Recent changes in the status of classification and taxonomy have shifted taxonomy away from being seen as a purely subjective and descriptive science. It is no longer defined solely by similarities in a few phenotypic traits. Unlike traditional classification, computers perform a significant portion of modern taxonomy without direct human input (Vinarski, 2022). This approach theoretically seeks the concept of evolutionary species, which is based on the 'unified' concept defined by De Queiroz (2007), describing species as separately evolving metapopulation lineages. According to this theory, species are independent of phylogenetic lineages and are identified by analyzing the genetic history of individual allele populations (Vinarski, 2022).

Over a decade ago, the first reference genome of cannabis was published (van Bakel et al., 2011), and contemporary research has sought to use genetic features to differentiate various clades, primarily focusing on distinguishing drug-type from hemp-type cannabis. The consensus from these studies is that classification cannot be based solely on cannabinoid content; the entire genome and metabolome must be considered in their differentiation. Additionally, recent studies have employed advanced genomic approaches, including whole-genome sequencing (WGS), to reassess the classification of *sativa*, *indica*, and *ruderalis* (Balant et al., 2024; Jin et al., 2021; Lynch et al., 2016; Vergara et al., 2016, 2021). These studies examined different accessions labeled as each of these types based on phenotype and emphasized the monotypic nature of *Cannabis*.

A Genotyping-by-sequencing (GBS) study including 137 drug-type *Cannabis* plants, using 116,000 SNPs to perform PCA analysis revealed that *sativa* and *indica* samples were genetically indistinguishable genome wide (Watts et al., 2021). The authors concluded that the commonly used *sativa/indica* classification does not accurately reflect genetic variation across *Cannabis* plants (Watts et al., 2021). Similarly, Ren et al. (2021) employed WGS and identified 12 million SNPs in 110 cannabis accessions, concluding that *Cannabis* should be considered a single species due to the genetic overlap between hemp and drug-type cultivars. Another study utilized WGS and identified 344 multiallelic SNPs to investigate the genetic structure of 23 cannabis genotypes. Using methods such as Discriminant analysis of principal components (DAPC), PCA, neighbor-joining (NJ) tree, and hierarchical clustering, they found that these genotypes could be divided into five clusters, with distinct groups for CBD dominant, THC-dominant, and balanced cultivars. This clustering revealed closer genetic proximity between CBD-dominant and balanced cultivars, likely due to recent breeding interests focusing on non-psychoactive, CBD-rich cultivars. The identified structural SNPs, particularly those on chromosomes containing *THCAS* and *CBDAS*, offer potential insights into the production of other cannabinoids and morphological traits (Jin et al., 2021).

In recent years, artificial intelligence (AI) has incorporated a wide range of tools and techniques, including machine learning (ML), deep learning (DL), neural networks (NN), computer vision, pattern recognition, clustering, and natural language processing (NLP). AI has converged with the field of big data, offering extensive datasets for model training and significantly advancing big data analytics and reasoning capabilities (Janbi et al., 2023; Riza et al., 2023). ML allows systems to enhance their performance and adapt with minimal human oversight by learning from data. DL, a subset of ML, utilizes neural networks to emulate the brain's learning mechanisms, enabling sophisticated pattern recognition and adaptation to new data (Taye, 2023). By applying ML to large datasets, including morphological, phytochemical, genetic data, and image data, researchers can uncover patterns and features that differentiate between various classes that a human alone would not be able to identify. This process facilitates the identification of biomarkers that can effectively separate and classify different categories, leading to improved accuracy in classification and a deeper understanding of the underlying biological differences (Niazian and Niedbała, 2020; Singh et al., 2016).

A recent study highlighted the practical application of these techniques in the *Cannabis* industry. Researchers utilized hyperspectral imaging (HSI) combined with advanced multivariate data analysis methods to classify different cannabis chemotypes during cultivation. Their approach demonstrated high accuracy in identifying and differentiating three chemotypes, achieving an overall classification trueness of 94.7%. This non-invasive method offers a promising alternative to traditional analytical techniques, providing enhanced control and monitoring capabilities for cannabis production (San Nicolas et al., 2024). In another approach, a new method was developed to overcome the challenge of nonhomologous landmarks in palmate, pinnate, and lobed leaves using cannabis. Researchers created continuous polynomial models by modeling corresponding pseudo-landmarks for each leaflet as angle-radius coordinates, bypassing issues related to variable leaflet numbers. Analyzing 341 leaves from 24 individuals across nine *Cannabis* accessions, they accurately predicted accession identity, leaflet number, and relative node number. This intra-leaf modeling offers a rapid, cost-effective means of identifying *Cannabis* accessions, making it valuable for taxonomic studies, cultivar recognition, and potentially for analyzing chemical content and sex identification (Balant et al., 2024).

The potential of computer analyses to reveal or identify new species is only fully realized when a formal description is carried out by a taxonomist (Vinarski, 2020). However, this poses a challenge, as many research teams are composed primarily of geneticists and bioinformaticians, often lacking taxonomic experts familiar with the nomenclature and procedures for describing new taxa (Vinarski,

2022). This highlights the crucial role of taxonomists and underscores the need for modern taxonomy to be conducted through interdisciplinary collaboration to effectively address this gap.

6 Conclusion

In recent decades, biological systematics has seen a convergence of once-opposing methodologies, largely driven by technological advances in computing and molecular biology rather than theoretical shifts. These advancements have provided the necessary tools for unifying classification methods, particularly at the species level, and have enabled the school of cladistics (phylogenetics) to leverage genomic tools and techniques to identify genetic changes and differentiate between species and subspecies.

Classifying cannabis through traditional taxonomy is particularly challenging due to factors like wind pollination, heterozygosity, modern cultivation practices, incomplete domestication, and poorly documented breeding histories. By considering the genus *Cannabis* to be made up of a single species, *C. sativa*, we can define the primary phenotypic ancestral traits of *sativa*, *indica*, and *ruderalis* as distinct 'varieties.' Numerous hybrids among these 'varieties' can be further classified into sub-varieties ('forma') based on distinctive features. This multi-tiered system underscores the complexity and diversity of cannabis, highlighting the importance of advanced genomic and molecular tools in refining our understanding of its taxonomy. Integrating these technologies with traditional taxonomy allows for a more precise and comprehensive classification that truly reflects the diversity and evolutionary relationships within the *Cannabis* genus.

7 Where to look for further information

For a comprehensive foundation in plant systematics, Sambamurty (2013) provides an essential academic overview. Regarding the specific classification of the genus, McPartland (2018) offers a detailed review of *Cannabis* systematics at the family, genus, and species levels. For readers interested in modern genomic approaches, Lapierre et al. (2023b) demonstrates how genomics-based taxonomy is currently being used to clarify Cannabis classification. Additionally, researchers can follow the latest industry standards and regulatory updates through the International Hemp Association (IHA) (http://www.internationalhempassociation.org) and the European Industrial Hemp Association (EIHA) (https://eiha.org).

8 References

Ahmed SA, et al. (2008) Cannabinoid ester constituents from high-potency Cannabis sativa. *Journal of Natural Products* 71(4): 536–542.

Alamgir ANM (2017) Pharmacognostical botany: classification of medicinal and aromatic plants (MAPs), botanical taxonomy, morphology, and anatomy of drug plants. In: Alamgir ANM (ed.), *Therapeutic Use of Medicinal Plants and Their Extracts: Volume 1: Pharmacognosy*. Springer: 177–293.

Amaducci S, et al. (2005) Fibre development in hemp (Cannabis sativa L.) as affected by agrotechnique: preliminary results of a microscopic study. *Journal of Industrial Hemp* 10(1): 31–48.

Anwar F, et al. (2006) Analytical characterization of hemp (Cannabis sativa) seed oil from different agro-ecological zones of Pakistan. *Journal of the American Oil Chemists' Society* 83(4): 323–329.

Babaei M and Ajdanian L (2020) Screening of different Iranian ecotypes of cannabis under water deficit stress. *Scientia Horticulturae* 260: 108904.

Babaei M, et al. (2022) Morphological and phytochemical changes of Cannabis sativa L. affected by light spectra. In: Singh HB and Vaishnav A (eds) *New and Future Developments in Microbial Biotechnology and Bioengineering*. Elsevier, Amsterdam: 119–133.

Babaei M, et al. (2024) Characterization of indigenous populations of cannabis in Iran: a morphological and phenological study. *BMC Plant Biology* 24(1): 151.

Bachir F, et al. (2022) Origin, early history, cultivation, and characteristics of the traditional varieties of moroccan cannabis sativa L. *Cannabis and cannabinoid research* 7(5): 603–615.

Baker PB, et al. (1982) The physical and chemical features of cannabis plants grown in the United Kingdom of Great Britain and Northern Ireland from seeds of known origin. *Bulletin on Narcotics* 34(1): 27–36.

Balant M, et al. (2024) Intra-leaf modeling of Cannabis leaflet shape produces leaf models that predict genetic and developmental identities. *New Phytologist*. Epub ahead of print 2024.

Barbaro D, et al. (1516) *In Hoc Volumin Haec Continentur, Ioannis Baptistae Egnatii in Dioscoridem Ab Hermolao Barbaro Tralatum Annotamenta* (ed. Aloisius Barbari & Franciscus Barbari). Venice.

Barcaccia G, et al. (2020) Potentials and challenges of genomics for breeding cannabis cultivars. *Frontiers in Plant Science* 11: 573299.

Block JS (2023) 1 The history of cannabis: a story of coevolution and human discovery. In: A ElSohly (ed.) *Fundamentals*. De Gruyter, pp. 1–26. https://doi.org/10.1515/9783110718362-001.

Braich S, et al. (2020) A new and improved genome sequence of Cannabis sativa. *Gigabyte*.

Burbrink FT, et al. (2022) Empirical and philosophical problems with the subspecies rank. *Ecology and Evolution* 12(7): e9069.

Canada G of (2015) Industrial hemp regulations Règlement sur le chanvre industriel. *Ministry of Justice*. Epub ahead of print 2015.

Carlson CH, et al. (2021) Morphometric relationships and their contribution to biomass and cannabinoid yield in hybrids of hemp (Cannabis sativa). *Journal of Experimental Botany* 72(22): 7694–7709.

Carpentier C, et al. (2012) *Cannabis Production and Markets in Europe*. Publ. Office of the Europ. Union.

Cascini F, et al. (2012) Increasing delta-9-tetrahydrocannabinol (Δ-9-THC) content in herbal cannabis over time: systematic review and meta-analysis. *Current Drug Abuse Reviews* 5(1): 32–40.

Catalogue of Life (COL) (2024) Available at: https://www.catalogueoflife.org/ (accessed 11 August 2024).

Cherney JH and Small E (2016) Industrial hemp in North America: Production, politics and potential. *Agronomy* 6(4): 58.

Christodoulou MD, et al. (2020) The Cinderella discipline: morphometrics and their use in botanical classification. *Botanical Journal of the Linnean Society* 194(4): 385–396.

Clarke RC (1998) *Hashish!* Red Eye Press

Clarke RC and Merlin MD (2013) *Cannabis*. Evolution and Ethnobotany, University of California Press.

Cook A (2010) Linnaeus and Chinese plants: a test of the linguistic imperialism thesis. *Notes and Records of the Royal Society* 64(2): 121–138.

Crocq M-A (2020) History of cannabis and the endocannabinoid system. *Dialogues in Clinical Neuroscience* 22(3): 223–228.

Cuperus JT (2022) Single-cell genomics in plants: current state, future directions, and hurdles to overcome. *Plant Physiology* 188(2): 749–755.

Dach J, et al. (2015) *Cannabis Extracts in Medicine: The Promise of Benefits in Seizure Disorders, Cancer and Other Conditions*. McFarland.

Dawidowicz AL, et al. (2024) Cannabigerol (CBG) signal enhancement in its analysis by gas chromatography coupled with tandem mass spectrometry. *Forensic Toxicology* 42(1): 31–44.

de Candolle A (1867) *Lois de La Nomenclature Botanique Adoptées Par Le Congrès International de Botanique Tenu à Paris En Août 1867: Suivies d'une 2e Édition de l'introduction Historique et Du Commentaire Qui Accompagnaient La Rédaction Préparatoire Présentée Au Congrès*. H. Georg.

de Lamarck J-B (1786) *Encyclopédie Méthodique: Botanique*. chez Panckoucke.

De Meijer EPM and Hammond KM (2005) The inheritance of chemical phenotype in Cannabis sativa L.(II): cannabigerol predominant plants. *Euphytica* 145(1): 189–198.

De Meijer EPM and Keizer LCP (1996) Patterns of diversity in Cannabis. *Genetic Resources and Crop Evolution* 43: 41–52.

De Meijer EPM, et al. (2003) The inheritance of chemical phenotype in Cannabis sativa L. *Genetics* 163(1): 335–346.

De Queiroz K (2007) Species concepts and species delimitation. *Systematic Biology* 56(6): 879–886.

de Ronne M, et al. (2023) Genetic Insights into Agronomic and Morphological Traits of Drug-Type Cannabis Revealed by Genome-Wide Association Studies. *bioRxiv*.

Dowling CA, et al. (2024) A Flowering Locus T ortholog is associated with photoperiod-insensitive flowering in hemp (Cannabis sativa L.). *The Plant Journal* 119(1): 383–403.

Doyle JA (2018) Phylogenetic analyses and morphological innovations in land plants. *Annual Plant Reviews Online*: 1–50.

Erdtman G (2023) *Pollen Morphology and Plant Taxonomy: Angiosperms (an Introduction to Palynology)*. Brill.

Erkelens JL and Hazekamp A (2014) That which we call Indica, by any other name would smell as sweet. *Cannabinoids* 9(1): 9–15.

FDA US (2020) *FDA Regulation of Cannabis and Cannabis-Derived Products, Including Cannabidiol (CBD)*. US Food & Drug Administration.

Fetterman PS, et al. (1971) Mississippi-grown Cannabis sativa L.: preliminary observation on chemical definition of phenotype and variations in tetrahydrocannabinol content versus age, sex, and plant part. *Journal of Pharmaceutical Sciences* 60(8): 1246–1249.

Flores-Sanchez IJ and Verpoorte R (2008) Secondary metabolism in cannabis. *Phytochemistry Reviews* 7: 615–639.

Fonseca BM, et al. (2018) Inspired by nature: modern medicines derived from Cannabis. *Frontiers in Natural Product Chemistry* 4: 119–155.

Fordjour E, et al. (2023) Cannabis: a multifaceted plant with endless potentials. *Frontiers in Pharmacology* 14: 1200269.

Freedman AH and Wayne RK (2017) Deciphering the origin of dogs: from fossils to genomes. *Annual Review of Animal Biosciences* 5(1): 281–307.

Fu X, et al. (2023) Phylogenomic analysis of the hemp family (Cannabaceae) reveals deep cyto-nuclear discordance and provides new insights into generic relationships. *Journal of Systematics and Evolution* 61(5): 806–826.

Fulvio F, et al. (2021) Analysis of sequence variability and transcriptional profile of cannabinoid synthase genes in cannabis sativa L. chemotypes with a focus on cannabichromenic acid synthase. *Plants* 10(9): 1857.

Geweda MM, et al. (2024) Evaluation of dispensaries' cannabis flowers for accuracy of labeling of cannabinoids content. *Journal of Cannabis Research* 6(1): 11.

Gilmore S, et al. (2007) Organelle DNA haplotypes reflect crop-use characteristics and geographic origins of Cannabis sativa. *Forensic Science International* 172(2): 179–190.

Gould J (2015) The cannabis crop. *Nature* 525: S2–S3.

Grant V (2003) Incongruence between cladistic and taxonomic systems. *American Journal of Botany* 90(9): 1263–1270.

Hajizadeh M (2016) Legalizing and regulating marijuana in Canada: review of potential economic, social, and health impacts. *International Journal of Health Policy and Management* 5(8): 453.

Halpin-McCormick A, et al. (2024) Examining population structure across multiple collections of Cannabis. *Genetic Resources and Crop Evolution*. Epub ahead of print 2024. https://doi.org/10.1007/s10722-024-01928-1

Hazekamp A and Fischedick JT (2012) Cannabis-from cultivar to chemovar. *Drug Testing and Analysis* 4(7–8): 660–667.

Hillig KW (2005) Genetic evidence for speciation in Cannabis (Cannabaceae). *Genetic Resources and Crop Evolution* 52: 161–180.

II APG (2003) An update of the Angiosperm Phylogeny Group classification for the orders and families of flowering plants: APG II. *Botanical journal of the Linnean Society* 141(4): 399–436.

Janbi N, et al. (2023) Distributed artificial intelligence: taxonomy, review, framework, and reference architecture. *Intelligent Systems with Applications* 18: 200231.

Jin D, et al. (2021) Classification of cannabis strains in the Canadian market with discriminant analysis of principal components using genome-wide single nucleotide polymorphisms. *PLoS One* 16(6): e0253387.

Koren A, et al. (2020) Controversial taxonomy of hemp. *Genetika* 52(1): 1–13.

Kovalchuk I, et al. (2020) The genomics of Cannabis and its close relatives. *Annual Review of Plant Biology* 71: 713–739.

Lapierre É, et al. (2023a) Comprehensive phenotypic characterization of diverse drug-type cannabis varieties from the Canadian legal market. *Plants* 12(21): 3756.

Lapierre É, et al. (2023b) Genomics-based taxonomy to clarify cannabis classification. *Genome* 66(8): 202–211.

Lata H, et al. (2023) 2 Cannabis: botany and biomass production. In: ElSohly MA (ed.), *Cannabis Chemistry and Biology: Fundamentals*. Walter de Gruyter GmbH & Co KG: 27.

Lindley J (2011) *Flora Medica: A Botanical Account of All the More Important Plants Used in Medicine, in Different Parts of the World*. Cambridge University Press.

Linnaeus C (1938) The 'Critica Botanica' of Linnaeus. Translated by Sir Arthur Hort. London: Ray Society.

Luisa BG (2012) *An Introduction to Numerical Classification*. Elsevier.

Lynch RC, et al. (2016) Genomic and chemical diversity in Cannabis. *Critical Reviews in Plant Sciences* 35(5–6): 349–363.

Mandolino G and Carboni A (2004) Potential of marker-assisted selection in hemp genetic improvement. *Euphytica* 140: 107–120.

Mark T, et al. (2020) Economic viability of industrial hemp in the United States: a review of state pilot programs. Epub ahead of print 2020.

McKelvey B (2022) *Organizational Systematics: Taxonomy, Evolution, Classification*. University of California Press.

McNeill J (1996) The BioCode: integrated biological nomenclature for the 21st century? In: *Proceedings of a mini-symposium on Biological Nomenclature in the 21st Century*, 1996.

McPartland JM (2017) Cannabis sativa and Cannabis indica versus "Sativa" and "Indica". In: S Chandra, et al. (eds) *Cannabis sativa L.-botany and biotechnology*. Springer, pp. 101–121.

McPartland JM (2018) Cannabis systematics at the levels of family, genus, and species. *Cannabis and Cannabinoid Research* 3(1): 203–212.

McPartland JM and Guy GW (2017) Models of Cannabis taxonomy, cultural bias, and conflicts between scientific and vernacular names. *The Botanical Review* 83: 327–381.

McPartland JM, et al. (2018) Cannabis is indigenous to Europe and cultivation began during the Copper or Bronze age: a probabilistic synthesis of fossil pollen studies. *Vegetation History and Archaeobotany* 27(4): 635–648.

McPartland JM, et al. (2019) Cannabis in Asia: its center of origin and early cultivation, based on a synthesis of subfossil pollen and archaeobotanical studies. *Vegetation History and Archaeobotany* 28(6): 691–702.

Michael T, et al. (2024) Domesticated cannabinoid synthases amid a wild mosaic cannabis pangenome. Epub ahead of print 2024.

Moher M, et al. (2021) Photoperiodic response of in vitro Cannabis sativa plants. *HortScience* 56(1): 108–113.

Monthony AS, et al. (2021) The past, present and future of Cannabis sativa tissue culture. *Plants* 10(1): 185.

Mudge EM, et al. (2018) Chemometric analysis of cannabinoids: chemotaxonomy and domestication syndrome. *Scientific reports* 8(1): 13090.

Murovec J, et al. (2022) Analysis of morphological traits, cannabinoid profiles, THCAS gene sequences, and photosynthesis in wide and narrow leaflet high-cannabidiol breeding populations of medical cannabis. *Frontiers in Plant Science* 13: 786161.

Muyle A, et al. (2016) SEX-DETector: a probabilistic approach to study sex chromosomes in non-model organisms. *Genome Biology and Evolution* 8(8): 2530–2543.

Naim-Feil E, et al. (2021) The characterization of key physiological traits of medicinal cannabis (Cannabis sativa L.) as a tool for precision breeding. *BMC Plant Biology* 21(1): 294.

Naim-Feil E, et al. (2022) Empirical evaluation of inflorescences' morphological attributes for yield optimization of medicinal cannabis cultivars. *Frontiers in Plant Science* 13: 858519.

Naim-Feil E, et al. (2023) The cannabis plant as a complex system: interrelationships between cannabinoid compositions, morphological, physiological and phenological traits. *Plants* 12(3): 493.

Niazian M and Niedbała G (2020) Machine learning for plant breeding and biotechnology. *Agriculture* 10(10): 436.

Orgogozo V, et al. (2015) The differential view of genotype–phenotype relationships. *Frontiers in Genetics* 6: 179.

Pacifico D, et al. (2008) Time course of cannabinoid accumulation and chemotype development during the growth of Cannabis sativa L. *Euphytica* 160: 231–240.

Pant SP, et al. (2024) Exploring the historical, botanical, and taxonomical foundations of cannabis: a review. *Cannabis and Derivatives* (1): 3–36.

Patten MA (2015) Subspecies and the philosophy of science. *The Auk: Ornithological Advances* 132(2): 481–485.

Peng H and Shahidi F (2021) Cannabis and cannabis edibles: a review. *Journal of Agricultural and Food Chemistry* 69(6): 1751–1774.

Petit J, et al. (2020) Genetic architecture of flowering time and sex determination in hemp (Cannabis sativa L.): a genome-wide association study. *Frontiers in plant Science* 11: 569958.

Prentout D, et al. (2021) Plant genera Cannabis and Humulus share the same pair of well-differentiated sex chromosomes. *New Phytologist* 231(4): 1599–1611.

Raman V, et al. (2017) Morpho-anatomy of marijuana (Cannabis sativa L.). In: S Chandra, et al. (eds) *Cannabis sativa L.-botany and biotechnology*. Springer, pp. 123–136.

Ramzy V and Priefer R (2021) THC detection in the breath. *Talanta* 222: 121528.

Razumova OV, et al. (2023) GISH painting of the Y chromosomes suggests advanced phases of sex chromosome evolution in three dioecious Cannabaceae species (Humulus lupulus, H. japonicus, and Cannabis sativa). *Protoplasma* 260(1): 249–256.

Ren G, et al. (2021) Large-scale whole-genome resequencing unravels the domestication history of Cannabis sativa. *Science advances* 7(29): eabg2286.

Riza LS, et al. (2023) Implementation of machine learning in DNA barcoding for determining the plant family taxonomy. *Heliyon* 9(10): e20161.

Rouhan G and Gaudeul M (2021) Plant taxonomy: a historical perspective, current challenges, and perspectives. In: Besse P (ed.), *Molecular plant taxonomy: Methods and protocols*. Springer: 1–38.

Russo EB, et al. (2008) Phytochemical and genetic analyses of ancient cannabis from Central Asia. *Journal of Experimental Botany* 59(15): 4171–4182.

Salentijn EMJ, et al. (2015) New developments in fiber hemp (Cannabis sativa L.) breeding. *Industrial Crops and Products* 68: 32–41.

Sambamurty A (2013) *Taxonomy of Angiosperms*. IK International Pvt Ltd.

San Nicolas M, et al. (2024) NIR-hyperspectral imaging and machine learning for non-invasive chemotype classification in Cannabis sativa L. *Computers and Electronics in Agriculture* 217: 108551.

Sawler J, et al. (2015) The genetic structure of marijuana and hemp. *PLoS One* 10(8): e0133292.

Schilling S, et al. (2020) The cream of the crop: Biology, breeding and applications of Cannabis sativa. *Authorea Preprints*. Authorea. Epub ahead of print 2020.

Schwabe AL and McGlaughlin ME (2019) Genetic tools weed out misconceptions of strain reliability in Cannabis sativa: implications for a budding industry. *Journal of Cannabis Research* 1(1): 1–16.

Shephard H, et al. (2004) Sex expression in hop (Humulus lupulus L. and H. japonicus Sieb. et Zucc.) floral morphology and sex chromosomes. In: Ainsworth CC (ed.), *Sex Determination in Plants*. Garland Science, pp. 139–150.

Silver RJ (2019) The endocannabinoid system of animals. *Animals* 9(9): 686.

Singh A, et al. (2016) Machine learning for high-throughput stress phenotyping in plants. *Trends in Plant Science* 21(2): 110–124.

Small E (2015) Evolution and classification of Cannabis sativa (marijuana, hemp) in relation to human utilization. *The Botanical Review* 81: 189–294.

Small E (2017) Classification of Cannabis sativa L. in relation to agricultural, biotechnological, medical and recreational utilization. In: S Chandra, et al. (eds) *Cannabis sativa L.-botany and biotechnology*. Springer, pp. 1–62.

Small E and Beckstead HD (1973) Cannabinoid phenotypes in Cannabis sativa. *Nature* 245(5421): 147–148.

Small E and Cronquist A (1976) A practical and natural taxonomy for Cannabis. *Taxon* 25(4): 405–435.

Small E and Marcus D (2002) Hemp: a new crop with new uses for North America. *Trends in New Crops and New Uses* 24(5): 284–326.

Stack GM, et al. (2021) Season-long characterization of high-cannabinoid hemp (Cannabis sativa L.) reveals variation in cannabinoid accumulation, flowering time, and disease resistance. *GCB Bioenergy* 13(4): 546–561.

Struik PC, et al. (2000) Agronomy of fibre hemp (Cannabis sativa L.) in Europe. *Industrial Crops and Products* 11(2–3): 107–118.

Sun X (2023) Research progress on cannabinoids in cannabis (Cannabis sativa L.) in China. *Molecules* 28(9): 3806.

Taye MM (2023) Understanding of machine learning with deep learning: architectures, workflow, applications and future directions. *Computers* 12(5): 91.

Torkamaneh D and Jones AMP (2021) Cannabis, the multibillion dollar plant that no genebank wanted. *Genome* 65(1): 1–5.

Turner CE, et al. (1979) Constituents of cannabis sativa L., XIV: intrinsic problems in classifying cannabis based on a single cannabinoid analysis. *Journal of Natural Products* 42(3): 317–319.

Upton R (2023) Nomenclature: herbal taxonomy in the global commerce of botanicals. In: Kinghorn AD, et al. (eds), *Progress in the Chemistry of Organic Natural Products 122: Botanical Dietary Supplements and Herbal Medicines*. Springer, pp. 221–260.

van Bakel H, et al. (2011) The draft genome and transcriptome of Cannabis sativa. *Genome Biology* 12(10): 1–17.

Van der Gulik PTS, et al. (2017) In defence of the three-domains of life paradigm. *BMC Evolutionary Biology* 17: 1–5.

Van der Gulik PTS, et al. (2023) Renewing L innaean taxonomy: a proposal to restructure the highest levels of the N atural S ystem. *Biological Reviews* 98(2): 584–602.
Vergara D, et al. (2016) Genetic and genomic tools for Cannabis sativa. *Critical Reviews in Plant Sciences* 35(5–6): 364–377.
Vergara D, et al. (2020) Modeling cannabinoids from a large-scale sample of Cannabis sativa chemotypes. *PloS one* 15(9): e0236878.
Vergara D, et al. (2021) Widely assumed phenotypic associations in Cannabis sativa lack a shared genetic basis. *PeerJ* 9: e10672.
Verma BK (2010) *Introduction to Taxonomy of Angiosperms*. PHI Learning Pvt. Ltd.
Vinarski MV (2020) Roots of the taxonomic impediment: is the "integrativeness" a remedy? *Integrative Zoology* 15(1): 2–15.
Vinarski MV (2022) Today's taxonomy: a way to a new synthesis or an endless deadlock? *Biology Bulletin Reviews* 12(6): 561–569.
Von Linné C (2005) *Linnaeus' Philosophia Botanica*. Oxford University Press.
Warf B (2014) High points: an historical geography of cannabis. *Geographical Review* 104(4): 414–438.
Watts S, et al. (2021) Cannabis labelling is associated with genetic variation in terpene synthase genes. *Nature plants* 7(10): 1330–1334.
Wiley EO and Lieberman BS (2011) *Phylogenetics: Theory and Practice of Phylogenetic Systematics*. John Wiley & Sons.
Yang M-Q, et al. (2013) Molecular phylogenetics and character evolution of Cannabaceae. *Taxon* 62(3): 473–485.
Zhukovskii PM (1971) Cultivated plants and their wild relatives. Epub ahead of print 1971.

Chapter 2

Advances in understanding the nutrient and water requirements of hemp

Xinlin Zhao, Caisheng Qiu, Songhua Long and Huajiao Qiu, Institute of Bast Fiber Crops – Chinese Academy of Agricultural Sciences, China

1 Introduction

Like us plants need 'food' in order to grow and reproduce. 'Food' refers to all the matter that plants must obtain from the environment to complete their life cycle. This mainly includes O_2, CO_2, water, inorganic nutrients (e.g. N, P, K), and some low-molecular-weight organic matter (e.g. amino acid, urea) present in the environment.

O_2 and CO_2 are easy for plants to obtain from the air, but both compounds are not usually limiting factors in agricultural production. Water and inorganic nutrients are paid the most attention in agriculture since both factors can greatly affect crop growth and yields, with many studies focusing on them (e.g. Vamerali et al., 2003, Bahador et al., 2023, Wise et al., 2024). Plants can continually absorb water and inorganic nutrients from the soil. However, to

http://dx.doi.org/10.19103/AS.2025.0161.02

achieve higher and more sustainable yields, extra water and nutrients are often supplied by farmers to avoid water stress and nutrient deficiency.

In particular, the use of high-yield crop varieties and multiple cropping patterns accelerates exhaustion of soil-available nutrients, which makes fertilization more necessary. Studies also show that some low molecular weight organic matter can be absorbed by plants directly (Näsholm et al., 1998, Jones et al., 2005), which provides a new approach to applying fertilizers. 'Food' for crops can thus be supplied using fertilization and irrigation. With scientific management of water and nutrients, crops can achieve higher and sustainable yields.

Industrial hemp or hemp (*Cannabis sativa* L.) has attracted increasing attention in recent years due to diverse applications in various industries, including textiles, construction materials, food, medicine, fodder, phytoremediation, and energy (Zhao et al., 2021, Mohammadi et al., 2024). These various applications have increased the economic value of hemp and promoted its cultivation. It is estimated that the global industrial hemp market will reach $18.6 billion by 2027, with more than 40 countries actively involved in growing hemp. Therefore, there is a great need for farmers to (1) understand the nutrient and water requirements of hemp and (2) optimize its growth, yield, and quality.

This chapter examines current knowledge on the nutrient and water needs of hemp, highlighting recent advances in research and identifying areas for future investigation. The chapter discusses the primary macronutrients (N, P, K), secondary macronutrients (needed in smaller quantities to boost growth) (Ca, Mg, S), micronutrients (Fe, Zn, Cu, etc.), and the role of water in hemp growth and development. It also explores the interactions between nutrients and water, as well as the impact of environmental factors on hemp's nutrient and water requirements. Finally, the chapter examines the various methods used to assess and manage hemp's nutrient and water status, including soil testing, fertilization strategies, and irrigation techniques. By providing a comprehensive overview of the nutrient and water requirements of hemp, this chapter aims to contribute to the sustainable production and utilization of this valuable crop (Fig. 1).

Common terms used in this chapter are:

- **CBD**: Cannabidiol, one of the main metabolites of hemp;
- **Available nutrients**: The nutrients that plants can absorb from the environment in a relatively short time;
- **Essential nutrient**: An element that is indispensable for the plant and cannot be replaced by others during growth; and
- **Beneficial nutrient**: An element that is not essential for the plant but can improve growth.

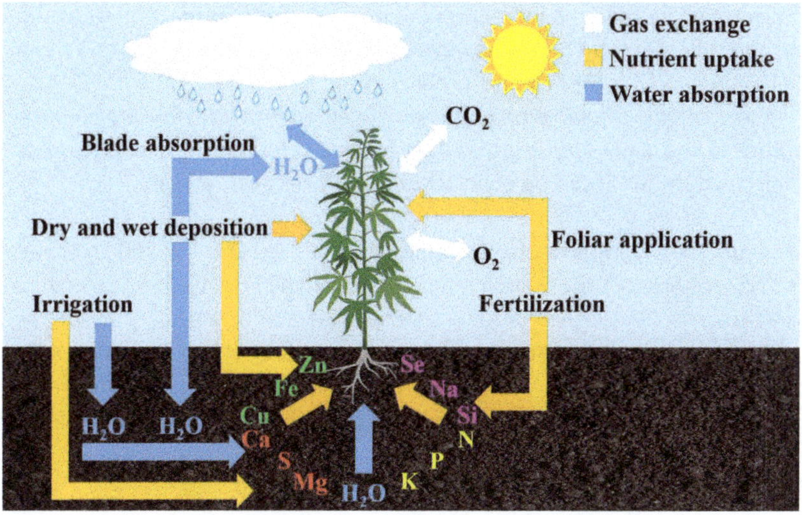

Figure 1 Nutrient types and possible sources for hemp.

2 Nutrient requirements of hemp: an overview

It is generally accepted that there are 17 essential elements for higher plants: C, H, O, N, P, K, S, Ca, Mg, Fe, Mn, Cu, Zn, B, Cl, Ni, and Mo. These elements play important roles in plant growth and reproduction and cannot be replaced by other elements. As soon as some of these elements are deficient, the growth of the plants will be inhibited. Apart from these essential elements, some elements are referred to as beneficial elements, such as Si, Na, Co, Se, and Al. These elements are not essential for growth, but a suitable supply enhances plant growth.

Although hemp is highly adaptable to abiotic stresses such as low soil fertility, moderate salinity, and concentrations of heavy metals (Mastinu et al., 2024), high-yield varieties and intensive cultivation make fertilization necessary, especially for hemp grown on marginal lands (Blandinières and Amaducci, 2022). This is because of a limited soil nutrient budget which is exhausted by continuous nutrient consumption by crops. To obtain higher yield, high quality, and sustainable hemp production of grain, fiber, or inflorescence, a scientific approach to fertilization is needed to meet the nutrient requirement of hemp.

3 Essential macronutrients: nitrogen, phosphorus and potassium

The macronutrients that hemp needs are: C, H, O, N, P, and K. In this chapter we focus on N, P, and K, since hemp mainly obtains C, H, and O from water and the atmosphere. N, P, and K are often treated as the three essential elements of fertilizers. While there have been many studies on the interaction between N, P, K, and different food crops, however, research related to N, P, K, and hemp is still limited, and further research is needed.

Based on the available hemp production literature from the late nineteenth century to the early 1990s, as well as recorded fertilizer application rates in different countries, it can be concluded that (Wylie et al., 2020):

- nitrogen (N) application rates range from 60 kg N ha^{-1} to 200 kg N ha^{-1};
- phosphorus (P) application rates range from 30 kg N ha^{-1} to 120 kg P ha^{-1}; and
- potassium (K) application rates range from 40 kg N ha^{-1} to 200 kg K ha^{-1}.

However, higher fertilizer rates have also been reported (Tang et al., 2022). We will now review N, P, and K for hemp.

3.1 Nitrogen

Nitrogen (N) is one of the most important macronutrients for hemp, playing an essential role in the synthesis of proteins, nucleic acids, chlorophyll, plant hormones, and other important plant compounds (Liu et al., 2023). It is generally thought that N affects hemp yields more than P and K under field conditions (Deng et al., 2019). The N requirement for hemp is relatively high, especially during the vegetative stage. An adequate N supply can promote growth, increase leaf area, and enhance photosynthetic activity. However, excessive N may lead to lush vegetative growth at the expense of flower and seed production. In contrast, an insufficient supply of N can lead to growth suspension, chlorosis, and delay of flowering. Therefore, it is a challenge to optimize the N rate in hemp cultivation.

Sufficient and timely N supply is essential to achieving high yield (Cai et al., 2023). Several studies have investigated the effects of N on hemp growth and yield (stem/fiber, seed, and inflorescence). Based on multisite field experiments in Europe, Tang et al. (2017) found that the stem yield of several dual-purpose hemp varieties at an N rate of 60 kg ha^{-1} increased by 32%, compared with the blank control, but a further increase of N fertilizer supply had an insignificant effect on stem yield. On the other hand, compared with stem yield, seed yield was not significantly affected by increasing N rate.

However, N fertilizer rate should be based on the N fertility of the soil. When the soil is rich in bioavailable N, the response of hemp to N fertilization can be insignificant, which might lead to the incorrect conclusion that hemp does not have a high N requirement (Wylie et al., 2020). Nitrogen management methods should also be based on the desired end product. For fiber hemp, high-stem biomass with good quality fibers is the goal. Based on the literature, stem biomass tends to increase as the N rate increases (Campiglia et al., 2017), but excessive N supply might negatively affect fiber quality (Wylie et al., 2020).

The planting density should also be taken into consideration for N usage since a higher density of hemp means a higher demand for N. One study has shown that a high N rate is necessary for high fiber yield when planting density is high (Werf et al., 1995). However, other researchers disagree (Wylie et al., 2020, Tang et al., 2022). For grain hemp, many studies have investigated the effects of N fertilization on the yield of hemp seed. Based on the literature, the optimal N rate for grain production ranged from 60 kg N ha^{-1} to 200 kg N ha^{-1} (Aubin et al., 2015, Tang et al., 2017). These different conclusions indicate that the response of hemp to N heavily relies on environmental conditions.

Recently, the cultivation of hemp for CBD and other phytochemicals has become popular. Most studies have focused on outdoor, irrigated fields, whereas some have focused on N management via greenhouse hydroponic cultivation. For greenhouse growers, Wei et al. (2023) investigated the effects of nutrient solution composition on hemp cannabinoid yield and found that 8 mmol N L^{-1} in solution is the optimum N level for CBD production via aeroponic cultivation. In a hydroponic experiment, it was found that the total dry weight, stem diameter, and plant height at 1.50 mmol N L^{-1} increased by more than 26.16%, 11.43%, and 4.51%, respectively, compared with that at 0.30 mmol N L^{-1} (Zhu et al., 2021). Table 1 shows some published studies about N fertilization in hemp cultivation (Fig. 2).

3.2 Phosphorus

Phosphorus (P) is an important component in nucleic acids (DNA and RNA), cell membranes, enzymes, and energy-containing molecules (Verlinden et al., 2022). It also plays an indispensable role in hemp growth and yield development (Gamuyao et al., 2012). P deficiency can cause stunted growth

Table 1 Nitrogen application rates and hemp yields in published studies

References	N rate (kg/ha)	Biomass yield (t/ha)	Dry stem yield (t/ha)
Papastylianou et al. (2018)	120	92.6[a]	6.36
	180	100[a]	7.41
	240	110[a]	9.54
Wogiatzi et al. (2019)	184	12.7~15.4	10.7~13.0
	244	12.2~12.8	9.99~10.6
Tang et al. (2022)	300	12.3	8.55
	600	12.7	8.95
	900	13.3	10.5
Giannoulis et al. (2024)	140	11.4	-

Note: [a] Fresh biomass.

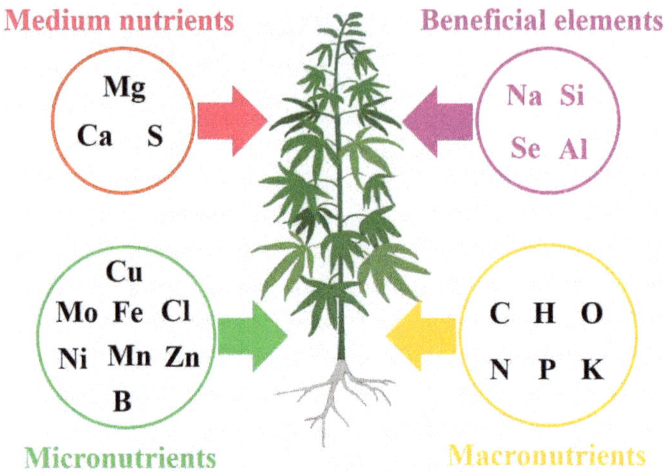

Figure 2 The nutrient elements required by hemp.

and eventually reduce hemp yield (Shiponi and Bernstein, 2021). However, similarly to N, there isn't much literature on the effects of phosphorus (P) on hemp. Islam et al. (2023) conducted a study to investigate the effects of P rates on growth, morpho-physiological adaptations, and nutrient uptake of hemp and concluded that the effects of P on hemp rely significantly on the particular variety.

Compared to N, the requirement of P for fiber hemp is less important. Jordan et al. (1946) found that 30 kg P ha^{-1} was sufficient for fiber hemp. However, hemp cultivated for grains is thought to need more P, which is related to the accumulation of P in hemp seeds. To date, researchers have not drawn any consistent conclusions about P rate and hemp grain yield. For phytochemical hemp, Wei et al. (2023) suggest that 1 mmol P L^{-1} in solution is adequate for CBD production via aeroponic cultivation. Table 2 shows some published studies about K fertilization in hemp cultivation.

Table 2 Phosphorous fertilizer application and hemp yield in published studies

References	P rate (kg/ha)	Dry stem yield (t/ha)	Fiber yield (t/ha)	Seed yield (kg/ha)
Vera et al. (2010)	20~80	-	-	865~1028
Deng et al. (2019)	30~150	-	1.70~3.29[a]	-
Wogiatzi et al. (2019)	60	9.99~13.0	-	-
Tang et al. (2022)	300	8.55	2.43	-
	600	8.95	2.44	-
	900	10.5	2.57	-

3.3 Potassium

Potassium (K) plays important roles in maintaining cell turgor, regulating water balance, and enhancing plant resistance to stress (Wang et al., 2013, De Prato et al., 2022, Liu et al., 2023). Hemp requires a relatively high amount of K, especially during the flowering and fruiting stages. K deficiency can lead to weak stems, yellowing of leaves, and reduced yield. However, compared with N, the K requirements of hemp are not as well defined. Finnan and Burke (2013) found that there was no significant relationship between hemp yield and either K rate or soil K concentration. However, there are guidelines that exist for K fertilization of hemp, with K rates recommended between 140 kg K ha^{-1} and 230 kg K ha^{-1} according to soil properties (Finnan and Burke, 2013). Table 3 shows some published studies about K fertilization in hemp cultivation.

4 Secondary/beneficial macronutrients, micronutrients and other nutrients

4.1 Secondary or beneficial macronutrients

Secondary or beneficial macronutrients usually refer to calcium (Ca), magnesium (Mg), and sulfur (S) which play an important role in the growth and development of plants (Valikhanova, 1968, Monshausen, 2012, Bouranis et al., 2020). The demand for these elements of a crop is less than N, P, and K but is greater than for trace elements.

Ca is an important component of plant cell walls, which can stabilize cell membrane structure and participate in physiological processes such as cell division and elongation (Wang et al., 2024). It is important for root development of plants, resistance to pests and diseases, and prevention of fruit softening.

Mg is the central atom of the chlorophyll molecule and is essential for photosynthesis. It also participates in the activation of many enzymes in the plant body, affecting the physiological functions of plants such as respiration and carbohydrate metabolism (Zhu et al., 2025).

Table 3 Potassium fertilizer application and hemp yield in published studies

References	K rate (kg/ha)	Dry stem yield (t/ha)	Fiber yield (t/ha)	Seed yield (kg/ha)
Iványi and Izsáki (2009)	290~490	9.45~12.51	-	-
Deng et al. (2019)	75~225	-	1.70~3.29	-
Wogiatzi et al. (2019)	60	9.99~13.0	-	-
Tang et al. (2022)	300	8.55	2.43	-
	600	8.95	2.44	-
	900	10.5	2.57	-

S is a component of proteins and some bioactive substances. It participates in the redox reaction in plants and plays a role in the regulation of plant growth and the formation of flavor substances (Ausma et al., 2025).

To date, few studies have focused on hemp requirements for Ca, Mg, and S. Due to the high values placed on nutrients supplied by hemp grains, the need for research on Ca and Mg is more important than ever (Alonso-Esteban et al., 2022). More studies are needed on the response of hemp to these nutrients to optimize hemp production.

4.2 Micronutrients

Micronutrients refer to a class of elements with extremely low content in plants but which play an indispensable role in plant growth and development. Commonly known trace elements closely related to plant growth include iron (Fe), copper (Cu), and boron (B), among others.

Iron is a component of many important enzymes in plants and participates in electron transfer in important physiological processes such as photosynthesis and respiration (Xu et al., 2025). It is crucial for the synthesis of chlorophyll. If plants lack Fe, the synthesis of chlorophyll will be blocked, and leaf chlorosis will occur.

Cu is a component of some redox enzymes in plants and participates in electron transfer in processes such as respiration and photosynthesis. It also has an effect on the lignification of cell walls (Wang et al., 2025).

B mainly participates in the reproductive growth process of plants and has important influences on pollen germination, pollen tube elongation, fruit and seed development, etc. (Liu et al., 2024). At the same time, B also plays a role in the division and elongation of plant cells.

There are a few studies on micronutrients requirements of hemp. After searching the literature, we found the nutrient value for Fe supplementation in hemp seed (Smułek and Jarzębski, 2023). However, most of the studies about the other micronutrients and hemp, such as Zn, Cu, and Mn, are related to phytoremediation rather than fertilization (Guo et al., 2024).

4.3 Other nutrients

Apart from the essential elements, beneficial elements also play a positive role in promoting hemp growth and development, yield, and quality when present in appropriate amounts, such as silicon (Si). Si is an element abundant in the Earth's crust. Many plants can absorb Si and deposit it in parts such as cell walls. Si can enhance the mechanical strength of plants; improve the resistance of plants to biotic and abiotic stresses such as pests, lodging, drought, and high temperature; and affect physiological processes such as photosynthesis and nutrient absorption of plants (Yamaji et al., 2024). To date, few studies have focused on the requirement of beneficial elements of hemp.

5 Water requirements of hemp

5.1 Water uptake and transport in hemp plants

Water is an important component of plant cells, and generally the water content of the protoplast of plant cells can reach 80–90%. Thus, water can significantly affect plant survival and growth (Feng et al., 2024). Hemp plants absorb water through their roots and transport it to various parts of the plant through the xylem. The rate of water uptake and transport in hemp plants is influenced by various factors, including soil moisture content, root morphology, and environmental conditions. Adequate soil moisture is crucial for ensuring optimal water uptake by hemp roots.

5.2 Effects of water stress on hemp growth and development

Compared with cotton, hemp is considered a water-efficient crop with a lower water requirement (Duque Schumacher et al., 2020). By analyzing published studies, Wise et al. (2023) found that hemp has a 38% lower crop water requirement and 84% lower crop irrigation requirement compared to cotton. However, water stress can have a significant impact on hemp growth and development. Mild water stress can lead to reduced growth, wilting of leaves, and decreased photosynthetic activity. Severe water stress can result in plant death.

During the seedling stage, sufficient moisture is key to the healthy growth of hemp seedlings. After seed germination, seedlings need to absorb enough moisture to activate the internal enzyme system so that the nutrients in the endosperm or cotyledons can be decomposed and utilized, thereby starting the normal growth process. At this time, the soil should be kept moderately moist but not waterlogged in order to avoid causing root hypoxia or rot.

Industrial hemp grows rapidly during the vegetative growth stage and has a large demand for moisture. Sufficient moisture can support the growth and development of organs such as stems and leaves, making the stems stronger, the leaves larger, and the root system more developed. If there is a lack of water, it will lead to thinner stems, smaller and thinner leaves, poor root development, and even stunted plants.

During the flowering and fruiting stage, the supply of water is also crucial. Lack of water may lead to difficulties in pollen germination, poor fruit development, and other issues, affecting the yield and quality of industrial hemp. However, if there is too much water, it may also cause problems such as flower and fruit drop, so a moderate supply of water should be maintained.

Research has shown that hemp plants can adapt to water stress to some extent by adjusting their physiological and morphological characteristics. For example, plants may develop deeper roots to access water from deeper soil layers or reduce leaf area to conserve water (Gill et al., 2022, Morgan et al., 2023).

5.3 Irrigation strategies for hemp cultivation

By analyzing 27 prior publications, Wise et al. (2023) found that the crop water requirement (CWR) – total liters of water required per area of hemp growth – was 5024 m^3/ha, while the CWR of cotton was 8050 m^3/ha. They also found that the crop irrigation requirement (CIR) – the volume of irrigated water required per area of hemp crop growth – was 761 m^3/ha, while the CIR of cotton was 4840 m^3/ha. Through a field experiment, Wogiatzi et al. (2019) found that, when irrigation was equal to local evapotranspiration (500 mm), the dry stem yield of hemp varied between 9.69 t/ha and 13.0 t/ha. However, when irrigation was equal to 60% of local evapotranspiration (300 mm), the dry stem yield of hemp varied between 10.6 t/ha and 10.7 t/ha.

Proper irrigation is essential for maximizing hemp yield and quality (Wogiatzi et al., 2019). Different irrigation methods, such as drip, sprinkler, and flood irrigation, can be used depending on the soil type, climate, and available resources. Drip irrigation is often considered the most efficient irrigation method for hemp cultivation, as it delivers water directly to the root zone, minimizing water loss through evaporation and runoff (Giannoulis et al., 2024). Sprinkler irrigation can be used in areas with limited water resources or where drip irrigation is not feasible. Flood irrigation is less commonly used due to its high water consumption and potential for soil erosion. The timing and amount of irrigation should be carefully controlled to avoid over- or under-irrigation. Soil moisture sensors and weather data can be used to determine the optimal irrigation schedule.

6 Interactions between nutrients and water in hemp

6.1 Nutrient uptake and transport under different water conditions

The availability and uptake of nutrients by hemp plants can be influenced by soil moisture content (Li et al., 2009). Adequate soil moisture is essential for ensuring the solubility and mobility of nutrients in the soil, facilitating their uptake by plant roots.

Water stress in soil can significantly reduce the bioavailability of most nutrients such as P, K, Ca, and Mg. Water stress can also affect root growth and function, as well as the activity of nutrient transporters in plant roots. As a result, uptake of nutrients will be limited and nutrient deficiency symptoms will occur. On the other hand, excessive soil water content can lead to leaching of nutrients from the soil, reducing their availability to plants and restraining hemp growth. Therefore, maintaining an appropriate balance between water and nutrients is crucial for optimal hemp growth.

6.2 Effects of nutrient deficiencies and excesses on water use efficiency

Nutrient deficiencies and excesses can also affect water use efficiency of hemp. For example, N deficiency can lead to reduced photosynthetic activity and increased water use efficiency, while excessive N can result in lush vegetative growth and decreased water use efficiency. Similarly, K deficiency can reduce plant resistance to stress and increase water loss through transpiration, while excessive K can interfere with the uptake of other nutrients and affect plant growth and development. Understanding the relationship between nutrients and water use efficiency can help farmers optimize fertilization and irrigation practices to improve productivity and sustainability.

7 Methods for assessing and managing hemp nutrient and water status

7.1 Soil testing

Soils in different areas have different levels of fertility. It is important to measure the soil nutrient status of a hemp field prior to planting in planning a fertilization program. After learning about soil nutrient status, we can apply fertilization based on soil nutrient concentration data. In this way, nutrient deficiency can be avoided, and the nutrient-utilization efficiency can also be improved. Finally, both the planting cost and the environmental risk can be reduced. Similarly, based on the soil test results, appropriate irrigation strategies can be developed to meet the specific water requirements of hemp plants.

7.2 Fertilization strategies

Appropriate fertilization is essential for providing hemp plants with the necessary nutrients for optimal growth. Fertilizers can be applied in different forms, such as organic fertilizers, inorganic fertilizers, and slow-release fertilizers. The choice of fertilizers should be based on the soil type, nutrient requirements of hemp plants, and environmental considerations. Organic fertilizers can improve soil fertility and water-holding capacity, while inorganic fertilizers can provide quick-release nutrients. Slow-release fertilizers, which can be either organic or inorganic, can release nutrients gradually over time, reducing the risk of nutrient leaching and environmental pollution.

Fertilization rates should be determined based on soil test results, planting density, hemp variety, and the growth stage of hemp plants. Over-fertilization can lead to nutrient imbalances, environmental pollution, and reduced plant quality, while under-fertilization can result in stunted growth and low yield.

7.3 Irrigation management

Proper irrigation management is crucial for ensuring optimal water supply to hemp plants. Irrigation methods should be selected based on the soil type, climate, and available resources. Drip irrigation and other efficient irrigation methods can be used to minimize water loss and improve water use efficiency. The timing and amount of irrigation should be determined based on soil moisture, weather data, and the growth stage of hemp plants. Over-irrigation can lead to waterlogging, nutrient leaching, and increased disease incidence, while under-irrigation can result in water stress and reduced yield.

8 Current challenges and future trends

8.1 Synchronous supply of water and nutrients

Global crop yield variability is heavily affected by fertilization, irrigation, and climate (Mueller et al., 2012). The absorption of water and nutrients in soil or any other substrates goes on simultaneously during the growth of the crops. Proper water content can promote the bioavailability of nutrients and then enhance the growth of crops. Further studies are needed on the synchronous supply of both water and nutrients in hemp cultivation.

8.2 Understanding the genetic basis of nutrient and water use efficiency in hemp

Research on the genetic basis of nutrient and water use efficiency in hemp can help identify genes and molecular markers associated with these traits. This knowledge can be used to develop improved hemp varieties with enhanced nutrient and water use efficiency through breeding and genetic engineering.

8.3 Investigating the interaction between hemp and beneficial soil microorganisms

Beneficial soil microorganisms, such as mycorrhizal fungi and nitrogen-fixing bacteria, can play an important role in improving hemp's nutrient and water uptake. Further research is needed to understand the interaction between hemp and these microorganisms and develop strategies to enhance their beneficial effects.

8.4 Developing sustainable fertilization and irrigation practices for hemp cultivation

As the demand for hemp continues to grow, there is a need to develop sustainable fertilization and irrigation practices that minimize environmental impacts and ensure long-term productivity. This includes exploring the use of organic fertilizers, precision fertilization techniques, and water-saving irrigation methods.

8.5 Studying the impact of climate change on hemp's nutrient and water requirements

Climate change is expected to have a significant impact on hemp cultivation by altering temperature, precipitation patterns, and soil moisture conditions. Future research should focus on understanding how climate change will affect hemp's nutrient and water requirements and developing adaptation strategies to ensure sustainable production.

9 Conclusion

Understanding the nutrient and water requirements of hemp is essential for optimizing its growth, yield, and quality. This chapter has provided a comprehensive overview of the current knowledge on the nutrient and water needs of hemp, highlighting recent research advances and identifying areas for future investigation. Adequate supply of nutrients and water is crucial for ensuring healthy plant growth, maximizing productivity, and minimizing environmental impacts. By implementing appropriate fertilization and irrigation strategies based on soil testing and other assessment methods, farmers can improve the sustainability and profitability of hemp cultivation. Future research should focus on understanding the genetic basis of nutrient and water use efficiency, investigating the interaction between hemp and beneficial soil microorganisms, developing sustainable fertilization and irrigation practices, and studying the impact of climate change on hemp's nutrient and water requirements. This will contribute to the continued growth and development of the hemp industry and the sustainable use of this valuable crop.

10 Acknowledgments

I sincerely appreciate several researchers who offered their help in providing me with information about hemp cultivation and giving me advice in preparing this book chapter. The writing of this chapter was supported by the Agricultural Science and Technology Innovation Program [ASTIP-IBFC], the Key Research and Development Program Project of Hechi City [Heke AB240708], the Key Research and Development Program of Heilongjiang Province [2024ZXJ02B21-01], and the Central Public-Interest Scientific Institution Basal Research Fund [1610242024010].

11 Where to look for further information

More references to research and development on hemp's agronomy, ecophysiology, opportunities, and challenges of hemp cultivation in different regions of the world: Agrawalet al (2022) "Cannabis/hemp for sustainable agriculture and materials" https://link.springer.com/book/10.1007/978-981-16-8778-5.

For fertilization and weed management for hemp: Kousta et al (2023) "Effect of fertilization and weedmanagement practices on weed diversity and hemp agronomic performance" in *Agronomy-Basel*.

For hemp cultivation inlow-fertility soil: Greiner et al (2025) "Fiberhemp biomass yield and quality on shallow stony soil in southwest Germany" in *Land*.

12 References

Alonso-Esteban, J. I., et al. (2022), Mineral elements and related antinutrients, in whole and hulled hemp (*Cannabis sativa* L.) seeds, *Journal of Food Composition and Analysis*, 109, 104516.

Aubin, M. P., et al. (2015), Industrial hemp response to nitrogen, phosphorus, and potassium fertilization, *Crop, Forage & Turfgrass Management*, 1 (1), 1–10.

Ausma, T., et al. (2025), Regulation of sulfur metabolism in seedlings of the C_4 plant maize upon sulfate deprivation and atmospheric H_2S exposure, *Environmental and Experimental Botany*, 232, 106121.

Bahador, M., et al. (2023), Radiation and water use efficiencies of mycorrhizal inoculated hemp under water-deficit stress, *Journal of Soil Science and Plant Nutrition*, 23 (2), 2202–2214.

Blandinières, H. and Amaducci, S. (2022), Adapting the cultivation of industrial hemp (*Cannabis sativa* L.) to marginal lands: a review, *GCB Bioenergy*, 14 (9), 1004–1022.

Bouranis, D. L., et al. (2020), Advances in plant sulfur research, *Plants*, 9 (2), 256.

Cai, S., et al. (2023), Optimal nitrogen rate strategy for sustainable rice production in China, *Nature*, 615, 73–79.

Campiglia, E., et al. (2017), Plant density and nitrogen fertilization affect agronomic performance of industrial hemp (*Cannabis sativa* L.) in Mediterranean environment, *Industrial Crops and Products*, 100, 246–254.

Deng, G., et al. (2019), Planting density and fertilization evidently influence the fiber yield of hemp (*Cannabis sativa* L.), *Agronomy*, 9 (7), 368.

De Prato, L., et al. (2022), Morpho-physiology and cannabinoid concentrations of hemp (*Cannabis sativa* L.) are affected by potassium fertilisers and microbes under tropical conditions, *Industrial Crops and Products*, 182, 114907.

Duque Schumacher, A. G., et al. (2020), Industrial hemp fiber: a sustainable and economical alternative to cotton, *Journal of Cleaner Production*, 268, 122180.

Feng, M., et al. (2024), Variation in water uptake pattern of the constructive species Pinus koraiensis: Evidence from water stable isotopes, *Journal of Hydrology*, 636, 131224.

Finnan, J. and Burke, B. (2013), Potassium fertilization of hemp (*Cannabis sativa*), *Industrial Crops and Products*, 41, 419–422.

Gamuyao, R., et al. (2012), The protein kinase Pstol1 from traditional rice confers tolerance of phosphorus deficiency, *Nature*, 488 (7412), 535–539.

Giannoulis, K. D., et al. (2024), Cultivation practices affect biomass yield and quality of "Felina 32", an industrial hemp variety, *Agronomy*, 14 (11), 2743.

Gill, A. R., et al. (2022), Physiological and morphological responses of industrial hemp (*Cannabis sativa* L.) to water deficit, *Industrial Crops and Products*, 187, 115331.

Guo, Y., et al. (2024), Industrial hemp (*Cannabis sativa* L.) can utilize and remediate soil strongly contaminated with Cu, As, Cd, and Pb by phytoattenuation, *Chemosphere*, 358, 142199.

Islam, M. M., et al. (2023), Phosphorus fertilisation differentially influences growth, morpho-physiological adaptations and nutrient uptake of industrial hemp (*Cannabis sativa* L.), *Plant and Soil*, 492 (1–2), 301–314.

Iványi, I. and Izsáki, Z. (2009), Effect of nitrogen, phosphorus, and potassium fertilization on nutrional status of fiber hemp, *Communications in Soil Science and Plant Analysis*, 40 (1–6), 974–986.

Jones, D. L., et al. (2005), Dissolved organic nitrogen uptake by plants–an important N uptake pathway?, *Soil Biology and Biochemistry*, 37 (3), 413–423.

Jordan, H. V., et al. (1946), Effects of fertilizers on yields and breaking strengths of american hemp, *Cannabis Sativa*, *Agronomy Journal*, 38 (6), 551–562.

Li, S.-X., et al. (2009), *Chapter 7 Nutrient and water management effects on crop production, and nutrient and water use efficiency in dryland areas of China, Academic Press*. Pittsburgh. pp. 223–265.

Liu, B., et al. (2023), The anion channel SLAH3 interacts with potassium channels to regulate nitrogen-potassium homeostasis and the membrane potential in Arabidopsis, *Plant Cell*, 35 (4), 1259–1280.

Liu, W., et al. (2024), *BnaC4.BOR2* mediates boron uptake and translocation in *Brassica napus* under boron deficiency, *Plant, Cell & Environment*, 47 (10), 3732–3748.

Mastinu, A., et al. (2024), Hemp cultivation opportunities for marginal lands development, *PLoS One*, 19 (3), e0299981.

Mohammadi, M., et al. (2024), Recent progress in natural fiber reinforced composite as sound absorber material, *Journal of Building Engineering*, 84, 108514.

Monshausen, G. B. (2012), Visualizing Ca^{2+} signatures in plants, *Current Opinion in Plant Biology*, 15 (6), 677–682.

Morgan, W., et al. (2023), Identifying physiological traits related with drought tolerance and water-use efficiency in floral hemp (*Cannabis sativa* L.), *Crop Science*, 64 (1), 354–372.

Mueller, N. D., et al. (2012), Closing yield gaps through nutrient and water management, *Nature*, 490 (7419), 254–257.

Näsholm, T., et al. (1998), Boreal forest plants take up organic nitrogen, *Nature*, 392, 914–916.

Papastylianou, P., et al. (2018), Effect of nitrogen fertilization on growth and yield of industrial hemp (*Cannabis sativa* L.), *Notulae Botanicae Horti Agrobotanici Cluj-Napoca*, 46 (1), 197–201.

Shiponi, S. and Bernstein, N. (2021), Response of medical cannabis (*Cannabis sativa* L.) genotypes to P supply under long photoperiod: functional phenotyping and the ionome, *Industrial Crops and Products*, 161, 113154.

Smułek, W. and Jarzębski, M. (2023), Hemp seed oil nanoemulsion with *Sapindus* saponins as a potential carrier for iron supplement and vitamin D, *Reviews on Advanced Materials Science*, 62 (1), 20220317.

Tang, K., et al. (2017), A comprehensive study of planting density and nitrogen fertilization effect on dual-purpose hemp (*Cannabis sativa* L.) cultivation, *Industrial Crops and Products*, 107, 427–438.

Tang, K., et al. (2022), Fiber hemp (*Cannabis sativa* L.) yield and its response to fertilization and planting density in China, *Industrial Crops and Products*, 177, 114542.

Valikhanova, G. Z. (1968), Physiological role of Mg in plants, *Agrokhimiya*, 12, 121–129.

Vamerali, T., et al. (2003), A comparison of root characteristics in relation to nutrient and water stress in two maize hybrids, *Plant and Soil*, 255, 157–167.

Vera, C. L., et al. (2010), N, P, and S fertilization effects on industrial hemp in Saskatchewan, *Canadian Journal of Plant Science*, 90:179-184.

Verlinden, M. S., et al. (2022), Phosphorus stress strongly reduced plant physiological activity, but only temporarily, in a mesocosm experiment with *Zea mays* colonized by arbuscular mycorrhizal fungi, *Biogeosciences*, 19 (9), 2353–2364.

Wang, C., et al. (2024), Mechanisms of calcium homeostasis orchestrate plant growth and immunity, *Nature*, 627 (8003), 382–388.

Wang, M., et al. (2013), The critical role of potassium in plant stress response, *International Journal of Molecular Sciences*, 14 (4), 7370–7390.

Wang, Y., et al. (2025), Advances of the mechanism for copper tolerance in plants, *Plant Science*, 350, 112299.

Wei, X., et al. (2023), Effects of different N, P, and K rates on the growth and cannabinoid content of industrial hemp, *Journal of Natural Fibers*, 20 (1), 2159605.

Werf, H. M. G., et al. (1995), Nitrogen fertilization and row width affect self-thinning and productivity of fibre hemp (*Cannabis sativa* L.), *Field Crops Research*, 42 (1), 27–37.

Wise, K., et al. (2023), Comparative study of water requirements and water footprints of fibre crops hemp (*Cannabis sativa*) and cotton (*Gossypium hirsutum* L.), *Journal of Agrometeorology*, 25 (3), 392–396.

Wise, K., et al. (2024), Fertiliser supplementation with a biostimulant complex of fish hydrolysate, Aloe vera extract, and kelp alters cannabis root architecture to enhance nutrient uptake, *Scientia Horticulturae*, 323, 112483.

Wogiatzi, E., et al. (2019), Effect of irrigation and fertilization levels on mineral composition of *Cannabis sativa* L. leaves, *Notulae Botanicae Horti Agrobotanici Cluj-Napoca*, 47 (4), 1073–1080.

Wylie, S. E., et al. (2020), Fertility management for industrial hemp production: current knowledge and future research needs, *GCB Bioenergy*, 13 (4), 517–524.

Xu, L., et al. (2025), Iron deficiency and toxicity trigger divergent metabolic responses and adaptive plasticity in Ulmus pumila: Insights from integrated transcriptomic and metabolomic analyses, *Plant Physiology and Biochemistry*, 221, 109601.

Yamaji, N., et al. (2024), Shoot-Silicon-Signal protein to regulate root silicon uptake in rice, *Nature Communications*, 15, 10712.

Zhao, X., et al. (2021), Industrial hemp–an old but versatile bast fiber crop, *Journal of Natural Fibers*, 19 (13), 6269–6282

Zhu, R., et al. (2021), Morpho-physiological traits, antioxidant capacity and nutrient accumulation in hemp (*Cannabis sativa* L.) under varying levels of nitrogen nutrition, *Journal of Plant Nutrition*, 45 (6), 854–865.

Zhu, Z., et al. (2025), Comprehensive analysis of the effects on photosynthesis and energy balance in tomato leaves under magnesium deficiency, *Plant Physiology and Biochemistry*, 222, 109671.

Chapter 3

Advances in hemp breeding

Stephen Baluch, Oregon State University, USA

1 Introduction

Due to recently becoming legal to grow again in a number of countries, hemp (*Cannabis sativa* L.) has had a resurgence of interest both in cultivation and research. Industrial hemp has multiple uses, the main categories being fiber, grain, and cannabinoid oil. However, due to the previous illegality of the crop, hemp has not benefited from the Green Revolution innovations like other crops have. As the crop's cultivation increases in acreage, biotic stresses such as insects and pathogens will likely (re)emerge to challenge defense mechanisms within *Cannabis sativa*. In addition, the yields of grain, fiber, and cannabinoids need to rise to maximize profitability. All of these make excellent targets for breeding programs, both public and private.

Within this chapter, we will examine these targets and the strategies to achieve the goals of a breeding program. After defining some common terms, the chapter focuses initially on the challenging but promising potential use of marijuana lines to develop new hemp varieties. It then looks at the range of breeding techniques, from basic use of clones to advances in hybrid breeding.

http://dx.doi.org/10.19103/AS.2025.0161.04

The chapter also discusses issues in seed and grain production before concluding by looking at ways advances in breeding can improve resistance to biotic and other stresses.

2 Common terms used in this chapter

Cultivar: A compound word created from cultivated and variety denoting a seed-derived group of plants exhibiting certain characteristics and available to farmers on the open market.

Variety: Synonymous with cultivar, a genetically distinct, seed-derived group of plants grown by farmers for the open market.

Hybrid: A seed-derived variety formed from the mating of two genetically distant inbred lines usually exhibiting a performance advantage over the parental lines.

Genotype: The collective genetic make-up of an organism.

Phenotype: The observable expression of genes molded by gene expression, inter-gene action and the environment.

Population: A group of individuals sharing common ancestry.

Selection: The act of choosing certain individuals to constitute the next generation.

Selection pressure: The intensity of selection used to cull individuals from a population. It can be applied naturally by allowing the environmental stresses to select the individuals, or artificially where the breeder determines what and how many individuals will constitute the next generation.

3 Setting the scene: issues in defining cannabis and hemp

Unfortunately, the legal definition of hemp can be debated on scientific grounds. In a publication from 1976, Small and Cronquist note that the definition of <0.3% D^9-THC (THC) is arbitrary and can be discerned from young, fresh leaves (Small and Cronquist, 1976). This appears to be the source of the definition adopted by governments around the world. Some researchers have suggested a threshold of 1% would be more appropriate and still be below the threshold of impairment (Sliwoski, 2022). The author is not aware of any type III chemotype variety that will not exceed the 0.3% threshold if left to grow to full maturity. This is a direct result of the leaky nature of cannabidiolic acid synthase (CBDAS) (Zirpel et al., 2018), where the enzyme catalyzes cannabigerolic acid into THC acid (THCA) at an unpredictable level.

This characteristic creates a serious risk in growing the crop and maintaining legal THC levels, as well as difficulty in breeding varieties that will lead to a legal THC level. Within every step of the breeding program, it is important to test and

Published by Burleigh Dodds Science Publishing Limited, 2026.

select the plants with the lowest levels of THC possible while still delivering a level of cannabidiol (CBD) production for the crop to be profitable if one is breeding for the cannabinoid market.

Cannabis in general exhibits different chemical make ups, or chemotypes, based on the alleles at two tightly linked co-dominant loci (Grassa et al., 2018). Consequently, varying alleles at these loci give rise to five different types of cannabis:

- Type I is THC dominant with very little CBD produced if any at all detected.
- Type II is a mixed ratio chemotype consisting of both THC and CBD with CBD typically produced at higher levels than THC.
- Type III is CBD dominant with some THC.
- Type IV is CBG dominant and makes very little THC and CBD.
- Type V produces no cannabinoids.

Within the legal definition, only types III, IV, and V are considered hemp.

It is important to consider the underlying genetics of the germplasm you are working with to anticipate the expected chemotype of the crosses you are making. Originally, a single locus known as the B locus was hypothesized with two co-dominant alleles for THCA and CBDA synthases with a less-often observed null allele leading to the CBG-dominant type IV (de Meijer et al., 2003). Later research has shown that the chemotype is caused by two separate loci tightly linked on chromosome seven (Grassa et al., 2018). The two synthases are so close on the chromosome that they can be considered a single locus with two co-dominant alleles with the active synthase alleles in repulsion and is still referred to as the B locus.

An interesting case study regarding the B locus is evident in the prevalence of the B_D allele in farmers' landrace germplasm in traditional hash-producing regions. Although the desired product of hashish is a psychoactive one, farmers saving their seed from season to season cannot apply selection pressure on the B locus due to a lack of equipment for testing (de Meijer, 1999). The resulting populations grown for hash production contain the active CBDAS expressed in some members of these populations. Farmers save the seed for planting next season is selected mainly based on the yield. These populations would be very interesting to study the potentially multiple different active CBDAS encoding alleles for the development of new hemp varieties. These alleles could encode synthases with different levels of production of THCA, some of which could produce CBD/THC ratios advantageous to breeders and farmers alike.

Looking more broadly, hemp and marijuana have a strained relationship. Cannabinoid hemp is typically feminized and grown very similarly to marijuana with wide spacing to allow for growth. However, the situation is very different when you consider grain or fiber hemp which are either monoecious or dioecious with males intentionally left in the field and planted at a very high density. Pollen from these sources would most certainly destroy a marijuana producers' crop, and it is important for all in the industry to be good neighbors. This growing animosity is seen greatest in Humboldt County, California, where hemp production is strictly prohibited (Hemp Industry Daily Staff, 2021). Humboldt County is part of the famous Emerald Triangle where some of the country's highest quality marijuana is grown. The vote among the county board of supervisors unanimously supported the ban in a 5-0 vote. This situation is likely to get more complicated as more grain and fiber hemp producers come online.

4 Use of marijuana lines to develop new hemp varieties

Depending on jurisdiction and appropriate licensing, marijuana can be a great source of new variation in the development of new hemp varieties. Lines derived from these crosses can have quite different types of terpenes than are normally found in CBD-type hemp. Further, CBD types and those derived from marijuana have flowers that are much denser than what is normally observed in current hemp germplasm. These kinds of flower structures could be quite beneficial for grain-type hemp with a hypothesized reduction in shattering. Shattering has been an observed detrimental trait for both fiber and grain-type hemp, leading to significant yield losses. These losses affect both seed production for CBD types and fiber and grain varieties in general and has apparently not had much breeding focus to reduce or eliminate that phenotype. Currently, these crosses have been made and observed to have a reduced but not eliminated shattering in the Oregon State University breeding program.

While there are benefits to using marijuana varieties in hemp breeding, there are also some issues. A pitfall of using marijuana types is they were often bred for indoor cultivation. This characteristic can be detrimental to the final product since these plants and their progeny have not experienced outdoor cultivation to eliminate harmful traits such as fungal, bacterial, and general stress susceptibility. Repeated backcrossing could potentially be used to transfer the advantageous terpene and shattering traits while reintroducing the relatively robust nature of outdoor-bred hemp varieties.

Another potential pitfall is that marijuana varieties have been repeatedly crossed to other varieties with little attention to fixing some traits in the lines, leading to a high degree of heterozygosity in those populations (Sawler et al., 2015). This aspect could make transferring traits a little more difficult than if the lines had been further bred to reduce this variation. If the gene of interest is in the heterozygous state, the number of crosses required and plants tested will have to increase to account for the segregation of the trait. Moreover, the identification of these genes becomes more complicated in a heterozygous state.

4.1 Reproduction and sex manipulation

Cannabis is generally a dioecious, cross-pollinated species. However, some populations show varying degrees of monoecy and both dioecious (females) and monoecious individuals can readily self-pollinate. Self-pollination in dioecious populations requires the application of masculinization chemicals or the application of specific stress to cause male flowers to form on female individuals. These gynoecious matings lead to 'feminized' seed, where every individual from the cross will be genetically XX female. In general, there are five methods to produce feminized seed:

1. Gibberellic acid;
2. Daylength manipulation;
3. Colloidal silver;
4. Silver nitrate; and
5. Silver thiosulfate.

Each method has strengths and weaknesses. Gibberellic acid, although effective, leads to very long internodes and strange growth habits. Colloidal silver is easy to produce and highly effective, but requires application every day until male flower formation. Silver nitrate is an improvement over colloidal silver, but does not persist for great lengths of time.

Daylength manipulation is only applicable to indoor or light-depravation greenhouses but can be highly effective. For instance, after 2 weeks of short-day conditions, the light schedule is changed to 24 hours of light with no dark period. After 1 week of this regimen, the schedule is changed back to short-day conditions, and the plants are allowed to flower. Male flowers will form on these plants at relatively high frequency (depending on genetic propensity). However, not all individuals are susceptible to this manipulation, and this causes a significant level of stress, leading to potentially other undesired changes in physiology.

The gold standard is silver thiosulfate, which is produced from a reaction of silver nitrate with sodium thiosulfate. Silver thiosulfate is by far the most effective method; however, it must be made fresh each time it is to be used since the solution is unstable even under refrigerated conditions (Fig. 1).

Figure 1 An XX female induced by silver thiosulphate to produce male flowers. Note the existence of female flowers exhibiting incomplete transformation.

4.2 Pollination control

As with breeding any crop, maintaining pollen control and strict line purity is extremely important. With cannabis, this is quite difficult since dioecious cannabis is obligately outcrossing (Punja et al., 2017). Spatially separated males and females, which could be separate individuals or clones from the same mother plant, is one method of control. In this case, the males and females could be separated by being on different properties or through controlled rooms or greenhouses with positive pressure to exclude other sources of pollen. We have employed the use of a laboratory vacuum pump (developed by Irina Orlova of Scotts MiracleGro) fitted with a hose that accepts a 1 mL filter pipette to suck pollen from selected males (Fig. 2a and b). These labeled tips are then transported to the females and the end is cut off so that a paintbrush can be inserted to apply the pollen. The second method is to separate the individuals

Figure 2 Use of a vacuum pump to isolate pollen from masculinized XX plants.

temporally and use a method to store pollen. There are numerous protocols online; however, the general protocol is to dry and chill the pollen and store it in a refrigerator or a freezer if the moisture level is dropped low enough. A third method is useful with monoecious individuals and involves using 10 mm mesh pollinating bags from Midco (https://midcoglobal.com/product/cannalock-pollination-bag/) covering the whole plant. Utilizing this method allows for the self-pollination of many individuals in close proximity. This can be done within a controlled environment or in the field if the bags can be adequately secured against wind.

4.3 Growth in target environment

Two of the most important considerations for any plant breeding program for any trait of interest are selection environment and available variation. Nursery location is a significant consideration because the trait may not be expressed or adaptations to the local environment could alter the plant breeder's behavior. Breeding a crop for outdoor field production will certainly yield different results if each year during the winter there would be an indoor nursery. Time to homogeneity may be decreased but other attributes, such as disease susceptibility, could be a significant setback. Variation for any given trait is necessary to make gains from selection and must be assessed before one is able to set goals for their breeding program. Traits with a high level of environmental or management interaction will be difficult to breed since the heritability is low. Conversely, traits with high heritability can be bred for using simple methods and gains from selection will lead to fixing beneficial alleles quickly. For example, if seed size is relatively similar across all lines in a germplasm collection, it could be nearly impossible to breed a variety with large seeds. The variation simply doesn't exist for selection to act upon.

4.4 Latitude adaptation

Cannabis, especially hemp, has mostly been bred for temperate latitudes and in general, south of 35 degrees north and north of 35 degrees south, varieties bred for more poleward latitudes do not perform well. Typically, once a temperate variety becomes sexually mature enough to flower, flowering proceeds immediately. Some varieties will flower with as few as 8 hours of darkness. The result of this biological reality is farmers in the sub-tropical and tropical regions of the planet do not achieve the same yields as their temperate counterparts.

Although suitable varieties exist in fiber hemp, mostly from Yunnan province in China, grain and cannabinoid hemp varieties that perform well at these latitudes are very hard to find if at all. Introgressing late flowering genes into temperate germplasm is of high importance. To breed for this type of variety, the two paths are either to grow so fast they reach a greater biomass than temperate germplasm or to maximally delay flowering.

Consider the following scheme (Fig. 3). Tropical and sub-tropical germplasm can be sourced online from companies such as The Landrace Team (tltseeds.com). These farmer's varieties are typically type I, therefore a Schedule I license from the Drug Enforcement Agency (DEA) is required to enact this scheme or be sourced from another company that has already made the necessary crosses. These landraces are pollinated by a type III line to form a type II population. In general, it is important to use the type I as a female since using it as a male could spread potentially detrimental type I pollen to other hemp lines accidentally. The type II population is allowed to open pollinate to create an F2 population segregating for chemotype in a 1:2:1 ratio for type I, type II, and type III, respectively. Analyzing for active THCA synthases, the resultant population can ensure only type III individuals form the next generation. The

Figure 3 Schematic of the development of late maturing breeding lines for sub-tropical and tropical regions.

type III hemp identified by the previous step can then be randomly intermated to generate more diversity and increase the seed for the next phases of the breeding process implementing the breeder's preferred method. Once this seed is obtained, it can then be distributed to breeders in locations suitable for selection and continue the breeding process until variety development is complete.

5 Breeding methods for hemp: clones and mass selection

5.1 Clones

Before the introduction of improved seed varieties, CBD-type hemp farmers relied on clones for their seedstock. Typically, high-value individuals were selected from a population derived from a specific cross or highly variable seed variety. These plants would be cloned and evaluated in multiple environments before being available on the market. Once a high-value individual was identified, further crossing with other established clone-only varieties occurred and the process started over again. Until stable, uniform seeds were available, clones were the only way for farmers to achieve uniformity of their crop. Clones offered stable, predictable harvest times and since the varieties were widely tested, they were able to be deployed in multiple environments with predictable performance.

To satisfy demand, large cloning operations developed which required both a large amount of space to house all the mother plants required to supply thousands of clones and a large number of specialized environments for the clones to develop root systems before being sold on the open market. Unfortunately, this kind of variety production was not scalable to broad acreage. Tissue culture operations have come online to deliver increasingly larger numbers of plants to satisfy this demand; however, the cost per clone is still substantial. With decreasing margins per plant since legalization, this cost is often prohibitive. Further, agronomic practices have evolved where growing 1500 plants per acre was common, to now over 5000 plants per acre are often observed. Another major problem with tissue culture is that it can also introduce unwanted phenotypic differences between individuals (Debnath and Ghosh, 2022). These somaclonal variations could be genetic, epigenetic, or both and although these differences could be interesting to a plant breeder, they could be detrimental to the marketable crop due to the loss of homogeneity.

5.2 Mass selection

Mass selection can be a very productive method and can achieve multiple goals. In general, populations developed from crossing clone-only varieties or other crosses between target individuals are used as first-generation starting material (Fig. 4). A cycle is developed in which populations are repeatedly selected for the traits of interest ensuring that only plants expressing the desired phenotype contribute to the next generation. Over time, the population becomes more homogenous and can be ready for broad-acre testing and be maintained through open pollination.

Unfortunately, this method is best suited to highly heritable traits which can differ between individual populations and aren't necessarily the same between crops of different species (Fehr, 1991). Care must be taken to select only in environments where the desired phenotype is expressed which places constraints on time to variety release. The popular monoecious variety, Henola, was developed over 11 years of negative selection for male plants

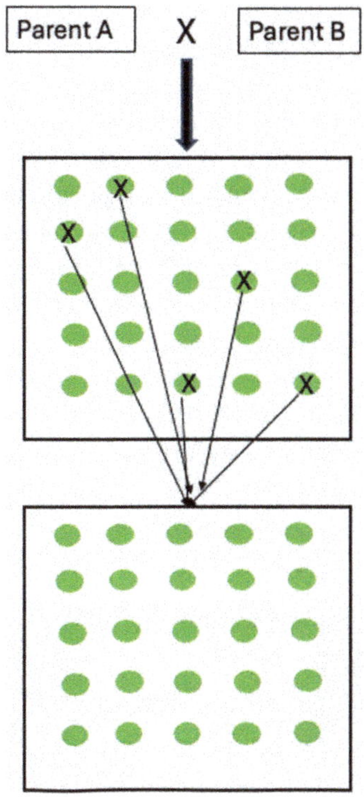

Figure 4 From an initial cross between two selected parents, the source population is evaluated and only the selected plants contribute to the next generation.

as well as high THC content, while positive selection occurred on yield (PVP# 202100192). Since cannabis is a cross-pollinated species, this method requires nurseries to occur in relative isolation from other pollen-shedding hemp. This illustrates the time and resource commitment to develop a variety through this method and the care must be taken to ensure that the variety performs as expected.

6 Breeding methods for hemp: single-seed descent and pedigree breeding

6.1 Single-seed descent

Single-seed descent (SSD) can be deployed for any of the main categories of hemp. A single randomly selected seed from each female produces the next generation. With regards to hemp, it is likely best suited for early generations of the breeding program to quickly reduce variation. For instance, after a gynoecious self-pollination or pollination between females of the same population, a single seed is selected from the S_1 generation to carry on the next generation. This reduces the variation observed and can allow the breeder to focus on later stages, closer to commercialization, when they are grown as $S_{1:2}$ or $S_{2:3}$ families. Growing large numbers of families from this scheme ensures that much of the variation in the population is still captured and subject to selection. A great benefit of this method is that the original S_0 individuals can be grown in a nursery in the target environment for initial selection, ensuring beneficial alleles are present for the challenges of that environment. After which the rapid inbreeding can occur in off-season nurseries either at another outdoor location or within a greenhouse or indoor facility.

Successful use of this method can involve direct self-pollination if the plants are monoecious, or the plants can be sprayed with silver thiosulfate in only one location on the target plant, allowing both female and male inflorescences to be present on the same plant. Ten-micron pollen bags, such as those shown in Fig. 5, can be used to ensure only self-pollination occurs.

Figure 5 A group of grain breeding lines undergoing self-pollination.

6.2 Pedigree breeding

Like SSD methods, the pedigree method is designed to rapidly inbreed individuals of a selected population. This is typically done in an 'ear-to-row' manner where a single selected female's seed will be grown in a family row. Selection can occur on both a family level and an individual level and multiple individuals within a family can be selected or just one. Consider the following breeding scheme. Prior to flowering, selections can be made based on growth, growth habit, and plant structure. From the plants that are selected, four clones will be taken to serve various purposes (Fig. 6). The first clone is to be a female, while the second clone is designed to be a male upon which one of the selected masculinizing protocols to induce male flowers on an otherwise female plant. This allows the line to be self-pollinated to support the ear-to-row process. A third clone will be pollinated by an inbred tester line serving as a male.

It is important to use an inbred line to reduce any variation that would occur due to segregation in the resultant population. This testcross serves two purposes. The first is to observe the variation held within the line upon which breeding is occurring. This resulting population can be observed to determine how close the line is to true breeding and give an indication of how soon the line can be submitted to further testing. Second, this cross allows the breeder to observe what the result of the breeding effort may appear in all phenotypes. This tester can be chosen based on its overall general combining ability and for its own commercial potential. Finally, the fourth clone, if it exists, is to make a new developmental cross to start the breeding process again. This process has two major considerations. First, record keeping must be complete and meticulous. Second, there is a significant issue of space to hold all these clones in an organized fashion. Once pollination needs to occur, a further level of organization must be employed to make sure pollination happens on the correct plants in an efficient manner.

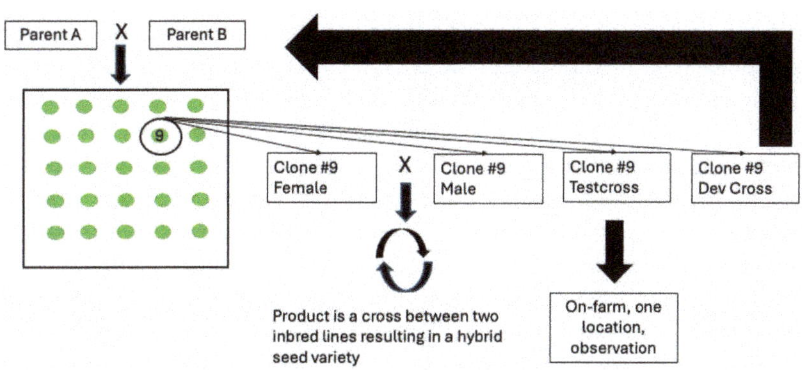

Figure 6 Schematic of the four-clone method of breeding.

Published by Burleigh Dodds Science Publishing Limited, 2026.

True males are of little value in this breeding scheme and are a hindrance. If the final product is a feminized crop, it is inappropriate to use a true male since their presence will have to be removed from the process at some point anyway, which adds complexity and cost. Importantly, the observations fueling the development of female cannabis allow the breeder to control both sides of the cross which improves gain from selection. Further, a true male inbred line still requires a progeny test to assess breeding value and general combining ability, and their male progeny will still have to be removed wasting further resources. These types of matings slow the breeding process since all the individuals are not self-pollinators but sib crossers, therefore the inbreeding process could take twice the time. The advent of gynoecious matings is a tool that cannot be understated in its value to any successful breeding program.

It has been reported that cannabis is quite sensitive to inbreeding depression. Fertility is usually a significant issue that must be mitigated. After three rounds of self-pollination, it may be most efficient to maintain the line by maintaining the inbred line as a clone. However, this is impractical for fiber and grain hemp since the seed is typically sold by the kilogram, whereas seeds for cannabinoid production are sold individually or a higher seed cost comparatively. Thus, producing significant quantities of seed becomes expensive and maintaining the clone lines requires large efforts to protect against a total loss. Since there is still some variation left in a line that has been selfed three times, it may be most practical to maintain the line at this point by open pollination and partial mass selection. This will mitigate any fertility issues that have a direct impact on the cost of goods sold (COGS).

7 Breeding methods for hemp: hybrid breeding, fast plants and synthetics

7.1 Hybrid breeding

Since the limited legalization in 2014 in the US, hybrid hemp has become much more prevalent. The products of the pedigree method are inbred lines that can exploit heterosis and be crossed together to form a very uniform crop. These hybrid lines add a great layer of uniformity in many traits such as maturity, chemotypic stability, and growth habit. Unlike varieties developed through mass selection, these types of products also afford Intellectual property protection to the breeder since the inbred lines used to make them will be impossible to recreate.

Cytoplasmic male sterility (CMS) would enable hybrids to be made cheaply and could work with both monoecious and dioecious individuals or combinations thereof. If sufficient isolation is achieved, both the hybrid seed

Published by Burleigh Dodds Science Publishing Limited, 2026.

and the male parent line could be harvested and is a bonus to the production of the hybrid.

We have begun experiments to determine the potential of a non-genetic system. We have been studying Ethephon (Bayer Crop Science), 2-chloroethylphosphonic acid, to control the sexual expression of monoecious populations. The use of Ethephon in cucurbit hybrid seed production is well documented and can serve as a guide to the development of the system in cannabis (Robinson, 2008). For cucurbits, the range of concentration is between 100 ppm and 750 ppm. Our initial experiment tested 250 ppm, 500 ppm, and 1000 ppm (Figs 7 and 8). Plant size and architecture were drastically altered (Fig. 8) with plants from all treatments 1/3 of the size of the control plants with increased branching. Male flower structures were suppressed in all treatments. The 250 ppm treatment did not eliminate male flowers; however, the 1000 ppm treatment led to deformed flower structures with leaves at a

Figure 7 Effect of Ethephon on plant growth. From left concentrations are control (water), 250 ppm, 500 ppm, and 1000 ppm.

Figure 8 Malformed growth of an individual from the 1000 ppm treatment.

steeper angle (Fig. 7). The 500 ppm plants showed the best reaction in regard to plant health and flower suppression; however, a few male flower structures were observed. We repeated the experiment using 550 ppm, 650 ppm, and 750 ppm and found that the 550 ppm treatment suppressed male flowers and had a satisfactory response in overall phenotype. These results were obtained on a single population and more research is needed using other unrelated populations to understand genotype by management interactions. Further, we have not tested this system on dioecious, XY males. Future experiments will also test the system in the field.

7.2 Fast plants

Fast plants, as they are sometimes referred to, are plants that have one parent photosensitive and the other day-neutral (also known as autoflower) (Dowling et al., 2024). Hybrid hemp can exploit this feature to extend the range of adaptation of individual inbred lines. The main benefits of these varieties are that they achieve a level of biomass that day-neutral plants cannot obtain, while flowering time appears at a mid-parent level. These varieties are sometimes a month earlier than their full photoperiod counterparts, which is highly beneficial to avoid the poor weather observed in the fall in some locations. In the Pacific Northwest, summer conditions are ideal for growing cannabis; however, west of the Cascade Range the weather turns quite wet around general harvest time and risks the crop to loss from diseases such as botrytis or powdery mildew. These varieties allow the farmer to avoid this risk and take advantage of the excellent environmental conditions during the summer months.

7.3 Synthetics

Synthetics are populations or varieties that are generated from random intermating of three or more inbred clone lines. For fiber and grain types, a variety derived from this mating scheme is impractical due to the columnar growth habit making producing enough clones to generate the Syn1 generation a nearly impossible task. Moreover, if one were to produce enough clones, COGS would far outweigh any profits. However, these kinds of varieties may work well for the extract market. Matching maturities in these varieties is paramount to producing a uniformly harvestable crop.

A modified approach to synthetics in the fiber and grain market could be employed to generate highly productive varieties (Fig. 9). One or more independent inbred populations could serve as the initial starting material to

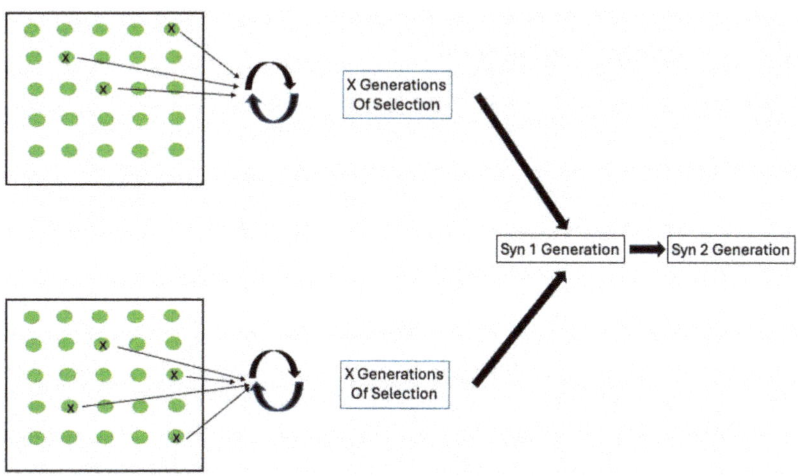

Figure 9 Schematic of the modified synthetic method.

generate the Syn1. With this strategy, the only limiting step to creating the Syn1 is the number of seed available of each inbred population.

8 Seed and grain production

8.1 Seed production

Location of seed production is also a major consideration for the industry. A study from the year 2000 found that in the latter part of August, 36% of the pollen counts were from cannabis (Stokes et al., 2000). These results were found in the US Midwest where marijuana remains illegal suggesting that the source of the pollen is due to feral populations or the result of clandestine grows or breeding operations. This alarming statistic suggests that genetically pure seed may be impossible in regions where feral hemp is known to exist. In Canada, a minimum distance of 4800 m between a registered seed production field and any other pollen-bearing cannabis is required (Canadian Seed Growers' Association, 2019).

These findings indicate that seed production fields need great isolation and may not be possible in some locations. Through personal correspondence, it has been suggested that a 'zone' system across the country needs to be established. Certain regions of the country would be designated as either cannabinoid hemp only, fiber and/or grain production or seed production could occur. In the Pacific Northwest, cannabinoid hemp and marijuana could be designated west of the Cascade Mountains, while fiber and grain could occupy the eastern side. Unfortunately, a nationwide agreement such as this would require amazing cooperation and administration between producers and the political will likely does not exist.

8.2 Grain variety development

The US grain market at present is quite small. Grain varieties can be supplied from both Canada and Europe, but few dedicated grain variety are of US origin and available on the market. The downside of Canadian and European sources is that the varieties can be day-neutral or otherwise very early maturing which when considering mainland US these varieties are too early and therefore do not yield up to the full potential that a photoperiod sensitive variety could obtain.

We believe we have a novel approach to develop later maturing varieties and address the issue of shattering. We have crossed the shortest cannabinoid types in our germplasm with monoecious fiber types. This achieves multiple aims. First, this should influence plant height with the goal of the resultant lines to be somewhere around the mid-parent value. Second, these should be later maturing and potentially higher yielding than the currently available day-neutral types. Third, the flower structure of the cannabinoid types is certainly more dense than is typical of grain and fiber types and may be useful to address the issue of shattering currently observed in fiber and grain types. Fourth, the columnar plant architecture of the fiber types would be beneficial for high-density planting. Finally, the addition of the monoecious trait would eliminate the need for XY males to pollinate the crop and therefore adding to overall yield since non-seed-producing individuals will not be taking precious space.

9 Breeding for resistance to biotic and abiotic stress

It is often claimed that hemp is a low-input crop and does not suffer from the disease or abiotic stress pressures that devastate other fiber crops such as cotton (Lawson et al., 2022). This may be true in the infancy of the cannabis industry but is likely to change as fungal, bacterial, and insect pests adapt to the new host. It is important to diligently monitor what stresses are present and to forecast new challenges. The field is developing quickly and identifying a gene-for-gene resistance system first described by Flor (1956), for a multitude of pests would allow breeders to begin pyramiding these sources of resistance. However, transferring multiple resistance genes into a single variety is a long and costly process and gene-for-gene resistance is the easiest for pests to overcome.

There are three forms of host resistance: infection resistance, resistance to spread, and tolerance. Resistance to initial infection is highly desirable since there would not be damage to the final product. Resistance to disease spread minimizes the impact of the pathogen, but some crop damage is observed. Finally, tolerance to the pathogen, such as viruses, is observed when the pathogen is detectable but does not cause economic loss.

Powdery mildew is an important disease as it is common and reduces the quality of the flower. *PM1* is the first resistance gene found in cannabis and is of the *NBS-LRR* class of resistance genes (Mihalyov and Garfinkel, 2021), while other researchers have found and characterized *MLO* genes and have found two potential sources of PM susceptibility (Pépin et al., 2021). Recently, building on the Pépin et al. research, an MLO gene, *CsMLO1*, has been examined and found to confer susceptibility to PM (Stack et al., 2024). These potential breeding targets are of great interest. However, breeding for disease resistance can be challenging since some diseases, such as powdery mildew, cannot be cultured and therefore inoculant material and standardization is very difficult to obtain (Pérez-Garcia et al., 2009). This, in turn, makes finding the genes responsible for susceptibility and resistance to PM reliant on other methods of providing inoculum to screen for resistance.

Bud rot, caused by *Botrytis cinerea*, is a devastating disease affecting the flowers of the cannabis plant rendering them unusable. Botrytis typically begins to cause economic losses under cool conditions with relatively high humidity (Mahmoud et al., 2023). Unfortunately, this kind of environment is often observed in late flowering in many locations, causing rapid onset of the disease. In the Pacific Northwest, these conditions typically begin in October which is in the middle of harvest season. Botrytis can completely ruin a crop in under a week (author's personal observation). No specific genes have been identified conferring resistance to date. However, five putative defense genes have been found to be upregulated in the presence of a Botrytis infection (Balthazar et al., 2020). Finding and characterizing potential sources of genetic resistance are paramount to controlling this disease.

Hop latent viroid (HLVd) is becoming a serious pathogen to the cannabis industry. A study done by Dark Heart Nursery in California with hundreds of thousands of samples analyzed found that 90% of cannabis farms in California are infected by HLVd (Cannabis Business Times, 2021). This great infection rate suggests that it is likely already nationwide. The practice of sharing clones on the marijuana side of the cannabis industry is likely a major route of infection.

Viroids are low-molecular-weight, RNA molecules without proteins which are pathogenic (Diener, 1971). Infected plants can be asymptomatic; however, symptoms include stunting, chlorosis, tight internodes, and weak, malformed branches (Warren et al., 2019). Currently, the only known control is meristem culture (Adkar-Purushothama et al., 2023), but without some resistance mechanism, the variety will remain prone to reinfection. Further compounding the problem, HLVd is also seed transmitted (Atallah et al., 2024). For short-term breeding goals, it is likely that tolerance to the viroid will be a primary method of control, and in the long-term, genetic resistance must be sought. However, this may be difficult as natural resistance across viroid–host interactions are elusive (Flores et al., 2017)

To breed for a multitude of abiotic stresses, extensive testing in multiple target environments ensures the variety being developed is exposed to as many stresses as possible to understand genotype by environment interactions and for product placement. Stresses such as salinity, heat, nutrient availability, and drought are all breeding targets that great strides could be made. Suitable environments exist throughout the US to test and breed for these traits.

10 Conclusion

Despite missing the benefits of the Green Revolution, thanks to the tools developed therein and the rapid increase in molecular techniques and abilities, cannabis is quickly catching up with other major crops. This has been enabled by dedicated researchers and forward-looking companies pushing the plant

Published by Burleigh Dodds Science Publishing Limited, 2026.

toward its full potential. It will be interesting to observe the developments of the next few decades. Further information can be obtained through specific sources in the NCBI database (https://www.ncbi.nlm.nih.gov/guide/literature/).

11 References

Adkar-Purushothama, C.R., et al. (2023). Hop latent viroid: a hidden threat to the cannabis industry. *Viruses*, 15(3), 681. https://doi.org/10.3390/v15030681.

Atallah, O.O., et al. (2024). New insights into hop latent viroid detection, infectivity, host range, and transmission. *Viruses*, 16(1), 30. https://doi.org/10.3390/v16010030.

Balthazar, C., et al. (2020). Expression of putative defense responses in cannabis primed by pseudomonas and/or bacillus strains and infected by Botrytis cinera. *Frontiers in Plant Science*, 11, 572112. https://doi.org/10.3389/fpls.2020.572112.

Canadian Seed Growers Association. (2019). https://seedgrowers.ca/wp-content/uploads/2019/01/Circ6-SECTION-10-ENGLISH_Rev-02.01-2019_20181218.pdf.

Cannabis Business Times. Eric Sandy. (2021). Dark heart nursery research finds 90% of California facilities test positive for hop latent viroid. https://www.cannabisbusinesstimes.com/disease/hop-latent-viroid/news/15693417/dark-heart-nursery-research-finds-90-of-california-facilities-test-positive-for-hop-latent-viroid.

Debnath, S.C. and Ghosh, A. (2022). Phenotypic variation and epigenetic insight into tissue culture berry crops. *Frontiers in Plant Science*, 2022, 13. https://doi.org/10.3389/fpls.2022.1042726.

de Meijer, E.P.M. (1999). Cannabis germplasm resources. In: Ranalli, P. (ed.) *Advances in Hemp Research*. Haworth Press Inc. p. 138.

de Meijer, E.P.M., et al. (2003). The inheritance of chemical phenotype in cannabis sativa L. genetics. https://doi.org/10.1093/genetics/163.1.335.

Diener, T.O. (1971). Potato spindle tuber "virus": IV. A replicating, low molecular weight RNA. *Virology*, 45(2), 411–428. https://doi.org/10.1016/0042-6822(71)90342-4.

Dowling C.A., et al. (2024). A Flowering Locus T ortholog is associated with photoperiod-insensitive flowering in hemp (Cannabis sativa L.). *The Plant Journal*. 119(1): 383–403. https://doi.org/10.1111/tpj.16769.

Fehr, W.R. (1991). *Principles of Cultivar Development*. Macmillian Publishing Company.

Flor, H.H. (1956). The complementary genic systems in flax and flax rust. *Advanced Genetics*, 8, 29–54.

Flores, R., et al. (2017). Engineering resistance against viroids. *Current Opinion in Virology*, 26, 1–7. http://dx.doi.org/10.1016/j.coviro.2017.07.003.

Grassa, C.J., et al. (2018). A complete Cannabis chromosome assembly and adaptive admixture for Elevated Cannabidiol (CBD) Content. bioRxiv. https://doi.org/10.1101/458083.

Hemp Industry Daily Staff. (2021). Legendary California marijuana growing region humboldt permanently bans hemp cultivation. *Hemp Industry Daily*. https://hempindustrydaily.com/legendary-california-marijuana-growing-region-humboldt-permanently-bans-hemp-cultivation/.

Lawson, L., et al. (2022). Cellulose textiles from hemp biomass: opportunities and challenges. *Sustainability*, 14(22), 15337. https://doi.org/10.3390/su142215337.

Mahmoud, M., et al. (2023). Understanding bud rot development, caused by botrytis cinerea, on Cannabis (Cannabis sativa L.) plants grown under greenhouse conditions. *Botany*, 101(7), 230–231. https://doi.org/10.1139/cjb-2022-0139.

Mihalyov, P.D. and Garfinkel, A.R. (2021). Discovery and genetic mapping of PM1, a powdery mildew resistance gene in cannabis sativa L. *Frontiers in Agronomy | Disease Management*, 3, 2021. https://doi.org/10.3389/fagro.2021.720215.

Pépin, N., et al. (2021). Genome-wide characterization of the MLO gene family in cannabis sativa reveals two genes as strong candidates for powdery mildew susceptibility. *Frontiers in Plant Science*, 12, 1268706. https://doi.org/10.3389/fpls.2021.729261.

Pérez-Garcia, A., et al. (2009). The powdery mildew fungus *Podosphaera fusca* (synonym *Podosphaera xanthii*), a constant threat to cucurbits. *Molecular Plant Pathology*, 10(2), 153–160.

Punja, Z.K., et al. (2017). Assessing genetic diversity in *Cannabis sativa* using molecular approaches. In: Chandra, S., et al. (eds) *Cannabis sativa L. - Botany and Biotechnology*. Springer. https://doi.org/10.1007/978-3-319-54564-6_19.

Robinson, R.W. (2008). Rationale and methods for producing hybrid cucurbit seed. *Journal of N ew Seeds*, 1(3–4), 1–47. https://doi.org/10.1300/J153v01n03_01.

Sawler, J., et al. (2015). The genetic structure of marijuana and hemp. *PLoS One*, 10(8), e0133292. https://doi.org10.1371/journal.pone.0133292.

Sliwoski, H. (2022). Federal bill would raise THC limit in hemp to 1%. Canna Law Blog. https://www.lexology.com/library/detail.aspx?g=07f7de73-6c26-420c-8270-520def54962b.

Small, E. and Cronquist, A. (1976). A practical and natural taxonomy for cannabis. *Taxon*, 25(4), 405–435.

Stack, G.M., et al. (2024) Genetic mapping, identification, and characterization of a candidate susceptibility gene for powdery mildew in cannabis sativa. *Molecular Plant-Microbe Interactions*, 37(1), 1543229. https://doi.org/10.1094/MPMI-04-23-0043-R.

Stokes, J.R., et al. (2000). Cannabis (hemp) positive skin tests and respiratory symptoms. *Annals of Allergy, Asthma & Immunology*, 85(3), 238–240. https://doi.org/10.1016/S1081-1206(10)62473-8.

The Landrace Team. (2024). https://tltseeds.com (Accessed: June 2, 2024).

Warren, J.G., et al. (2019). Occurrence of hop latent viroid causing disease in cannabis sativa in California. *Plant Disease*, 103(10), 2699. https://doi.org/10.1094/PDIS-03-19-0530-PDN.

Zirpel, B., et al. (2018). Elucidation of structure-function relationship of THCA and CBDA synthase from *Cannabis sativa* L. *Journal of Biotechnology*, 284, 17–26.

Chapter 4

Current state of the art in outdoor cultivation of hemp (*Cannabis sativa* L.)

David R. Gang and Fatemeh Sefidkon, Washington State University, USA; and Bonny Jo Peterson, Industrial Hemp Association of Washington, USA

1 Introduction

The cultivation of hemp (*Cannabis sativa* L.), particularly its low-tetrahydrocannabinol (THC) varieties, has seen a remarkable resurgence in recent years. This resurgence is driven by the multifaceted applications of hemp across various industries, ranging from pharmaceuticals to textiles, and from food products to biofuels. The versatility of hemp is unparalleled, making it a crop of significant economic and environmental importance.

In this chapter, we describe the current state of the art in outdoor cultivation of hemp, encompassing all its primary uses: cannabinoid/chemical/essential oil production, fiber production, and grain (seed) production. Each type of hemp cultivation demands specific agronomic practices to optimize yield and quality, addressing diverse requirements such as planting methods, soil conditions, irrigation, nutrient management, and pest control.

Outdoor cultivation of hemp presents unique challenges and opportunities compared to indoor cultivation. It requires a thorough understanding of

http://dx.doi.org/10.19103/AS.2025.0161.05

environmental variables and adaptive management practices to ensure successful growth and consistent production. This chapter provides a comprehensive overview of these practices, informed by the latest research and field studies.

We will explore the various agronomic strategies tailored to different types of hemp and highlight the critical factors that influence their successful cultivation. From selecting the right soil and irrigation techniques to managing pests and diseases, our discussion aims to equip growers with the knowledge needed to maximize their crops' potential while adhering to sustainable and regenerative practices when possible and desired.

Furthermore, this chapter will examine the impact of regional and climatic variations on hemp cultivation, describing studies that illustrate how different environments can influence cultivation practices and outcomes. By understanding these factors, growers can better adapt their strategies to local conditions, ultimately enhancing the resilience and productivity of their crops.

The future of hemp cultivation is bright, with ongoing advancements in genetics, technology, and agronomy promising to further enhance its viability and profitability, thereby supporting rural communities and economies. As we navigate the complexities of modern hemp farming, this chapter will serve as a guide, offering insights and practical advice to help growers achieve success in the dynamic and evolving landscape of outdoor hemp cultivation.

1.1 Overview of hemp: definitions and classifications

Cannabis sativa L. is a versatile plant species with two main subtypes arbitrarily distinguished by their cannabinoid content.

1.1.1 Hemp

Defined by law in Europe, the United States, and several other nations, hemp possesses a low concentration of Δ9-tetrahydrocannabinol (Δ9-THC or just THC). The specific level of Δ9-THC that separates hemp from high-THC cannabis varies but is typically around 0.2–0.3% by dry weight in most countries. Hemp is grown primarily for industrial uses such as fiber, seeds, and non-THC cannabinoids, such as cannabidiol (CBD).

1.1.2 High-THC cannabis

This form of the plant, also called Marijuana or Marihuana in statute in many countries, contains a higher Δ9-THC content than hemp (anything above the limit that defines hemp). High-THC cannabis is cultivated primarily for recreational or medicinal use where the psychoactive properties of THC are desired.

The distinction between hemp and marijuana under legal and agricultural classifications has evolved globally, with varying Δ9-THC thresholds currently being applied to define non-psychoactive hemp (Adesina et al. 2020). The levels of other tetrahydrocannabinols, such as the Δ8-THC and Δ10-THC isomers, are present in the plant at much lower levels than Δ9-THC and thus have so far been largely ignored in policy and legal determinations, but with the rise in production and sale of hemp-based Δ8-THC in particular in the last few years, these compounds are now part of conversations around how hemp versus high-THC cannabis should be defined relative to each other. This may have implications for outdoor growth as changing regulations may develop in response.

1.1.3 Legal classifications and thresholds

The exact levels of THC in hemp that separates it from high-THC cannabis varies worldwide, leading to a range of legal definitions:

US Definition: The 2018 Farm Bill in the United States classified hemp as any part of the plant *C. sativa* that contains less than or is equal to 0.3% THC on a dry weight basis, distinguishing it from marijuana under federal law.

European Union Standards: In the EU, hemp is legally defined as cannabis with a THC concentration of 0.2% or lower in most nations, but this definition is likely to change in 2025.

Different countries have set their own thresholds for THC content, with some regions allowing up to 1% THC in hemp (e.g. Czechia). These variations influence breeding, cultivation, and trade practices internationally (Small and Marcus 2019).

1.1.4 Types of hemp based on end-use

Hemp cultivars are often classified based on the intended use, as each use type is bred for specific traits (Salentijn et al. 2015). Phenotypes are targeted, and plants with corresponding traits are planted accordingly. Each type of hemp has distinct genetic and agronomic profiles. Fiber varieties prioritize height and fiber quality. Seed varieties optimize seed yield per hectare, and some are now being bred for oil composition, seed size, and oil yield. CBD/essential oil varieties focus on maximizing resinous flower production with high cannabinoid/terpene profiles. Modern breeding efforts aim to enhance these traits while also improving resistance to pests, diseases, and environmental stressors.

Fiber Hemp: These are high-biomass varieties grown for fiber production to support the textile, paper, and building materials industries, among other end products. These plants tend to be tall with dense stalks and are harvested typically prior to flowering for their strong outer fiber (bast fiber) and inner woody core (hurd/shives).

Essential Oil, Cannabinoid, or CBD Hemp: These varieties are grown specifically for floral material to be harvested prior to seed production (or are produced as feminized types or triploids that do not produce much, if any, seed). They are high in CBD, other non-psychoactive cannabinoids, or terpenoids. Floral biomass is predominantly used for wellness and medicinal applications (cannabinoid types and some terpene types). However, recently, floral biomass has been utilized in emerging industrial applications; for example, terpenes are being developed for use as fuel additives or in specialty solvents. These plants are grown outside in large fields if they are to be used as biomass for large-scale extraction. If grown for craft smokable bud applications, they are commonly grown indoors (see respective chapter on indoor growth).

Grain Hemp: These varieties are bred for their seeds to be eaten, used as animal feed, or as feedstocks for industrial applications, such as glue/resin production. Their seeds tend to be rich in omega fatty acids, proteins, and vitamins. Seed hemp varieties are shorter and somewhat bushier than fiber types, to maximize seed production.

Hemp's adaptability and wide range of uses make it unique among agricultural crops. Each part of the plant (fiber, seed, flower, root) can be harvested and processed for various industries, including textiles, building materials, food, cosmetics, biofuels, and wellness products, to name some of the largest, most mature sectors. This multipurpose nature adds complexity to its classification, as different regions might grow hemp for entirely different industries based on local market demand.

Since each country's regulations around THC content and hemp cultivation vary, classifications of hemp cultivars are often tailored to meet specific legal and market requirements. This creates a unique challenge in establishing global standards for hemp cultivation and trade, impacting everything from seed choice to export possibilities. All those factors have an impact on how hemp is grown, particularly outdoors, as agronomic practices can influence THC levels.

1.2 Historical context of outdoor hemp cultivation

Hemp is one of the oldest cultivated crops, with evidence of its use dating back over 10,000 years in regions like ancient Japan and China. Early civilizations cultivated hemp outdoors for fiber, which was essential for making ropes, textiles, and even early forms of paper (Clarke and Merlin 2016). Hemp quickly spread from Asia to Europe and Africa, with ancient Egyptians, Greeks, and Romans all incorporating it into their agriculture for use in textiles and sails.

Ceremonial and medicinal applications of the plant were also known early on. Outdoor cultivation was the primary method, as these societies relied on the natural environment to produce the hardy crops at scale. During the Middle Ages, hemp continued to be grown extensively outdoors in Europe, particularly in regions like France, Germany, Italy, and England, where it was integral to making ropes, ship sails, and other materials essential for exploration and trade. Historically, the region that is now Ukraine, with centuries of hemp cultivation, grew upwards of 150 000 ha pre-1950 (Goloborod'ko 1995).

Hemp was one of the first crops grown by settlers in colonial America, encouraged by European governments to ensure a steady supply of rope and sail material. States like Virginia and Massachusetts required farmers to grow hemp as part of their contribution to the colonial economy, and it was exclusively cultivated outdoors in open fields. Industrial hemp production expanded significantly during the early industrial era due to its use in textiles and naval supplies.

With the advent of industrialization in the eighteenth and nineteenth centuries, hemp was a critical resource, especially for the navy and shipping industries. It was grown outdoors in large fields to meet the demand for strong, durable fiber in rope, canvas, and uniforms. Despite this demand, the rise of cheaper cotton fibers and shifts in political power and influence, particularly of major industrial interests associated with timber, oil, and steel production, led to a decline in outdoor hemp cultivation. The development of steam-powered machines in textile mills further favored cotton over hemp, which was more challenging to process at scale (Cherney and Small 2016).

1.2.1 Twentieth-century decline and legal restrictions

The US Marihuana Tax Act of 1937, which grouped hemp with high-THC cannabis (marijuana), led to a dramatic decline in outdoor cultivation. Even though hemp had minimal psychoactive content, its production was nearly eradicated in the United States. The US newspaper industry's transition from hemp to timber pulp is believed by many to be a major factor in this decline.

Similar trends happened in other countries around the world, with the exceptions being France and China. Later, the 'Hemp for Victory' campaign in World War II reversed this trend briefly and encouraged a temporary increase in outdoor hemp farming for the war effort (Clarke and Merlin 2016). Farmers were urged to grow hemp to supply fiber for the war effort, but after the war, hemp cultivation again declined due to regulatory pressures and competition from synthetic materials. The age of petroleum-based synthetic chemistry, with fibers such as nylon taking the world by storm, put further pressure on hemp production. Thenail in the coffin for outdoor hemp production was Title II of the Comprehensive Drug Abuse Prevention and Control Act of 1970, which was followed by similar legislation in other countries around the world. Hemp growth was effectively banned in most of the world. In the following four decades, the final dismantling and elimination of infrastructure for hemp growth and processing occurred, and much of the knowledge of how to grow hemp was lost to most of the world. There were of course exceptions, and hemp remained in production mainly in China, France, Ukraine, and Russia and to a lesser extent in a handful of other countries.

1.2.2 Recent resurgence and legalization

Despite its official legal status as a Schedule I controlled substance in the United States (and similar standing in other nations), hemp (and its sister version high-THC cannabis) was grown behind closed doors (literally) and hidden in corn and cane fields or under forest canopies. This affected what varieties and the kind of knowledge regarding hemp's production were retained. Recently, we have seen the resurgence of both the high-THC cannabis and hemp. Ironically, the resurgence began with the long-simmering effort to push acceptance of the high-THC versions of the cannabis plant, first through medical marijuana efforts beginning in the 1990s and then full legal marijuana/high-THC cannabis efforts in parts of the United States beginning in 2012 in Colorado and Washington. The resurgence of hemp began a few years later in 2018 with the growth of CBD-based markets but rapidly expanded to other types due to its environmental benefits, such as soil regeneration and low pesticide requirements. Outdoor cultivation became an attractive option once again as demand rose for sustainable, natural materials.

Many countries, including Canada, parts of Europe, and, eventually, the United States (with the 2018 Farm Bill), began to re-legalize hemp, specifically low-THC varieties for industrial purposes. This shift enabled a resurgence of outdoor hemp cultivation on an industrial scale, especially for CBD production. Processing facilities for CBD are the same as required for some other agricultural crops, including high-THC cannabis types, and so efforts to process hemp grown for CBD were able to utilize existing agricultural infrastructure or even in some regions to follow the growth in the legal THC markets. Fiber and grain varieties, although specifically legalized by the 2018 Farm Bill, have struggled to catch up due to very different processing requirements and the complete dismantling of previously existing processing facilities in the last century.

Today's outdoor hemp cultivation incorporates modern farming techniques, such as selective breeding, soil health management, and precision agriculture. These advancements are helping to optimize yields, improve plant quality, and reduce environmental impact, making hemp a viable crop for sustainable and regenerative agriculture, particularly when added to crop rotation systems. With the global trend of a growing market for CBD products (although not as large of a trend as touted back in 2018), outdoor hemp cultivation focused in the United States first heavily on cannabinoid production, with markets for other traditional uses like fiber and seed being much slower to develop. The first large-scale non-CBD processing facility in the United States (seed processing) was built in 2019, and the first large-scale fiber decorticator was completed on the same site in 2021. Since then, over two dozen processing facilities for hemp fiber have sprung up across the nation, which compares to hundreds of CBD processing facilities. This difference in processing capacity has impacted the rate of knowledge generation relative to best growing practices for the different hemp types. Countries such as France and China, which did not shut down their fiber hemp industries in the 1970s, established robust processing industries over the past 50 years, which have been adopted, mimicked, or used as inspiration for how the hemp industry can be built in other nations such as the United States. Breeding programs and cultivation practices are evolving to meet these diverse market demands.

Thus, hemp cultivation has evolved significantly over thousands of years, from ancient fiber production to a diverse modern industry. With current technological advancements and legal support, outdoor hemp is set to play an increasingly prominent role in global agriculture, balancing economic viability with environmental sustainability and regenerative agricultural practices. The rest of this chapter will discuss current knowledge of best practices for hemp production. Due to the large diversity in how hemp is grown around the world, this chapter cannot be a complete, comprehensive treatise on all aspects of hemp growth in all of those locations. Instead, we focus on major trends that appear to support the best practices that can be optimized to fit local market and environmental needs. A key factor that must be kept in mind when considering how hemp is grown is that the method applied for its growth must match the ultimate downstream purpose/use of the final product. Different uses demand different growth approaches, as outlined below.

1.3 Importance of outdoor cultivation in the hemp industry

1.3.1 Economic viability and cost-effectiveness

Several researchers have evaluated the feasibility of outdoor hemp production in different regions and for different purposes. Mark et al. (2020) discussed the economic feasibility for industrial hemp production across various US states. Outdoor cultivation can be more cost-effective than indoor and greenhouse methods, depending on the ultimate use as well as region. Outdoor hemp production does not require expensive infrastructure, such as artificial lighting, climate control (and associate large energy demands), or extensive pest and disease management systems. This makes it more accessible for a larger number of farmers and contributes to lower production costs for the hemp industry. The exception to this now is hemp grown for specific high-value markets, such as the smokable flower market that is likely best grown indoors, much like its high-THC cannabis sister. Smokable flower hemp has emerged as a boutique and craft segment of the hemp industry that, while important in its own right in certain regions, will most likely not be a long-term major component of industrial hemp industry. Indoor growth of hemp is discussed more extensively in a separate chapter.

Outdoor hemp fields allow for expansive planting areas, enabling farmers to grow larger volumes at a fraction of the cost. This scalability is crucial for hemp applications that require bulk production, such as grain for food, industrial fiber, building materials, plastics, biofuel, and animal bedding. Compared to indoor cultivation, outdoor hemp growth relies on natural sunlight and seasonal weather patterns, eliminating the carbon footprint associated with artificial lighting and climate control. Outdoor hemp cultivation aligns with sustainable and regenerative practices by promoting soil health improvement, reducing erosion, and facilitating phytoremediation. It can also enhance soil microbial diversity. Hemp is known for its ability to improve soil health by replenishing nutrients, reducing soil erosion, and suppressing weeds naturally. As a bioaccumulator, hemp absorbs toxins and heavy metals from the soil through an aggressive root system, making it beneficial in phytoremediation projects where soil restoration is needed. Outdoor hemp cultivation contributes to carbon sequestration, as the plant absorbs CO_2 during its growth cycle, and then can be used in applications, such as building construction using hemp lime (also called hempcrete), that can last for a hundred years or potentially millennia. This makes outdoor-grown hemp an environmentally advantageous crop, particularly in the context of climate change and the need for carbon reduction solutions.

1.3.2 Enhanced product quality for certain applications

Outdoor-grown hemp often exhibits a more diverse cannabinoid and terpene profile due to natural environmental stresses (Kaminski et al. 2024). Natural sun exposure promotes cannabinoid synthesis, often leading to higher CBD yields, as well as higher THC levels, which can be problematic for regulatory compliance for some varieties. Hemp grown outdoors often produces longer, stronger fibers due to its natural growth pattern from a robust root system with a long tap root, which is advantageous for textile, building materials, composites, and paper industries. The strength and length of outdoor-grown hemp fibers can surpass those grown indoors, making it ideal for durable, high-quality products. The scale needed for large-scale industrial applications such as the building or automotive industries can only be achieved through outdoor growth. Industrial applications for hurd, the inner woody core of the stalk covered by fiber, require the same large-scale cultivation as for fiber. Grain crops for food or industrial purposes grown outdoors have higher yield capacity with larger seed heads.

1.3.3 Alignment with sustainable and organic markets

Hemp's natural resistance to many pests often allows for minimal pesticide use, making it a viable candidate for organic cultivation under suitable conditions (Kaminski et al. 2024). For instance, in France, hemp is promoted as an eco-friendly crop, requiring 'no irrigation, no fertilizer, and no pesticides,' as has been stated to the first author many times in visits with French hemp industry members (growers, processors, and seed suppliers, all state the same). However, while hemp faces relatively few pest and disease pressures as a minor crop within larger agricultural systems, it is not entirely immune to such challenges. In regions like the Pacific Northwest of the United States, diseases such as Beet Curly Top Virus (BCTV) and hemp latent viroid (HLV) pose significant threats, particularly to CBD hemp varieties. BCTV is transmitted by the prevalent endemic brown leafhopper, while the transmission method for HLV remains uncertain. When both pathogens infect a plant, they can cause severe stunting (Jarugula et al. 2023). To mitigate these issues, strategies such as managing BCTV vectors or developing virus-resistant germplasm are crucial.

As hemp acreage expands, pest and disease pressures are likely to increase, particularly without proper crop rotation practices. On the contrary, if proper crop rotation practices are followed rigorously, delay or potentially even prevention of pest problems may result. Adopting such practices supports sustainable and regenerative agriculture while aligning with organic certification standards, which meet growing consumer demand for chemical-free hemp products in wellness, food, and cosmetics sectors. Furthermore, outdoor hemp cultivation complements natural growth cycles, aligning with the market's shift toward organic, sustainable, and regenerative agriculture. By relying on fewer artificial inputs and natural resources, outdoor hemp reinforces its role as a sustainable/regenerative crop, appealing to environmentally conscious consumers.

1.3.4 Economic opportunities for farmers and rural communities

Hemp supports crop diversification and job creation in rural communities. The ability to grow hemp outdoors on a large scale makes it an appealing crop for economically challenged areas (Adesina et al. 2020). Hemp provides a new crop option for farmers, where it can be incorporated into rotation with other crops. This can contribute to improved soil health and economic stability for farmers by diversifying their income sources. Outdoor hemp cultivation, especially at larger scales, creates jobs in planting, maintaining, harvesting, and processing sectors, stimulating rural economies. Because a hemp-based circular bioeconomy requires hemp to be grown and processed at a smaller regional scale, rather than at a national level (processing facilities should be within less than 100 km from where the hemp was grown to meet sustainability and carbon negative/neutral goals), local industries that lead up to at least the fiber/seed processing facility level need to be in place, supporting local economies and rural community development. This is particularly important in regions transitioning from traditional crops to hemp, where hemp's diverse applications present new market opportunities.

1.3.5 Supporting industrial-scale applications

Outdoor hemp cultivation is not immune to challenges related to weather and disease impact. Unlike indoor cultivation, outdoor hemp farming is subject to seasonal changes, extreme weather events, and natural disasters, which can impact yield quality and consistency. Although generally resilient, outdoor hemp is still susceptible to specific pests and diseases (as mentioned above). However, advances in integrated pest management (IPM) and selective breeding can help overcome these challenges (Kaminski et al. 2024). This in turn will support hemp's integration into large-scale industrial applications. Indeed, hemp's biomass potential for industrial-scale applications like biofuels, bioplastics, construction materials, and textiles requires large-scale outdoor cultivation to be economically viable (Prade et al. 2011). Large, open fields enable the mass production needed for industrial uses, which is impractical in an indoor or greenhouse setting. As the demand for plant-based, renewable materials grows across industries, outdoor-grown hemp can serve as an essential raw material for other eco-friendly products. Outdoor cultivation allows the hemp industry to scale sustainably and meet demands for industrial hemp applications like biodegradable plastics and composite materials.

Outdoor hemp cultivation is indispensable for the future of the hemp industry due to its economic, environmental, and product-quality benefits (Mark et al. 2020). By leveraging natural resources and requiring fewer inputs, outdoor growth supports a sustainable or regenerative approach to agriculture that can meet the growing demand for hemp across diverse markets, from wellness and textiles to construction and biofuels. As the industry expands, outdoor cultivation is likely to remain a cornerstone of hemp production, balancing profitability with ecological stewardship.

2 Types of hemp for outdoor cultivation

Hemp is cultivated primarily for three categories of use: **essential oil hemp**, **fiber hemp**, and **grain hemp**. These types vary in their intended uses, growing conditions, and chemical composition, and understanding each type's specific requirements is crucial for optimizing outdoor cultivation. Figure 1 shows examples of each of the various hemp types being grown outdoors in the field. Table 1 outlines properties of the above types of hemp that will be described in more detail throughout the rest of this chapter.

2.1 Cannabinoid/chemical/essential oil hemp

Hemp under this broad classification is grown primarily for its specific chemical constituents, typically terpenoids or cannabinoids, both of which can be extracted by heat-applied distillation, hence the term 'essential oil' hemp. Cannabinoid hemp is grown specifically for its high concentrations of cannabinoids, most commonly CBD (cannabidiol) but also other minor cannabinoids like CBG (cannabigerol) and CBC (cannabichromene). Terpene essential oil hemp is cultivated specifically for its terpene-rich flowers, which are used in aromatherapy, personal care products, and even in food flavoring (Russo 2011). Terpenes such as myrcene, limonene, and pinene are extracted from the flowers, providing medicinal and aromatic properties (Abdollahi et al. 2020). Some of the predominate sesquiterpenes found in hemp are E-caryophyllene, caryophyllene oxide, E-β-farnesene, and β-caryophyllene (Verma et al. 2014). These plants are typically bred to produce minimal THC (<0.3%) to comply with legal standards in many regions (Mead 2017, Abdollahi et al. 2021).

Growing Conditions: Essential oil hemp requires growth conditions with adequate sunlight, well-drained soil, and consistent moisture. Essential oil

Figure 1 Examples of different types of hemp grown in the field. Left, essential oil hemp. Center, fiber hemp. Right, grain hemp.

Table 1 Properties and growth parameters for different classes of hemp

Characteristic	Fiber hemp	Grain hemp	Dual-purpose hemp	Essential oil/cannabinoid hemp
Primary use	Bast fibers for textiles, paper, and/or hurd/shives and/or full stalks for building materials	Seed/grain for oil, food, feed, resins, and plastics	Both fiber/hurd and seeds OR fiber/hurd and cannabinoids	CBD and other cannabinoids and/or terpenes for medicinal, wellness products, or industrial solvent additives
Row spacing	10–20 cm (closer for fiber, wider for hurd)	30–60 cm (for airflow and yield)	20–40 cm	Wide spacing (1 m × 2 m) for airflow, individual plant care OR higher density (0.9 m × 0.3 m) with proper mechanical weed control
Soil type	Well-drained, moderate fertility, loamy and/or sandy loam	Well-drained, moderate fertility, loamy and/or sandy loam	Well-drained, moderate fertility, loamy and/or sandy loam	Rich, well-drained soil
Water/irrigation needs	Moderate, rain-fed at the right time or irrigation early	Moderate, especially pre-flowering	Moderate, balanced for both	High: consistent water, especially during flowering
Planting depth	Prefers <1 cm, tolerates 1–2 cm	Prefers <1 cm, tolerates 1–2 cm	Prefers <1 cm, tolerates 1–2 cm	Prefers <1 cm, or use transplants
Fertilizer demand	50–150 units of N depending on variety, moderate P/K	50–150 units of N depending on variety, moderate K, high P at flowering	50–150 units of N depending on variety, moderate K, high P at flowering	50–150 units of N depending on variety, moderate K, high P at flowering
Planting density	200–300 plants/m^2	80–120 plants/m^2	150–200 plants/m^2	Low density: 1–3 plants/m^2
Weed suppression	Excellent due to dense canopy	Moderate (wider rows allow weeds)	Good	Low: manual weed management often required, or appropriate mechanical cultivation
Harvest time	Pre-flowering (fibers stronger, low lignin)	Just prior to seed maturity (~75% mature due to shattering)	Match fiber or grain plan depending on dual purpose	Late flowering (maximum cannabinoid content), prior to seed maturation if any seeds present (seeds are not desired)
Climatic needs	Warm summer growth season preferred	Moderate rainfall, sunny climates, dry fall	Match fiber or grain plan depending on dual purpose	Warm, sunny climates with controlled humidity
Yield focus	High fiber yield OR high hurd yield	High seed yield	Match fiber or grain plan depending on dual purpose	High cannabinoid yield (CBD, CBG, others) or high terpene yield
Market focus	Textiles, composites, paper, building materials	Oil, food products, feed	Textiles, oil, food, bio-materials	CBD oil, medicinal extracts, beauty products

hemp plants are typically planted at a density between 2500 and 8000 plants per hectare when grown using ground-covering mulches (see below) but can be grown at higher densities (~38,000/hectare, still lower than fiber or grain varieties) to achieve better ground cover and still maintain high oil yield per hectare. High-quality hemp cultivars are bred for increased cannabinoid or terpene production and are typically grown in temperate climates (Fike 2016). The plants should be spaced to reduce the risk of mold, which is common at high density due to the resinous nature of the flowers (Valizadehderakhshan et al. 2021). This leads to challenges related to weed pressure, which is controlled either by application of soil mulches/covering (often typically plastic, although biobased plastic mulches are being developed for this purpose) or by mechanical cultivation practices that remove weeds. Some growers are required to resort to manual labor force to remove weeds, which is only feasible in regions with inexpensive farm labor. More modern approaches include the deployment of automated weed control robots that use lasers to kill weeds before they grow large enough to be a problem. Modern approaches also employ densities that lead to canopy closure to shade out weeds, but at a density that supports the airflow required to prevent the mold growth mentioned above. Each variety must be evaluated individually to determine the proper density for its growth to allow for optimal yield, lowest weed pressure, and lowest growth-associated costs.

Harvesting: The timing of harvest is crucial for essential oil hemp. For optimal CBD yields, the plant should be harvested at full maturity (Tanney et al. 2021, Valizadehderakhshan et al. 2021). Farmers have typically focused on harvesting the flowering parts of the plant where cannabinoids or terpenes are most concentrated. However, many large-scale industrial applications now focus on total plant extraction, which simplifies harvest (no need to separate flower material from the stem) and downstream biomass management prior to extraction. Importantly, plants grown for essential oils need to be carefully managed to avoid excessive THC accumulation to remain compliant with current regulations. Stresses such as heat or lack of proper water can lead to elevated levels of THC, potentially rendering a crop non-compliant, a situation that all growers want to avoid.

2.2 Fiber hemp

Fiber hemp is grown primarily for its long fibers and/or its hurd (also called shives in some parts of the world, the short-fiber 'woody' section in the middle of the stem). The long fibers are used for textiles, bioplastics, composites, and building materials, among other applications, and the short fibers are used in building materials such as hemp lime, animal bedding, etc. This type of hemp is usually bred for robust growth and fiber quality and often (but not always) has lower cannabinoid concentrations. Some older fiber varieties have cannabinoid concentrations exceeding 5%.

Growing Conditions: Fiber hemp thrives in fertile soils with good irrigation. The plants are cultivated with higher plant densities compared to other types (typically over 2.2 million plants per hectare), which helps promote fiber growth (Karus and Kaup 2002). This type of hemp is resilient and can grow in a variety of climates, though it prefers temperate zones with well-distributed rainfall. It can also be grown effectively under irrigation.

Harvesting: The plant's height is key to maximizing fiber yield, and harvesting typically occurs before the plant reaches full maturity, before the flowers fully develop (Karus and Kaup 2002). The fibers are obtained from the stalks (which are also often called hemp straw) through a variety of processes, typically decortication or similar.

2.3 Grain hemp

Hemp under this classification is grown primarily for the seeds, which are valued for their high protein and fatty acid content (Callaway 2004). Hemp seed oil is typically 22–35% of the dry weight of the seed, comparable to or exceeding levels found in other oilseed crops. In addition, the seeds contain 20–40% protein, depending on the variety and the conditions under which they are grown. The hemp seeds by themselves can be used directly as animal feed, as novel 'seeds' in human food applications (multigrain breads) but are typically dehulled prior to human consumption (such as in the form of 'hemp hearts'), which is used as a high protein addition to numerous food dishes and bread products. The seeds can also be pressed or extracted to generate high-quality hemp seed oil or protein cake or powder. These products (oil and protein) are also used as food ingredients, although they also serve as high-quality sources for industrial feed stocks due to their unique properties.

Growing Conditions: As for fiber and essential oil hemp, grain hemp plants prefer well-drained soils with good nutritional properties. Hemp grown for grain is planted at a density lower than fiber hemp but higher than essential oil hemp, typically 0.5–1.7 million plants per hectare, depending on the variety and the location of growth.

Harvesting: When grown for grain, hemp is allowed to reach almost (but not complete) full maturity. Due to a large shattering problem still existing in all available varieties (as of this writing), the plants cannot be allowed to reach complete end of life maturity or a large fraction of the seed will fall to the ground prior to harvest, drastically reducing yields. Instead, plants are grown until around 70–85% (depending on the variety) of seeds reach maturity, with the remaining seeds still 'green.' This maintains the seeds on the stems long enough to allow the yield demand required to be met.

2.4 Hemp varieties and characteristics

2.4.1 Overview of hemp varieties

Essential oil hemp varieties are bred primarily for high concentrations of specific cannabinoids or terpenes, like CBD, CBG, or CBC, while keeping THC levels below 0.3% to comply with legal regulations in the United States and Canada (Mead 2017). These varieties are typically used in the production of CBD oils, edibles, and topicals in the traditional CBD marketplace but are also now of interest for inclusion in the mainstream cosmetic industry, or they are used for flavor/aroma production if bred for specific terpene profiles. There are dozens of essential oil hemp cultivars, and their availability varies by country. These plants tend to be bred for shorter height and maximum branching/ inflorescence ('cola') production to make their harvest easier and to maximize cannabinoid or terpene production per plant.

Industrial hemp varieties bred for fiber and grain production, on the other hand, typically have lower cannabinoid concentrations than cannabinoid-rich varieties, focusing instead on maximizing biomass, seed production, and fiber yield. However, some varieties, such as Carmagnola and other Italian and some eastern European strains, have higher CBD and CBG levels, exceeding 5%. Some Chinese fiber varieties also have high cannabinoid levels, which are often associated with THC levels exceeding regulatory compliance levels if the plants are allowed to mature too long, leading to problems of acceptance of some of these varieties in other regions. For example, many Chinese varieties are on the 'do not grow' list for the State of Kentucky because they were found to repeatedly exceed compliance levels in trials conducted across that state.

Fiber hemp plants are typically tall, with a strong stalk that is ideal for fiber extraction. The plants are grown such that they have minimal branching, due to their growth in dense rows (Karus and Kaup 2002). These varieties often exhibit a high level of adaptability to different climates and soil conditions, which makes them ideal for large-scale commercial cultivation. Modern approaches to fiber hemp growth tend to favor high density and plant harvest at a height of 2–3 m due to availability and suitability of harvesting equipment, even though many of the fiber hemp varieties can easily exceed 4, 5, or 6 m in height. Growing fiber hemp plants too tall or at lower densities leads to large, thick, highly lignified (compared to other hemp) stems that contain a lot of hurd but that can be exceptionally hard to harvest and difficult to process once harvested.

Ideal stem diameter for growth appears to be less than 2 cm for most locations and fiber hemp applications. For some applications, such as very high-quality textile production, as is found in Normandy and Belgium, plants are commonly targeted for only 2 m high, growth and stem diameters of less than 0.7 cm. The 'right' varieties grown like that can produce a very high bast fiber to hurd ratio, with exceptionally high-quality bast fiber quality and strength, making them ideal for textile production. Recent efforts with graphene production suggest that the best graphene can also be produced from the best quality bast fibers (Hossain et al. 2022). As mentioned above, fiber hemp plants are typically harvested prior to inflorescence development, before significant lignification begins in the stem. So, collection of seed is usually not an option. But leaf material and the initial inflorescence (early flowers) can be collected separately from the stem (if the right equipment is used) and utilized in applications such as animal bedding, animal feed (although there are concerns about cannabinoid levels in the tissues and what that might mean for the animals), and extraction for CBD/other cannabinoid collection. Many fiber hemp varieties can therefore be grown as a dual-purpose crop.

Grain hemp varieties can be short (around 1 m in height) or taller (2–3 m), but most are in the range of 1.5–2 m. These plants tend to have thicker stems than fiber hemp types due to the necessity to hold up a large grain head that can result at the top of these plants. Most grain types have been bred to possess a single large cola. These traits are affected by the method of growth of the plants, with wider spacing favoring branching and more dense spacing supporting single cola production.

Due to the need to harvest the plants at the end of life, the stems tend to be more lignified, and most leaves have dropped, so dual-purpose cropping can occur, but there are limits on specific downstream applications. For example, the stems from grain varieties can be harvested after (or at the same time with the right equipment) as the seeds, and then used in some building material applications, such as pressed stem-based boards. Even after proper retting, such stems do not produce fiber of the same quality as true fiber types do, but the hurd can be used, and the fiber can be used in short-fiber applications (can be cottonized and used in non-woven textiles such as paper towels, baby wipes, or even running shoes). Most grain varieties have been bred for more northern climates, and as such typically do not perform very well at lower latitudes. Advanced breeding efforts may change that trend in the future.

'Autoflower' hemp varieties have not been discussed up until now in this chapter. Autoflower varieties, named as such by the cannabis community, are in fact day neutral varieties of *C. sativa* that are highly susceptible to stress-induced flower initiation. They are exclusively essential oil hemp varieties. Other, more 'standard' varieties of hemp (essential oil, grain, and fiber types alike) are day length dependent, meaning that they initiate flowering not based on how long ago the seed was planted in the ground, but instead based on the length of the day (actually the night length matters – long nights are the actual signal, but since people are not typically nocturnal creatures, they develop terms to describe phenomena based on the environment they spend most of their time in). Most hemp varieties sense a change in the length of the night associated with the changing of the seasons.

As the seasons move from summer to fall, the days get shorter and shorter, the nights longer and longer. Once a certain night length is reached, a signal within the plants turns on a change in levels of plant-growth regulators (plant 'hormones') that turn on inflorescence initiation and subsequent flower development. This feature is important because various hemp varieties differ in when exactly they perceive and act on this signal, leading to different maturation dates (some earlier, some later). Autoflower varieties have a mis-regulated system that normally does not respond to the seasonal light change, but instead turns on flowering after a certain amount of time after plant emergence, much like most other crops that humans have developed. What is unusual about the autoflower hemp (and cannabis) varieties, compared to other crops, is that they are extremely susceptible to environmental stresses, all of which act as signals for the plant to initiate flowering immediately. This signal can be flood stress, drought stress, heat stress, pest pressure, mechanical damage, or switching day length too quickly. Some growers claim that saying a cross word at an autoflower plant will cause it to initiate flowering. Autoflower varieties tend to perform best in the lower latitudes, where day length changes are slow and where they can reach plant sizes comparable to other essential oil hemp varieties. In more northern latitudes, where significant changes in day length occur from day to day, autoflower varieties tend to perform poorly outdoors.

We and other private/commercial growers we have worked with in Washington State have evaluated such varieties in different locations across the state, and not a single one has performed at a level that anyone would consider acceptable. The plants tend to max out at less than a half meter in height. Often, they initiate flowering when only 15–30 cm in height and produce one or just a small handful of inflorescences. Any kind of stress, be it weed pressure, heat stress, or even wildfire smoke, has appeared to trigger these plants to immediately turn to flower production. Because they have not established a substantial root or leaf system to support that switch, the plants remain small and very unproductive. Such varieties appear to be much more suited to indoor growth facilities.

2.4.2 Some well-known varieties

2.4.2.1 Fiber and grain

Fédora 17: Fédora 17 is a French monoecious variety bred for high fiber yield and resilience. It is particularly known for its fiber quality, making it a popular choice for textile manufacturing. Under the right conditions it also produces a large hurd fraction.

Fiber Production: Fédora 17 produces strong, long fibers with high cellulose content, ideal for use in textiles, paper, and biocomposite materials.

Seed: It also produces seeds that can be used for hemp oil extraction or in food products.

Growing Conditions: Fédora 17 thrives in temperate climates and performs well in fertile, well-drained soils. We found Fédora 17 to outperform Férimon, Futura 83, and several other varieties when exposed to extreme conditions (high temperature, drought, and insect pressure). It does well in sandy soil.

Height: This variety is medium to tall in stature, allowing for a good biomass yield, with its fiber typically harvested at full maturity.

Férimon: Férimon (typically Férimon 12) is another industrial hemp variety that is primarily cultivated for fiber and seed production. It is well-regarded for its high fiber yield, as well as its resilience to disease and adaptability to a variety of soil types. Férimon is also known for its early maturation, which is a key characteristic for farmers who want a fast-growing, reliable variety.

Fiber Production: Férimon has a reputation for producing high-quality fiber that is suitable for textiles and composites. The plant has a high fiber-to-seed ratio, making it ideal for industrial hemp applications (Karus and Kaup 2002).

Seed: Férimon is also grown for its seeds, which are used in food products and for oil extraction.

Growing Requirements: Férimon is adaptable to a wide range of soil conditions, from loamy to clay soils. It requires a moderate amount of water and grows best in temperate climates. This variety is particularly resilient in regions with slightly harsher climates, making it suitable for diverse agricultural conditions (Valizadehderakhshan et al. 2021).

Early Maturation: Férimon's early flowering nature allows it to be harvested in a shorter growing period, which is beneficial for regions with shorter summers or those that want to rotate crops quickly (Fike 2016).

Industrial Fiber: Similar to Fédora, Férimon is used for its fiber in the production of biocomposites, textiles, and biodegradable products.

Nutritional Products: Its seeds are a rich source of hemp oil, used in cosmetic formulations and as a functional food ingredient.

In addition to Fédora 17 and Férimon, several other European hemp varieties are commonly cultivated, primarily for industrial purposes such as fiber, seed production, and sometimes for cannabinoid extraction. These varieties have been specifically bred for high yields, disease resistance, and adaptability to various growing conditions. Here are some of the notable European hemp varieties:

Futura 75: Futura 75 has been one of the most popular industrial hemp varieties in Europe. It is primarily grown for fiber and hurd production, but also for its seed production, being known for its strong stalks and high seed yield.

Fiber Production: Futura 75 is prized for its strong, long fibers, which are ideal for use in textiles, paper, and composite materials.

Seed: It produces high-quality seeds, which are rich in omega-3 fatty acids and are used in nutritional products.

Growing Conditions: This variety thrives in temperate climates and is resistant to many common hemp diseases, making it a reliable choice for large-scale farming (Karus and Kaup 2002).

Height: Futura 75 is typically tall, reaching heights of up to 3 meters or more, which makes it ideal for fiber production.

Futura 83: Futura 83 was derived from Futura 75. It matures later, which for the fiber/hurd-based industries is a good thing. It has higher fiber and hurd yields, has a higher hurd-to-fiber ratio, and better stress resistance. Otherwise, it is very similar to Futura 75.

USO 31: USO 31 is another popular variety in Europe, primarily used for fiber production but also cultivated for its seeds. It is well-regarded for its high biomass and early flowering, making it suitable for regions with shorter growing seasons.

Fiber Production: USO 31 produces strong fibers that are used in the textile and paper industries. The plant has a low lignin content, making the fibers easier to process (Fike 2016).

Seed: The seeds are of medium-to-high quality and are used in oil extraction for food and cosmetic products.

Growth Conditions: It performs well in a variety of soil types and is suitable for cultivation in many parts of Europe, including areas with slightly cooler climates. This variety is known for its early flowering, which allows farmers to harvest the plant in a shorter time span (Valizadehderakhshan et al. 2021). We have found that it does not perform as well as some of the other varieties on this list when exposed to drought or heat stress conditions.

Carmagnola: Carmagnola is a variety traditionally grown in Italy and has gained popularity across Europe and North America for its high fiber yield and quality seeds. It also produces about 5% CBD in its inflorescences, but not in the stems, and thus can be used for fiber, seed, or essential oil applications, although it is most commonly grown for fiber. It is a dioecious variety, with the males maturing and senescing earlier than the female plants. As a result, in some areas hand harvest has been used in its management, with the males harvested earlier than the females and then the females harvested a few weeks later. The males tend to have 'finer' fiber quality and are sought after for specialty paper applications, for example. It is possible to harvest both males and females at the same time, but the males will tend to be 'past their prime' and will yield a poorer fiber fraction as a result.

Fiber Production: Carmagnola is well-known for its long fibers that are perfect for use in textiles, ropes, and hurd production for building materials such as hemp lime (hempcrete).

Seed: The seeds are highly nutritious and used in the production of hempseed oil, which is rich in polyunsaturated fatty acids.

Growing Conditions: Carmagnola prefers fertile soils with good drainage and thrives in a mild climate. The variety also shows good resistance to disease and pests (Karus and Kaup 2002) and to heat stress (our experience in 2021).

Height: Carmagnola is known to grow quite tall, which maximizes its biomass and fiber potential.

Santhica 27: Santhica 27 is a French variety that is primarily used for fiber production and is also cultivated for its low THC content in compliance with European regulations.

Low-THC Content: Santhica 27 is specifically bred for its low-THC levels, making it ideal for markets that require low-THC industrial hemp varieties (Mead 2017).

Fiber Production: It produces long, strong fibers that are used for textile manufacturing, paper, and biocomposite materials.

Seed: The seeds of Santhica 27 are used to make hemp oil and have a relatively high protein content, making them a valuable commodity in the food industry.

Growing Conditions: This variety is early-maturing and performs well in moderate climates with consistent rainfall.

Kompolti: Kompolti is a Hungarian variety known for its resilience and high fiber yield. It has been used in Europe for many years, primarily for industrial purposes such as fiber extraction and seed production. It is one of our favorites for growth in Washington State.

Fiber Production: Kompolti produces long, high-quality fibers that are widely used in textiles and construction materials (Karus and Kaup 2002).

Seed: The seeds are nutritious, rich in essential fatty acids, and are often used in hempseed oil production for the food and cosmetic industries.

Growing Conditions: Kompolti is adaptable and can be grown in various soil types, though it performs best in well-drained, fertile soils with adequate water supply. It tolerates heat and drought stress better than most varieties we have evaluated.

Height: Kompolti plants typically grow tall and have a strong stalk, which is beneficial for fiber production.

2.4.2.2 Essential oil/cannabinoid/chemical

Harlequin: Harlequin is an essential oil hemp line bred for high CBD content. This variety was a poster child of the medical marijuana industry in the United States, known not for THC but for CBD instead.

Hot Blonde: Hot Blonde is an essential oil hemp line bred for high CBD content.

ACDC: ACDC is an essential oil hemp variety bred for CBD production that is a precursor to many current varieties. ACDC has shown variable regional results and seems to perform best indoors where a more controlled environment can be achieved.

Cherry Wine: Cherry Wine is an essential oil hemp variety bred in Colorado for terpene production. It was derived from Cherry, a high CBD-producing cannabis line, and was selected as having significantly reduced THC levels. Cherry wine has been shown to grow best at high elevations outdoors.

Hawaiian Haze: Hawaiian Haze is an essential oil hemp variety also bred for its terpene profile. It is frequently chosen for essential oil extraction due to its aromatic properties (Russo 2011).

Lifter: Lifter is an essential oil variety bred and cultivated for high CBD content and terpene profile production with large flowers. Lifter has become a prominent variety for extraction and smokable flower and has been shown to have good disease resistance. A triploid version of it was developed in the Western United States, leading to greatly reduced seed production in the harvestable fraction.

White CBG: White CBG is an essential oil hemp variety bred to target CBG specifically with low THC for extraction and smokable flower. The terpene profile is unique to this plant as well, and it appears to possess strong stress and disease resistance.

2.5 Uses and market demand

2.5.1 Uses of hemp

Essential Oil Hemp: Cannabinoids such as CBD (cannabidiol) and CBG (cannabigerol) are non-psychoactive compounds with potential health benefits, including anti-inflammatory, anxiolytic, and neuroprotective properties (Laprairie et al. 2015, Mead 2017). THC, within legal limits, has applications in pain management and nausea control (Rock et al. 2014, Serpell et al. 2014). Hemp extracts with CBD are used in skin-care formulations for their anti-inflammatory and antioxidant effects (Stella et al. 2019). CBD is being researched for use in animals, particularly for anxiety and chronic pain (Deabold et al. 2019). Terpenes from hemp, such as myrcene, pinene, and limonene, are bioactive compounds with therapeutic potential, including stress relief and antimicrobial properties (Pieracci et al. 2021, Russo 2011). These and other terpenes from hemp are increasingly used in flavoring, fragrances, and cleaning products (Weston-Green et al. 2021).

Fiber Hemp: Hemp fibers are valued for their tensile strength and are used in textiles, ropes, and composites (Karus and Kaup 2002). In construction, hemp lime/hempcrete offers thermal insulation and sustainability advantages (Stevulova et al. 2013).

Grain Hemp: Hemp seeds are a rich source of protein and essential fatty acids, beneficial for cardiovascular health (Callaway 2004). Hemp seed oil, known for its balanced omega-3 and omega-6 ratio, is used in nutrition and cosmetic products (Bakowska-Barczak et al. 2022, Mazzara 2023, Rizzo et al. 2023). Biodegradable plastics derived from hemp offer an eco-friendly alternative to petroleum-based products (Fike 2016). Epoxies and other glues/resins can be generated from hemp seed oil and hemp seed protein and are being developed for numerous industrial applications.

2.5.2 Market demand

Global Market Overview: The 2018 Farm Bill legalized hemp cultivation in the United States, defining hemp as cannabis with less than or equal to 0.3% THC content by dry weight (Agriculture 2020). This regulatory change spurred significant market growth, particularly in cannabinoid extraction. Similarly, Canada has permitted industrial hemp production since 1998, allowing for THC levels of up to 0.3% in flowering heads and leaves (Health Canada 2021). The North American hemp market was valued at approximately $2.8 billion in 2022, with projections estimating a compound annual growth rate (CAGR) of 16% through 2027 (Grand View Research 2024). The United States leads the market, driven by strong demand for CBD products, while Canada dominates seed and fiber production due to its favorable policies and established infrastructure (Health Canada 2021). The global hemp industry is projected to grow from $6.3 billion in 2024 to $16.3 billion by 2030, with a compound annual growth rate (CAGR) of over 17% (Grand View Research 2024). Major drivers include increasing legalization of hemp cultivation and cannabinoid use and rising demand for sustainable materials and plant-based wellness products.

Cannabinoid Products: The CBD segment dominates, with high consumer interest in wellness, pain relief, and sleep aids. Minor cannabinoids like CBG and CBN are gaining traction for specialized applications, such as neuroprotection and sleep regulation (Citti et al. 2018).

Industrial Applications: Hemp textiles are in demand for eco-friendly fashion, with countries like China and France leading production. Construction materials like hempcrete align with global trends toward sustainable building (Stevulova et al. 2013).

Nutritional Products: Hemp seeds and oil are gaining popularity as superfoods, particularly in North America and Europe, due to their health benefits and sustainability (Callaway 2004).

2.5.3 Key drivers of demand

Legalization and Policy Support: The United States and Canada have developed extensive programs to support hemp cultivation and processing. For example, New York's Industrial Hemp Program provides grants and technical assistance to growers (New York State Department of Agriculture and Markets 2025).

Consumer Awareness: Growing interest in sustainable and organic products drives demand for hemp-derived goods, particularly in wellness and construction sectors (Grand View Research 2024).

Technological Advancements: Innovations in cannabinoid extraction and bio-composite processing expand hemp's applications, creating opportunities in pharmaceuticals, bioplastics, and automotive industries (Fike 2016).

2.5.3.1 Key market trends

CBD Products: Rapid legalization of hemp-derived cannabinoids (e.g. CBD) in North America has significantly fueled demand, especially for wellness and pharmaceutical applications.

Fiber Applications: Rising interest in eco-friendly materials and sustainability has boosted demand for hemp textiles, paper, and composites.

Hemp Food Products: The growing popularity of plant-based diets and health-conscious consumers has driven demand for hemp seeds, oil, and protein.

Animal Feed: This is another emerging market with current allowances for dehulled and pressed seed meal for laying hens.

Hemp in Construction: Hempcrete, made from hurd and lime, is gaining traction as an alternative to traditional concrete due to its insulation properties and carbon sequestration potential.

Policy Support: Increased support for hemp cultivation, especially under the US 2018 Farm Bill and EU regulations, has expanded production and market penetration.

3 Agronomic practices

From proper seed selection through to harvesting techniques for specific end-uses, a number of factors are critical to successful cultivation of hemp. Purpose-driven choices on fiber, grain, and essential oil/chemical production along with efficiencies and best practices will be addressed in more depth throughout this section.

3.1 Seed selection

Choosing the appropriate seed for hemp cultivation is a fundamental agronomic decision that can significantly influence crop performance, yield, and quality. Seed selection must align with the desired end-use (essential oil, fiber, or grain production), local climate, soil type, and specific management practices. The importance of matching seed characteristics to the specific agronomic and climate conditions for desired hemp types – fiber, grain, or essential oil varieties – cannot be emphasized enough. Selecting high-quality, disease-free seeds with known genetic traits enhances yield predictability. Pre-sowing seed preparation, including scarification, pre-soaking, and temperature conditioning, can significantly boost germination rates and uniform seedling establishment (Wylie et al. 2021, Sunoj Valiaparambil Sebastian et al. 2023).

3.1.1 Genetic consistency and purity – worthwhile goals for hemp

Selecting genetically stable cultivars in crop plants ensures uniform plant growth, predictable yield, and consistent product profiles. That is the case for most crops today. But it is not the case for hemp, not yet. Most of what is currently available worldwide when it comes to suitable outdoor hemp 'varieties' are in reality mixed populations of genotypes that share a set of particular common traits that were used to select them. This can include an approximate plant height, falling within a range of stem diameters, approximate date of maturity, and overall approximate yield for fiber or grain. In reality, when these 'varieties' or 'cultivars' are planted in the field, particularly when any kind of stress is applied to them, a wide range of genotypes is observed. Seed oil composition varies, as does seed size (the latter widely in many cases). This feature of current hemp varieties is both a problem for many growers, who are keen to grow a uniform crop, and also a potential resource for breeders, because of the large amount of variability present in a given 'variety,' there is a lot of genetic space to play with for efforts to breed new varieties with desired characteristics. Certified seeds, which will result from such breeding programs, will have high genetic purity and are preferred because they minimize variability and maximize productivity. Genetic uniformity is especially critical for CBD production, where variations in cannabinoid concentration can affect market value and compliance with regulatory limits on tetrahydrocannabinol (THC) content (Amaducci et al. 2015). Over time, we will see certified varieties become more widely available.

3.1.2 Adaptation to local conditions

The choice of seed should be informed by the local growing environment, including climate, day length, and soil properties. Cultivars adapted to regional climates are better equipped to resist local pests, diseases, and abiotic stress factors such as drought or high humidity (Clarke and Merlin 2016). However, for many parts of the world that have not had continuous hemp breeding efforts over the past several decades, which is most of the world, varieties developed elsewhere are all that is available. In such cases, particular care should be placed on identification of those varieties to test out that come from areas that have conditions that are as close as possible to the conditions of the local site. There is a significant need for establishment of numerous hemp breeding programs around the world that focus on adapting and modifying hemp germplasm to meet the local needs.

Comparative studies assessed regionally adapted seed varieties of hemp across multiple European climatic zones. Germination rates, seedling growth, and yield outcomes were compared, leading to conclusions regarding the importance of selecting cultivars adapted to specific climates, leading to better resilience to local pests, optimized cannabinoid profiles, and improved yield stability. For instance, seeds bred for northern climates exhibited faster germination and canopy development, critical for overcoming shorter growing seasons . Teirumnieka et al. (2021) found that a fiber variety, 'Bialobrzeskie,' maintained consistent fiber yields regardless of temperature fluctuations, whereas 'Finola' and 'Purini' showed yield reductions with increased temperatures and enhanced yields with higher moisture levels when grown in Latvia. In contrast, 'USO-31' exhibited less sensitivity to temperature changes but slight reductions in yield with increased rainfall. The Estonian trials underscored similar regional climatic impacts, emphasizing the need to match cultivar characteristics with specific climate conditions to achieve optimal yields for fiber and grain hemp production.

As described (Stramkale et al. 2024, Teirumnieka et al. 2021), local climatic factors also significantly impacted hemp yield. Specifically, fiber hemp varieties showed in

creased sensitivity to temperature and moisture variations. In years with higher rainfall during the growing season, fiber yield was consistently higher. Meanwhile, dry spells negatively impacted early growth stages and overall yield. Seed hemp varieties also demonstrated variability, with moisture being a critical factor for optimal germination and seed establishment. These findings underscored the necessity of selecting region-adapted cultivars to maximize yield potential. These studies collectively highlight the value of certified, high-quality seeds and pre-sowing treatments in maximizing germination and stand uniformity while tailoring to local conditions and crop-specific end-uses.

3.1.3 Seed germination rates and viability

Seeds with a high germination rate (typically >80%, but preferably >90%) are crucial for successful plant establishment. Low germination rates can lead to uneven plant stands, higher input costs due to reseeding, and increased susceptibility to weeds and diseases. The seed lot's germination rate and viability can be assessed by germination tests performed under controlled conditions (Jovičić et al. 2019). This rate is then taken into account, and the amount of true live seed necessary to achieve the desired end density needs to be planted. For example, if a grower wants to plant a fiber variety at a final density of 3 million plants per hectare, but the seed lot obtained had a measured germination rate of just 85%, a total of 3 529 412 (3 000 000 ÷ 0.85) seeds per hectare would need to be sown. Based on the measured germination rate, the grower could conclude that about 3 000 000 seeds per hectare would germinate (true live seed) and that many plants would grow in the field. Shipping and storage conditions can affect seed viability and germination rates. Hemp appears to be particularly sensitive to that, particularly when seed is shipped around the world. Therefore, it is critical that growers measure the germination rate of each lot of seed obtained (see below).

3.1.3.1 How to measure rate

The germination rate of hemp seeds is a key agronomic parameter that determines plant population density and uniformity in the field. A high germination rate, typically above 80%, ensures a robust crop stand and reduces the need for reseeding, thus minimizing labor and input costs (Islam et al. 2021). Germination rates can be influenced by seed quality, storage conditions, and seed age, as well as environmental factors like temperature and moisture availability during the germination phase. The standard method for measuring germination rate involves placing seeds under optimal controlled conditions and recording the percentage of seeds that germinate within a specified period (e.g. 7–14 days). Testing is commonly conducted in laboratory settings or growth chambers to ensure consistent conditions of temperature, humidity, and light. Field-based tests also provide valuable insights but can be influenced by varying environmental conditions (Bouloc et al. 2013). Several recent reports emphasize the importance of following the above guidance for seed selection and preparation.

3.1.3.2 Seed quality impact

Using high-purity, disease-free seeds can lead to increases in germination rates and improved uniformity of plant stands. Genetic uniformity, particularly for cannabinoid hemp crops, due to the regulatory constraints on THC levels, is a significant factor to consider when choosing seed for planting. Clarke and Merlin found that genetic uniformity and regional adaptation in hemp seeds led to better germination rates and yield outcomes (Clarke and Merlin 2016). Seeds tailored for specific climates not only improved crop establishment but also reduced weed pressure and minimized THC variability, a factor critical for compliance in cannabinoid hemp production (Clarke and Merlin 2016, Kaminski et al. 2024). Although these studies were carried out on cannabinoid hemp types, it is reasonable to conclude that similar results will follow for fiber and grain hemp types as well.

3.1.4 Pathogen-free status or evaluation

Ensuring that the seed is free from pathogens (e.g. fungal spores, bacterial contamination) is essential to minimize disease pressure during early growth stages and to prevent contamination of the field by unwanted pathogens. Certified seed is often tested for the presence of pathogens, making it a safer option compared to non-certified sources (Islam et al. 2021), but see above for comments on the current availability of certified seed. It is possible to get seed evaluated for the presence of pathogens. For example, in the United States, the US Department of Agriculture has offices in each state or region that can perform such analyses – to look for pathogen spores. The cost is typically comparable to a soil test and should be considered as a good investment in the long-term maintenance of the farm.

3.2 Seed preparation techniques

Proper seed preparation enhances germination rates and establishes a robust plant stand, particularly under field conditions that may present challenges such as variable moisture or temperature extremes. Several methods have been shown to enhance seed germination, as follows.

Scarification: In some cases, hemp seeds may have a hard seed coat that inhibits water uptake. Scarification, which involves lightly scratching or weakening the seed coat using mechanical abrasion or chemical treatments, can improve water absorption and promote more rapid and uniform germination (Ranalli 1999).

Priming: Seed priming is a pre-sowing treatment involving soaking seeds in water or nutrient-enriched solutions and then drying them before planting. This treatment can enhance seed vigor, accelerate germination, and improve resistance to environmental stressors during early growth stages (Finnan and Burke 2013). Osmopriming, a variant that involves soaking seeds in osmotic solutions, may be effective for improving germination under suboptimal moisture conditions.

Coating and Pelleting: Seed coating involves applying a layer of protective material (e.g. clay, polymers, or a mix of nutrients and pesticides) to the seed surface. Coating can offer multiple benefits, including better handling and sowing uniformity, protection against soil-borne pathogens, and the incorporation of micronutrients or biostimulants (Amaducci et al. 2015). Pelleting, where seeds are coated to form uniform, rounded pellets, is especially useful for mechanical sowing, ensuring even distribution and improved seed-to-soil contact.

Pre-Soaking or Moisture Conditioning: Pre-soaking seeds in water for 12–24 hours before sowing can accelerate the initial stages of germination, particularly in dry environments. Care must be taken not to over-soak, as prolonged exposure can lead to seed decay or fungal growth. After soaking, seeds should be thoroughly dried to avoid clumping during sowing (Meijer 1995).

Disinfection Treatments: Treating seeds with fungicides, bactericides, or other antimicrobial agents reduces the risk of seed-borne and soil-borne infections during germination and early seedling establishment. Organic growers may opt for non-chemical treatments such as hot water or biological inoculants (Clarke and Merlin 2016).

3.3 Importance of seed quality testing

Before planting, it is advisable to conduct germination tests to evaluate seed viability. Standard germination tests involve placing a representative sample of seeds under controlled conditions (e.g. optimal temperature and moisture) and recording the percentage of seeds that successfully germinate within a predetermined period (Islam et al. 2021). This step ensures that only viable seeds are planted, improving plant density, uniformity, and yield potential.

3.4 Planting methods

Once seeds are selected, tested, and prepped, measuring the rate of seed germination to ensure the proper target seed count is the next crucial step pre-sowing. Sowing techniques with an emphasis on depth and plant density are as important to purpose-driven end-uses as seed selection. Essential oil plants have numerous specific differences *compared to* fiber and grain varieties.

3.4.1 Sowing techniques

Sowing methods for outdoor hemp cultivation vary depending on the end-use of the crop and local growing conditions. Broadcast sowing and row planting are the primary techniques used for establishing hemp fields. Broadcast sowing involves dispersing seeds uniformly across the field, often followed by harrowing to ensure soil contact. This method can result in uneven plant densities, though it may be economically advantageous for large-scale fiber production (Meijer 1995).

Row planting, in contrast, allows for precise control over seed depth, spacing, and plant density. This technique, typically performed using seed drills, is preferred for CBD and seed production as it facilitates better weed control, uniform growth, and targeted cultivation practices (Salentijn et al. 2015). Hemp prefers to be planted very shallow, the shallower the better. Most varieties display best germination and stand establishment rates when planted at a depth of less than 0.5 cm and when provided with substantial and solid moist soil contact. Planting into dry soil or planting too deep leads to very irregular germination, even with subsequent moisture application (e.g. rainfall or irrigation). The best practice is to plant into soil that is well hydrated but not muddy wet and to plant as shallow as is possible for the field conditions.

Hemp seeds need good soil contact for proper moisture transference and efficient germination, so soil compression over the planted seeds can significantly improve germination rates. Planting at a depth less than 0.5 cm is not always possible to achieve, however, depending on the soil conditions and equipment available. For example, when planting into very sandy soil, the seeds can easily end up at greater than 1 cm depth, even if targeted for shallower planting, particularly if the soil is not hydrated well enough. Reasonable germination can be achieved for many varieties when planted to a depth of 1–1.5 cm. Planting at depths greater than 3 cm often leads to greatly reduced germination and poor stand establishment, leading to issues related to weed pressure and overall yield. Breeding efforts could address this point and develop varieties that prefer or at least tolerate deeper sowing depths. Such a development would increase stand establishment success and decrease grower costs.

Another factor that can affect hemp establishment is soil crusting (Amaducci et al. 2015). When soil has high clay levels and is allowed to crust over after planting but before seedling emergence, that emergence can be severely affected, reducing stand establishment success. Keeping soil uniformly damp, but not soaking wet to the point of pooling, significantly improves seedling emergence and establishment.

Essential Oil hemp cultivators usually start seeds in a greenhouse and plant seedlings at 3–5 weeks of age. Planting in this manner provides better assurance of plant survival, while direct seeding of these varieties has shown in our field trials to have higher yields, but less plant counts in the field. One notable difference between transplants and direct seeded plants is that the former tend to have a more branched, shallower root system, which works well with drip irrigation, while the latter tend to have a single, deeper tap root, which leads to better resistance to drought stress and potentially could explain the higher yields observed with that method of planting.

3.5 Plant density and row spacing

Optimal plant density and row spacing for hemp depend on the intended crop purpose, as introduced above. Manipulating plant density and row spacing influences not only yield but also weed suppression, nutrient competition, and disease pressure. Row spacing in outdoor hemp cultivation has significant implications for weed suppression and final crop yield. Narrower row spacing effectively limits weed competition during the early growth stages of hemp due to rapid canopy closure (Finnan and Burke 2013). However, wider spacing may improve air circulation, reducing fungal disease incidence (Salentijn et al. 2015). Optimal configurations vary depending on the type of hemp being grown (whether for CBD, fiber, or grain) since each type exhibits unique growth habits and requires different management strategies. For fiber production, dense planting (300–400 plants per m^2) promotes taller, slender stalks, maximizing fiber yield and quality (Struik et al. 2000). Conversely, CBD production typically requires wider plant spacing (usually between 1 and 2 m between rows and between 0.5 and 1.5 m between plants within the row) to allow for extensive branching, light penetration, and airflow around each plant, which is critical for cannabinoid/terpene production and disease and pest control (Ranalli 1999). For grain production, hemp plants are typically planted at densities between 50 and 160 per m^2 (most often around 100 per m^2) so they have room to generate significant seed heads.

3.5.1 Fiber hemp

Fiber hemp is typically grown at high densities, with plant populations ranging from 200 to 400 plants per m^2. Studies have shown that narrower row spacing (e.g. 7–15 cm) for fiber hemp results in rapid canopy closure, which is highly effective in suppressing weed growth due to shading and competition for resources (Amaducci et al. 2015, Finnan and Burke 2013). This dense spacing encourages the development of tall, unbranched stalks, maximizing the fiber yield and quality by reducing lateral growth. In contrast, wider row spacing (e.g. 30 cm or more) has been found to reduce plant height and decrease overall fiber yield, primarily due to increased competition from weeds and the tendency of plants to branch more freely. An experiment conducted in Italy compared row spacings of 7.5 cm, 15 cm, and 30 cm for fiber hemp cultivation. It found that narrower spacings provided a weed suppression rate of over 80% and increased stalk yield by up to 25% relative to wider configurations (Amaducci et al. 2015).

3.5.2 Essential oil/cannabinoid/CBD hemp

Hemp grown for CBD or terpene production requires much wider spacing compared to fiber varieties. Optimal spacing ranges from 1 to 2 meters between rows and plants, promoting extensive lateral branching and increased exposure to light. Such configurations are critical for maximizing the number of flowering sites and enhancing resin production, which directly impacts CBD content and overall biomass yield (Ranalli 1999). Wider row spacing in essential oil/CBD hemp has been shown to reduce disease pressure, as increased airflow between plants minimizes the risk of fungal infections. However, these wider configurations pose challenges for weed management. An early experiment conducted across multiple North American sites suggested that narrower row spacing (0.5 to 1 meter) led to greater weed pressure due to inability of available equipment to cultivate effectively between rows and decreased CBD yield per hectare despite a larger plant count (Clarke and Merlin 2016). Effective weed suppression strategies in wider-spaced CBD crops often include mechanical cultivation or the use of cover crops and mulches. The largest CBD hemp grower in the United States (Columbia Basin Biosciences) has found the opposite to be the case due largely to its ability to manage weeds using effective control strategies that it employs closely planted rows with plants grown at higher densities per hectare (around 35,000 to 40,000).

3.5.3 Grain hemp

Grain hemp, typically cultivated for seed production, requires intermediate row spacing. Common configurations range from 15 to 60 cm row spacing with in-row plant density matching an overall end density sought after, depending on local conditions and desired outputs. Narrower row spacing (e.g. 15–30 cm) improves weed suppression by rapidly establishing a closed canopy, much like fiber hemp (Meijer 1995). However, excessively dense planting can lead to competition for light, nutrients, and moisture, ultimately decreasing seed yield and size.

3.5.4 Summary row spacing impacts

Fiber Hemp: Narrow row spacing (7–15 cm) maximizes fiber yield and suppresses weeds effectively; wider spacing reduces fiber yield due to increased weed competition.

CBD Hemp: Wider row spacing (1–2 m) maximizes flowering and resin production but may require additional weed management measures. Narrow spacing (0.5–0.9 m) can produce good yields if appropriate weed control equipment is available to match the spacing.

Grain Hemp: Intermediate spacing (15–60 cm) balances weed suppression and grain yield, with 30 cm often cited as optimal for seed production.

3.6 Soil requirements for outdoor hemp cultivation

Outdoor hemp cultivation relies heavily on soil quality and preparation to maximize yield and quality. The following sections focus on soil considerations specifically for growing hemp outdoors, addressing the natural challenges and opportunities unique to field conditions.

3.6.1 Soil types and preferences

For outdoor grown hemp, the soil type has a large impact on ultimate yield and success. Hemp performs best in well-drained loamy soils, which combine sand, silt, and clay in proportions that ensure proper aeration, water retention, and drainage. Field trials have shown that sandy loam soils provide the ideal structure for root development and nutrient availability, making them a top choice for outdoor cultivation (Small and Marcus 2019). Poorly drained soils, such as heavy clays, should be avoided or amended, as they can lead to root suffocation and increased susceptibility to diseases like root rot. This is particularly the case in areas with high rainfall, as hemp plants 'do not like wet feet.' Hemp roots need proper access to air and do not do well in constantly wet soils. Soil that gets very wet but that can drain well enough is suitable. Sandy soils, often found in outdoor fields, may require organic matter additions to improve water and nutrient retention, but with proper sowing (see comments above) and stand establishment, hemp grown in sandy soil can perform very well as long as plant nutrition and sufficient irrigation (either rainfall or human provided) is properly attended to.

3.6.2 Soil preparation and conditioning

Outdoor fields often require more extensive preparation than controlled environments. To prepare the soil, the following factors need to be properly addressed:

Tillage and Aeration: Tilling outdoor fields loosens compacted soil, promotes aeration, and allows roots to penetrate deeper. However, such practices go against soil conservation practices and are in opposition to sustainable and regenerative agriculture goals. Tilling can disrupt beneficial microorganisms (particularly mycorrhizal fungi) and lead to soil erosion. In conservation agriculture, minimal tillage practices or no till practices are recommended. Hemp, with its strong root system that can break up compressed soils, is an ideal addition to farming systems that seek to be sustainable or, better yet, regenerative in nature. Light and shallow surface discing, when appropriate, can provide a better soil bed for hemp, leading to better stand establishment and weed suppression.

Organic Amendments: Incorporating organic matter, such as compost or aged manure, improves soil fertility and structure. Outdoor fields can also benefit from using biochar to retain moisture and enhance microbial activity (McPartland et al. 2000).

Cover Crops: Planting cover crops such as clover, rye, barley, or vetch (among others) during the offseason enriches the soil by adding/maintaining nitrogen, suppressing weeds, and preventing erosion. For outdoor hemp, this is especially beneficial as hemp requires sufficient nitrogen levels for early growth.

Crop Rotation: Rotate hemp with legumes, cereals, or other crops to prevent nutrient depletion and reduce the risk of pests and diseases that thrive in monoculture systems. Repeat planting (planting in the same field in sequential years) is highly discouraged because it leads to disease and pest pressure increases over time. A three-to-four-year crop rotation schedule is recommended, but longer rotations may be required in some regions with particularly high disease pressure.

3.6.3 pH and nutrient balance

Outdoor soils often vary in pH and nutrient content depending on the location and prior use. Hemp grows best in soil with a pH range of 6.0 to 7.5, which allows for optimal nutrient availability. It can perform adequately with sub-optimal soil conditions, and different 'varieties' perform differently under these differing conditions.

Testing and Adjusting pH: Regular soil testing is critical. Lime is commonly used in outdoor fields to raise pH, while sulfur or gypsum can lower it. For large fields, amending the soil several months before planting ensures adjustments take effect prior to planting and stand establishment.

Nutrient Requirements: Outdoor hemp relies on naturally available and supplemented nutrients. Key elements include:

- **Nitrogen (N)**: Essential for vegetative growth, but excessive nitrogen can delay flowering. Outdoor growers may opt for slow-release fertilizers or green manure from legumes for a natural nitrogen boost. N levels for hemp are not the same as for many other crops. Most hemp varieties do best with N levels at 60–100 units (lbs/acre, kg/hectare) but can perform well between 50 and 150 units. Some varieties, such as Fibror, prefer N levels closer to 30 units to produce the best quality fiber. The exact N level applied depends on the purpose of the crop, whether it be for seed, fiber, or essential oil, with each purpose requiring different N levels.

- **Phosphorus (P)**: Supports root growth and seed production. Outdoor soils often need supplemental phosphorus early in the season.
- **Potassium (K)**: Enhances water regulation and plant strength, especially critical for hemp grown in areas prone to drought.
- **Micronutrients**: Elements like magnesium, zinc, and boron should not be overlooked, as deficiencies can hinder outdoor growth and cannabinoid production (Bocsa and Karus 1998).

3.6.4 Example of two fields in Prosser, WA

For one of the first years that we were able to grow hemp in Washington State, 2021, we grew in two fields:

Field 1: Pear Acres was a fiber and grain hemp trial field. The soil was a very sandy loam.

Field 2: Roza Farm was an essential oil/cannabinoid/CBD hemp trial field. The soil was a clay loam.

After performing soil test analyses for each field (Fig. 2), we received the following recommendations from the nutrient supplier field agronomist:

"Nitrogen: Rates depend on goal (whether the crop is for oil or grain/fiber). Hemp for oil can easily be over-fertilized with nitrogen.

- **Pear Acres**: I think we should target a top end of 100-lbs N. There is about 20 lbs residual N in the soil now and likely a very low contribution of N from organic matter as it reported very low. Therefore about 80 lbs N per

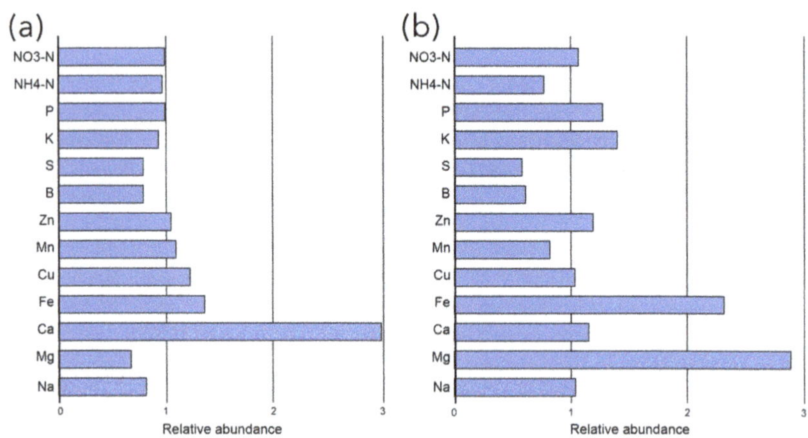

Figure 2 Soil nutrient test results from Roza Farm (a) and Pear Acres (b) in Prosser, WA, 2021, prior to crop planting.

acre should be top end rate considering the residual N, and that we don't want to overload salt.

- **Roza:** We can plan to use 40–50 lbs total N over the course of the season through regular weekly injections at Roza. This can be guided by tissue tests as well. Similar with hops and cannabis where the flower is the goal, about 75–80% of the N should be delivered to the crop by bloom time.

Phosphorous: Oil levels of hemp respond positively to phosphorous from what I have seen, and in the same way that hops, canola, and other oil crops do. I've researched this a lot out of curiosity over the years of working with oil crops. Apparently, all of the biochemical plant processes that produce oil, lipids, fatty acids, etc. require a lot of phosphate.

"Anyways, this is something that really fascinates me and is fun to watch actually work in the field. Some cannabis growers have speculated that phos directly increases THC, but so far I haven't seen data so it's all anecdotal. I also haven't been able to reproduce that result in field situations of industrial hemp receiving ample phos additions. It could be variety dependent as well.

- **Pear Acres**: Higher reported labile phos, so we could get by with a minimal addition in a dry blend of about 50 lbs P_2O_5 per acre.
- **Roza**: There will be a larger need for phosphorous because the reported labile levels are low, there is very little organic matter, there is a lot of free lime in the soil which prevents phos availability, and it matters so much to oil production. Liquid phos acid injections are recommended there because of the high pH and free lime. These will be the most efficient way to deliver P while also working to neutralize the calcium carbonate. We should target weekly applications of 3–5 lbs P_2O_5 per acre throughout the growing season.

Potassium:

- **Pear Acres**: Clint mentioned that grain hemp takes up incredible amounts of K. With incredible biomass and height production, K is not surprisingly a must (essential for cell elongation). Conveniently there is higher reported K2O at Pear acres where the grain will go. It would still be nice to see a target minimum level of 250 ppm. I recommend adding 75 lbs K as Sulfate of Potash 0-0-50 to the dry blend. We could go higher, but I believe in keeping salt levels as low as possible for successful seed germination. Recommending SOP instead of KCl also to keep the salts low.
- **Roza**: K would not be quite as essential here, but we should still supplement it. Adequate K timed correctly has a direct response in flower (cone) size.

Phosphorous aids with this as well, but of course has the bigger influence on oil content. I recommend light to moderate K early on, and transitioning toward heavier K at the same time we transition away from N at Roza. The "reproductive stage" would be the best timing to start reducing N and increasing K.

Calcium:

- **Pear Acres**: Calcium is slightly low at Pear Acres, and is an essential nutrient for cell division and cell wall structure. Therefore, it will be critical to ensure that an adequate amount is available to plants. Fortunately, the Pear Acres do not report any free lime (calcium carbonate), which is unavailable for plant uptake. To bring calcium to a more balanced level, I recommend adding 200 lbs gypsum (calcium sulfate, ~21% Ca) to the dry blend.
- **Roza**: Simply neutralizing the free lime in the soil will create more plant-available calcium. In the liquid program, we should opt for calcium complexed sources of nitrogen, such as calcium nitrate (CN9 9-0-0) or calcium ammonium nitrate (CAN17 17-0-0) to supply calcium as much as possible during the vegetative stage along with N.

Magnesium: Levels are overall slightly low. However, at Pear Acres, Magnesium takes up more of the CEC than it should because the other cations levels are also low. The dry gypsum will correct that. Magnesium as a proportion of the other cations is conversely relatively low in the soil test report for Roza, however when there is a lot of calcium carbonate in the soil it skews test results. As we neutralize the free lime with phos acid, we should see the cation ratio come back into balance naturally, and I don't think we need to bother with Mag additions unless tissue tests indicate that we should.

Sodium and Salts: All low and good results.

Micronutrients: Boron is low as it always is in our soils. I would recommend adding 1 lb of B in the dry blend at Pear Acres, and 0.5 lbs B in the liquid injections over the course of the vegetative growth stage at Roza. Manganese is also slightly low. Hop manganese removal is very high relative to other crops. Not sure if we have sufficient data on that for hemp. Budget depending, could consider an addition of 5 lbs Mn in the dry blend at Pear acres, and a target of 1 lb Mn over the course of the growing season at Roza in the liquid injections. Hops respond well to Mn and Cu applications all season long, as it helps with the lignification process and maintaining green cone color. Again, not sure if it would translate this way directly to hemp, but so far a similar approach has yielded very good results in the field where we copy those hops concepts. Zinc levels are also low and will be important to include if possible in the dry blend at Pear and the liquid at Roza. It is most important during root development, so I would recommend 5 lbs of it in the dry blend at Pear and 1 lb during the vegetative stage as a liquid at Roza. Cu and Fe levels appear adequate. It may be that a multimineral liquid injection is the simplest option for Roza' (pers. comm. Del Moro, 2021)."

3.7 Irrigation and water requirements

3.7.1 Water needs for different hemp types

Hemp water requirements depend on the type of cultivation (fiber, grain, or CBD) and local farm conditions. For fiber hemp, the seasonal water requirement typically ranges from 312 to 520 mm of water during the growing season, with significant variation depending on rainfall (Bajić et al. 2022) and the soil type. Essential oil/CBD hemp, which has higher water demands, requires 600–750 mm (24–30 inches) of precipitation or irrigation over the growing season, especially in semi-arid regions (Roseberg 2024). Grain hemp is similar to fiber hemp but requires extended water availability up until the last 2 weeks prior to harvest, at which time it is best (if possible) to cease irrigation, thereby inducing the plants to finish hemp seed filling and maturation.

The water demand for hemp also changes with growth stages. During germination and early vegetative stages, consistent soil moisture is critical. A general rule for optimal hemp establishment and best yield is that there are 5 weeks that are the most important in a hemp plant's life: the 2 weeks prior to planting, where the soil bed is being prepared and optimal soil moisture content for planting is established, and the 3 weeks after planting, where continued soil moisture maintenance is critical to get a uniform and robust plant stand established. After those initial weeks, occasional dry periods enhance root development, as the roots search deeper in the soil for moisture, thereby setting the stage for continued robust plant growth. In the mid-vegetative stage, hemp has the highest evapotranspiration (ET) rate, while the flowering stage requires slightly reduced irrigation to avoid overhydration and ensure optimal cannabinoid concentration (Bajić et al. 2022, Roseberg 2024), and as mentioned above, final cessation of water application is desired (again if possible; for some regions with fall rainfall, this may not be possible) during the final stages of seed maturation. The northern latitudes, where grain hemp has performed the best to date, tend to have good summer rains and a drier fall season, particularly at the end of September and into October, when the grain hemp is finishing up. In very dry climates, such as in central Washington State where our hemp research fields are located (and which gets only 150–280 mm (6–11 inches) of rain annually, mostly in the form of winter snows), human supplied irrigation is required but also adds the advantage of supporting optimization of water application precisely to the needs of the plant during its different growth stages. Ideally, sustainable and regenerative agriculture systems would not use human-applied irrigation systems, but for large areas of the world's agricultural lands, particularly those that are more arid, irrigation is a requirement.

3.7.2 Irrigation techniques and systems

Several different irrigation systems have been developed by humans over the millennia that farming has been the foundation of human civilization. These are not equal, with some approaches being much better for hemp and agricultural lands than others.

Drip Irrigation: This method delivers water directly to the root zone, minimizing evaporation and water loss. It is especially effective for essential oil/CBD hemp, where precise water control is essential to optimize yield and prevent diseases in humid climates (Bajić et al. 2022, Roseberg 2024). It is less desirable to be used for fiber or grain hemp applications due to the high cost associated with not just the drip line, which is not reusable after being buried underground, but also the rest of the equipment needed to run the drip line system at precise pressures and water application rates. It is not economically feasible in most regions to use drip irrigation systems in fields that are hundreds to thousands of acres in size. Perhaps in the future, hemp-based plastics will be used for dripline construction, at which point this approach may be more sustainable. For now, since drip lines are produced from petroleum-based materials, their use does not well meet the goals of sustainable/regenerative agricultural practices. However, this approach is the best for generating the highest yields of essential oil/cannabinoid/CBD hemp with the highest water use efficiency.

Overhead and Sprinkler Systems: Overhead irrigation is less efficient than drip lines due to higher evaporation rates, but it is more suitable for larger fiber or grain hemp fields, particularly when the irrigation system used is able to adjust sprinkler head height to best apply the water just above the tops of the plants as they continue to grow, thereby reducing water loss due to evaporation. Frequent irrigation through sprinklers can increase disease risks like powdery mildew if the canopy remains wet for extended periods (Roseberg 2024). This is especially the case for essential oil/cannabinoid/CBD hemp.

Flood Irrigation: Although significantly less efficient than other irrigation methods, flood irrigation is sometimes used for large-scale production where water availability is not yet a concern. Proper drainage is essential to avoid root diseases caused by waterlogging (Bajić et al. 2022). However, due to its tendency to salinate soils and its overall poor efficiency, this method of irrigation is losing favor in most parts of the world. Other methods, although requiring more specialized equipment, are simply better for long-term soil health, which should be the major factor considered when choosing the type of system to use for crop production.

Irrigation Monitoring: When an irrigation system is used, it is important to monitor the soil for moisture content and only apply water when the plants actually need it. As mentioned above, hemp roots grow and function best when not sitting in wet soil for extended periods of time. Applying water once the readily available water (RAW) in the soil is depleted improves irrigation water use efficiency (IWUE) (Bajić et al. 2022) and also stimulates hemp plants to produce the best quality end products. That RAW level depends on the soil type and needs to be evaluated for each location. Soil monitoring systems that measure soil moisture, soil conductivity, and soil temperature are ideally suited to aid in this monitoring, and they are getting less expensive each year. A mistake that new growers of fiber and grain hemp often make is a combined effect of adding too much nitrogen and too much water to their growing hemp crop. The plants appear to perform better early in the season when given high N rates and when irrigated such that the plants never want for water. However, the long-term quality of their end product will suffer. The young leaves will be happy, but the fiber will be weaker, and the seed oil levels will be lower. At the same time, extended and particularly severe water withholding can lead to elevated cannabinoid levels, particularly THC. For some cannabinoid varieties, this can be a good thing as long as the THC levels remain within regulatory limits. For some fiber types, particularly some Chinese fiber varieties, a drought or an irrigation malfunction at the wrong time of the growing season can turn a vigorous fiber crop into a regulatory disaster where THC levels exceeding 1% are not unprecedented. Proper water management and monitoring practices are a must for outdoor hemp production.

3.7.3 Water conservation practices

Mulching: Applying organic or plastic mulch helps retain soil moisture by reducing surface evaporation and suppressing weed growth (Roseberg 2024). This approach is very suitable for essential oil/cannabinoid/CBD hemp grown under drip tape with wide plant spacing. It is not suitable for large-scale fiber or grain hemp production. However, application of organic matter such as compost or of biochar can enhance water retention in large fields, leading to better water use efficiency and overall crop performance.

Crop Selection and Timing: Which type of hemp is most suitable to be grown at a particular location or within a particular region will depend on a number of factors, particularly processing facility availability. Nevertheless, the outdoor environment plays an important role in what type of hemp can be effectively grown where. Areas with shorter growing seasons require less water but require earlier maturing varieties to be successful. Adjusting planting density and timing irrigation based on local weather conditions can significantly conserve water while maintaining yield quality (Bajić et al. 2022).

Site Selection: Selecting well-drained soils with appropriate slope and depth improves water retention and reduces irrigation needs (Roseberg 2024). As mentioned above, proper soil evaluation to ensure correct nutrient and irrigation (if appropriate) application rates is critical to produce a quality crop.

3.8 Nutrient and soil amendment application for hemp cultivation

3.8.1 Essential nutrients for hemp growth

Hemp requires a balanced supply of the macronutrients nitrogen (N), phosphorus (P), and potassium (K) for optimal growth. Nitrogen promotes vegetative growth, phosphorus supports root and flower development, and potassium enhances overall stress tolerance. Secondary nutrients like calcium (Ca), magnesium (Mg), and sulfur (S) are also vital, as are micronutrients. As mentioned above, hemp's nitrogen needs range from 50 to 150 units (lbs/acre = 56–168 kg/ha), depending on soil fertility and hemp type (Bajić et al. 2022).

3.8.2 Organic vs. synthetic fertilizers

Organic fertilizers, such as compost and manure, improve soil organic matter and provide slow-releasing nutrients, enhancing microbial activity. Synthetic fertilizers offer immediate nutrient availability but risk leaching and soil degradation with overuse. A combined approach, integrating organic materials with minimal synthetic inputs, has shown promise for sustainable yield improvements (Cho et al. 2023).

3.8.3 Soil amendments and their applications

Biochar: A carbon-rich amendment produced through pyrolysis or other means, biochar improves water retention, enhances nutrient availability, and boosts microbial activity. It is often combined with compost to mitigate potential nutrient immobilization effects (Cho et al. 2023).

Compost: Compost enhances soil structure, increases nutrient availability, and promotes beneficial microbial populations. Application rates depend on soil testing but often range from 5 to 15 tons/ha.

Encapsulated Fertilizers: These slow-release fertilizers reduce nutrient leaching and ensure prolonged availability for hemp growth. They are particularly effective in sandy soil.

Granulated Products: Easy to apply, granulated organic or synthetic fertilizers provide a uniform nutrient distribution. Combining them with biochar has shown benefits in both yield and soil health.

3.9 Weed, pest and disease control in outdoor hemp cultivation

3.9.1 Common weeds, pests and diseases in outdoor hemp cultivation

Hemp is susceptible to a variety of weeds, pests, and diseases that can hinder growth and reduce yields. A number of resources are available that describe weed, pest, and disease issues in more detail (Punja et al. 2019, 2023, Buirs and Punja 2024, Gwinn et al. 2023, Punja 2021, Wang 2021, Bergstrom et al. 2020, Schwartz and Smart 2024), so this chapter will provide an overview. Common weeds in hemp cultivation include morning glory, pigweed, lambs quarters, Russian thistle, kochia, and bindweed, which compete for water, light, and nutrients (Bajić et al. 2022) and are cosmopolitan in nature. Common pests affecting hemp include aphids, spider mites, hemp russet mites, and flea beetles, which can damage plant tissues and potentially spread disease. A pest important for hemp cultivation in the Western United States is brown leaf hopper, endemic to Washington State but now present throughout the Western United States. As mentioned above, it spreads the beet curly top virus (BCTV), which, when combined with hemp latent viroid leads to significant stunting of some hemp varieties. Our observations over the past 4+ years of growing hemp in Washington State have indicated that different hemp varieties are more or less affected by this pest and these pathogens, with some CBD varieties being particularly susceptible (e.g. 'PhotoCBD'), whereas others are quite resistant ('Lifter'), and there appears to be enough genetic variability within each of the several dozen grain and fiber 'varieties' that we have tested to ensure that enough of the plants in a given plot/field are resistant enough that this pest and these pathogens, while endemic to our region, are not particularly great problems for grain or fiber hemp production in our region. Certain 'varieties' of fiber hemp have been developed that have very low to almost non-detectable levels of cannabinoids, and those were found by us and other researchers in the United States to be particularly susceptible to generalist pests like grasshoppers and Japanese beetles, where in both cases the plants were completely decimated by the pests. Apparently, cannabinoids play a critical role in the defense of hemp against such insect pests. In terms of other non-viral diseases, hemp cultivation is affected by numerous fungal and bacterial pathogens across the United States and globally, with some diseases being more prevalent depending on the local climate and agricultural practices.

3.9.2 Fungal pathogens

***Botrytis cinerea* (Gray Mold)**: Common in humid environments, Botrytis is a leading cause of flower and bud rot in hemp. It thrives in areas with high moisture and cool temperatures, making it a significant issue in the Northeast, Pacific Northwest, and regions with similar climates. It affects flower buds, causing disfigurement and browning of the cola tips, particularly of essential oil/cannabinoid/CBD hemp varieties, with visible gray spore masses and internal necrosis being present. It can ruin an essential oil crop.

Powdery Mildew: This disease, caused by species of *Golovinomyces*, is ubiquitous in warm, humid areas. It affects leaves and buds, forming a white, powdery coating on plant surfaces. It is prevalent in greenhouse operations and outdoor farms in regions with high humidity.

Septoria Leaf Spot: Found in many hemp-growing areas, this disease causes circular lesions with yellow margins, reducing photosynthetic capacity. It is particularly troublesome in regions with frequent rainfall and can lead to reduced yields.

***Fusarium* spp.**: There are a number of *Fusarium* strains that can cause root and stem rot, especially in poorly drained soils or areas with frequent flooding. It is prevalent in the Midwest and Southeast United States.

Pythium and Phytophthora (Root and Crown Rot): These pathogens cause significant losses in wet, poorly drained soils and are common in regions with heavy rainfall or irrigation issues. They lead to plant wilt, root discoloration, and cankers, significantly affecting yields.

3.9.3 Bacterial pathogens

***Xanthomonas* spp. (Bacterial Leaf Spot)**: This bacterial pathogen thrives in warm, moist conditions and causes water-soaked lesions on leaves, often progressing to necrosis. It has been reported in parts of the southeastern United States and in greenhouse settings.

Pseudomonas syringae: This ubiquitous opportunistic pathogen is known to affect hemp under high-humidity conditions, causing bacterial blights that result in leaf spotting and dieback in susceptible cultivars.

3.9.4 Geographical trends

In the United States, diseases such as Fusarium root rot and Botrytis are prevalent in areas like California and the Pacific Northwest due to conducive environmental conditions. In Canada, powdery mildew and Botrytis dominate in greenhouse hemp production due to controlled, humid environments. Globally, fungal pathogens like *Fusarium* and *Alternaria* are reported in hemp-growing regions of Europe, while bacterial diseases are less common but increasing due to changes in cultivation practices.

Preventative measures, such as crop rotation, resistant cultivars, and maintaining low humidity in growing environments, are crucial for managing these diseases. Improved scouting and pathogen-specific fungicides or biocontrol agents can also reduce disease impact. For detailed information on disease management, see the guidelines provided by agricultural extension publications generated by the University of Vermont, the Nevada Department of Agriculture, and Oregon State University.

3.9.5 Organic and conventional control methods for pests and diseases

Both organic and conventional control methods can be employed to combat these threats. Organic approaches typically involve using natural predators such as ladybugs for aphid control, predatory mites or neem oil for mite infestations, or applying compost or mulch for weed suppression (Bajić et al. 2022). On the other hand, conventional methods often involve the use of chemical pesticides or fungicides, which may be more effective in treating severe infestations but require careful application to avoid harming the crop or environment (Roseberg 2024). Few such chemical treatments have been approved for hemp, however.

3.9.6 Integrated pest management strategies

Integrated pest management (IPM) is a sustainable approach that combines various methods to manage pests while attempting to eliminate harm to the environment. IPM includes preventative strategies, such as crop rotation and maintaining optimal environmental conditions to reduce pest outbreaks. Monitoring pest populations and setting action thresholds allows for targeted interventions like introducing natural predators, using mechanical controls (e.g. hand weeding), or applying selective pesticides when necessary. A comprehensive IPM plan integrates multiple pest control strategies, ensuring long-term sustainability in hemp cultivation (Thweatt 2023). Costs for manual labor, however, are growing in various parts of the world, making that approach to weed control less viable going forward. Alternatively, mechanical approaches to weed control are being investigated for their applicability to hemp.

4 Growth for purpose-driven cultivation

Cultivation of hemp requires focusing from seed through harvest on the purpose it will be used for. Specification drives what seed is planted and how so, spacing matters for growth, e.g. fiber is planted at high density for stalk outputs, hurd for building materials at slightly lighter seed rates per acre, while grain is about half that of fiber. Essential oil hemp is grown for essential oils from the flower or biomass for specific chemical hemp end-uses and requires different approaches to cultivation and care throughout the process compared to fiber and grain hemp.

4.1 Cannabinoid/chemical/essential oil hemp

4.1.1 Cultivation practices for maximum cannabinoid production

Cannabinoids have different economic values with certain cannabinoids garnering hirer value. With growing demand in both the wellness and pharmaceutical industries, maximizing cannabinoid production is key to successful hemp cultivation for this purpose.

Genetics and Variety Selection: Selecting the right variety is crucial for high cannabinoid content. Hemp strains with high CBD or CBG concentrations are specifically bred for cannabinoid production. These varieties typically have low-THC content (below 0.3%) to comply with regulations.

Environmental Factors: Cannabinoid production is influenced by environmental factors such as light intensity, temperature, soil pH, and irrigation (Fiorentino et al. 2024). Plants grown in environments with stable temperatures (between 20 and 30 °C) and moderate light intensity (12–16 hours of daylight per day) produce the highest levels of cannabinoids. Furthermore, ensuring optimal soil conditions (pH 6.0–7.5) and avoiding excess nitrogen fertilizers are key to maintaining the quality and potency of cannabinoids.

4.1.2 Harvesting and processing techniques

Optimal Harvest Timing: To maximize cannabinoid yield, it is crucial to harvest at the right time. Harvesting too early may result in underdeveloped cannabinoids (low yield), while harvesting too late can lead to degradation, loss of potency, and potential regulatory non-compliance (if THC levels get too high). A typical indicator used by industry members to know when to harvest is the trichome maturity, which in many varieties turns from a 'milky' to an 'amber' appearance at prime maturity (Das et al. 2022). However, not all varieties follow this rule. Some have trichomes that turn pink or purple. Some never turn color but remain milky. Information about the particular variety, provided by the breeder/seed supplier, should be followed. Nevertheless, differences in maturity pattern can occur when a variety is grown in a location that is very different from those it has been grown in before, so proper chemical testing should be used when evaluating a new variety of this class.

Post-Harvest Processing: Post-harvest processing of cannabinoid hemp includes drying, curing, and extraction. Drying should be done in a controlled environment with low humidity (20–60%) to avoid mold growth. Extraction methods such as CO_2 extraction, ethanol extraction, and cold pressing are commonly used to isolate cannabinoids and terpenes from the plant material for use in oils, tinctures, and other products (Chacon et al. 2024). Older methods that used liquid butane or other highly explosive solvents have lost favor in recent years due to the dangers they present to the extraction team. And the CO_2- or ethanol-based methods have been shown to be safer and at least as cost-effective. Moreover, there is increased interest in recent years in extracting the natural carboxy forms of the cannabinoids in that form, which requires maintenance of low temperatures during the drying process and/or extraction when the material is still 'wet' due to the water-soluble nature of the cannabinoids in this form.

4.2 Fiber hemp

4.2.1 Cultivation practices for high-quality fiber production

Hemp has been traditionally cultivated for its strong and durable fiber, which is used in textiles, composites, and construction materials. Fiber hemp production focuses on optimizing fiber yield and quality.

Soil and Fertility Management: Hemp requires well-drained, fertile soil for optimal fiber production. High nitrogen levels promote vegetative growth, but excessive nitrogen can reduce fiber quality. A balanced fertilization strategy is essential, focusing on maintaining adequate phosphorus and potassium levels, especially during the vegetative phase.

Density and Spacing: Fiber hemp should be planted at a higher density (250–300 plants per square meter) to encourage tall, slender plants with high fiber content. This high-density planting strategy reduces branching and promotes the development of long, continuous fibers, which are ideal for textile production (Deng et al. 2019).

4.2.2 Harvesting and processing techniques

Timing of Harvest: For fiber hemp, harvesting is typically done when the plants are at full maturity, just before they begin to flower. At this stage, the cellulose in the stalks is at its highest, and the fibers are strongest, but lignin deposition is not yet high. Harvesting too early can lead to shorter, weaker fibers (Angulu and Gusovius 2024). Harvesting too late leads to higher levels of lignin, which makes separation of the fibers from the lignin a more challenging process, which again, weakens the fibers.

Processing Methods: Fiber extraction is typically done through retting, a process where the hemp stalks are soaked in water, allowed to lay on the field with morning dew or irrigation supplying water sufficient to encourage microbial activity that breaks down the pectin that binds the fiber to the rest of the plant. Mechanical processes such as decortication then separate the fibers from the woody core (hurd/shives). After retting, the fibers are dried, carded, and spun into yarn for textiles or further processed (such as by chopping or cottonization) for use in composite materials (Mariz et al. 2024) or non-woven fabrics. The hurd (called shives in Europe) is used for applications such as building construction (e.g. hemp lime/hempcrete). Large-scale processing equipment for decordication and fiber separation/purification has been around for 30+ years, but new approaches are evaluated continuously worldwide in efforts to decrease costs of processing while maintaining the desired properties of the resulting fiber/hurd.

4.3 Grain (seed) hemp

4.3.1 Cultivation practices for optimal seed yield

Hemp seeds are a valuable commodity, used in food products, oils, and as livestock feed. The focus of grain hemp production is maximizing seed yield and quality.

Soil Requirements and Fertilization: Grain hemp thrives in well-drained soils with a pH of 6.0–7.0. Nitrogen is critical during early growth stages for healthy plant development, but excessive nitrogen during seed development can lead to excessive vegetative growth at the expense of seed yield. Balanced fertilization, particularly with phosphorus and potassium during flowering and seed filling stages, is essential for maximizing seed yield (Wylie et al. 2021).

Planting Density and Irrigation: For high seed yield, hemp is generally planted at moderate density, around 100–150 plants per square meter. This allows for adequate airflow and light penetration, which promotes optimal seed development. Irrigation is crucial during the seed development stage to ensure consistent seed fill, especially in regions with unpredictable rainfall (Garcia Tejero et al. 2014). But, to stimulate final maturation of seeds, irrigation is typically withheld about 2 weeks prior to predicted end of seed maturity.

4.3.2 Harvesting and processing techniques

Harvest Timing: Seed hemp is harvested when the seeds are mostly mature, typically when the majority of the seeds have turned from green to brown. If harvested too early, seeds may be underdeveloped and low in oil content. Field monitoring of seed maturity is critical, with a typical harvest window occurring once about 75–80% of the seeds are ripe. If allowed to mature too much longer than that, the shattering problem of hemp becomes apparent, with significant losses occurring due to seed drop. The shattering gene for hemp has not yet been identified, and no seed variety has been reported, to date, that lacks the shattering trait. Identification of this gene(s) responsible for shattering and use of that information to eliminate shattering will be a significant advance in hemp breeding as it will lead to significantly higher quality in final seed products. Even with a significant green seed fraction in the harvested seed at present, hemp seeds are viewed as being among the highest quality oilseeds available.

Seed Harvesting and Processing: After harvest, hemp seeds are typically separated from the stalks using a combine harvester. The seeds are then cleaned, dried, and stored under controlled conditions (ideally at 10 °C) to maintain quality. Cold pressing or mechanical extraction is used to produce hemp seed oil, which is valued for its high omega-6 and omega-3 fatty acid content (Wylie et al. 2021). The remaining hemp seed cake is high in protein, which has other applications, such as a food protein additive, as an ingredient in glue production, or as a source for serum replacement for the purified edestin protein (the major seed storage protein of hemp). Dehulling prior to pressing is often performed now, which leads to a higher quality oil, with less chance of cannabinoid contamination. The seed coat does not naturally contain cannabinoids, but the method used to harvest hemp seed (combining in the field) can lead to contamination of the outside of the seeds due to residual cannabinoids present on the remaining dried flower parts of the mature seed heads, leading to trace levels of cannabinoids in some hemp seed oil products.

4.4 Outlook for hemp cultivation

Purpose-driven hemp cultivation is becoming increasingly sophisticated, with advancements in genetics, cultivation practices, and harvesting techniques designed to maximize the yield and quality of cannabinoid-rich flowers, fiber, and seeds. These innovations not only support the growing demand for hemp-based products but also position hemp as a sustainable crop with diverse applications. By refining cultivation and processing methods, farmers can achieve optimal results whether they are focusing on cannabinoid extraction, high-quality fiber production, or seed harvests for oil and protein production for food or industrial applications.

5 Regenerative and sustainability considerations in hemp cultivation

5.1 Healing the soil

Outdoor hemp cultivation is increasingly recognized for its environmental benefits, making it an attractive option in sustainable and regenerative agriculture. One of the most notable advantages of hemp is its relatively low environmental footprint compared to many other crops. Hemp requires minimal pesticide and herbicide use, at least at the present time, as the plant's natural properties – such as dense canopy coverage and rapid growth – help suppress weeds and deter many pests. This feature is particularly significant because it can reduce the chemical runoff into nearby water systems compared to other crops, thereby protecting aquatic ecosystems and maintaining soil health. Additionally, a reduced reliance on synthetic inputs lowers the overall energy and resource demands of hemp cultivation compared to other crops. When used in crop rotation systems, hemp can act as a 'break crop,' disrupting pest and disease cycles that affect other crops, thereby reducing the need for chemical interventions in subsequent planting seasons.

Another crucial environmental benefit of hemp is its role in carbon sequestration. During its growth cycle, hemp absorbs large amounts of CO_2 from the atmosphere, storing it within its biomass. Hemp can sequester approximately 1.6 tons of CO_2 per ton of harvested biomass, making it one of the most effective crops for carbon capture. This feature aligns hemp cultivation with global efforts to mitigate climate change by pulling carbon out of the atmosphere. Furthermore, when processed into durable products such as bioplastics, textiles, or construction materials, the carbon remains locked within these materials, extending the sequestration benefits over time. That is particularly the case for building materials, which can be expected to remain in place for a hundred years or more.

Hemp also contributes to soil health through its deep and extensive root systems, which reduce erosion, improve soil structure, and increase water infiltration. These roots can break up compacted soils and enhance the soil's organic matter content, fostering microbial activity that is vital for long-term fertility. In addition, hemp's roots help to stabilize the soil and prevent erosion, particularly on slopes or in areas with loose, sandy soils.

Hemp's role in improving soil nutrient levels is also linked to its ability to act as a 'phytoremediator.' The plant can take up heavy metals and other contaminants from the soil, potentially making degraded lands usable again for agricultural purposes. While this process requires careful management to avoid contaminating the hemp product itself, it illustrates hemp's broader utility in agricultural systems. For example, in an investigation that we have been carrying out this year, we have evaluated the ability of hemp varieties to remove the pesticide DTT and its derivatives (DDE and DDD) from contaminated soils and have found that the compounds are indeed efficiently removed from the soil and eventually end up in leaf and flower tissues of the mature plants, but not in the stems. Thus, proper processing of the plant used as a phytoremediator in this manner could lead to stem-based products free of the contaminating pesticide, whereas the dried leaves and flower components can be first selectively extracted for cannabinoids or other compounds and then utilized for biochar generation, a process that would completely destroy the carbon-based pesticide and derivatives.

Water use efficiency is another critical environmental advantage of hemp. Compared to water-intensive crops such as cotton, hemp requires significantly less irrigation. This trait is particularly advantageous in arid and semi-arid regions, where water scarcity is a pressing issue. Moreover, hemp's adaptability to diverse climatic and soil conditions makes it suitable for cultivation in marginal lands, reducing the pressure on fertile agricultural lands reserved for food production.

Beyond its direct agricultural benefits, hemp cultivation can support a circular economy by generating multiple by-products that can be repurposed in environmentally friendly ways. For instance, hemp fibers can be used in biodegradable composites and textiles, reducing reliance on petroleum-based materials and decreasing plastic pollution. Once those products reach the end of their life, they can be repurposed into biochar, a soil amendment that enhances carbon sequestration and nutrient retention in soils. Additionally, the seeds can be harvested for oil and protein-rich food products, contributing to a diverse and sustainable agricultural system that supports human health and nutrition.

Despite these advantages, it is essential to address challenges associated with outdoor hemp cultivation. Poor management practices, such as over-fertilization, can lead to nutrient leaching and waterway contamination. Moreover, the expansion of hemp farming into sensitive ecosystems could disrupt local biodiversity if not carefully managed. However, with responsible farming techniques and integrated pest management (IPM), the environmental benefits of outdoor hemp cultivation can be maximized while mitigating potential drawbacks and potentially leading to regenerative agricultural systems.

Thus, outdoor hemp cultivation presents a compelling case for sustainable and regenerative agriculture, with its low pesticide requirements, significant carbon-sequestering capacity, soil health improvement benefits, and water use efficiency. As hemp gains traction globally, continued research and sustainable practices will be vital to unlocking its full potential as a crop that supports both environmental and economic goals.

5.2 Impact on crop rotation systems

Hemp's versatility and unique agricultural traits make it an excellent candidate for crop rotation systems, benefiting both the crops that follow it in the rotation and the overall health of the farming ecosystem. Its ability to disrupt pest and disease cycles stems from its robust growth habits, extensive root system, and relatively low susceptibility to many common agricultural pathogens. Incorporating hemp into crop rotations can reduce the population of soil-borne pathogens and pests that often build up when the same crop is cultivated repeatedly, a practice that leads to monoculture-related problems like pest resistance and declining soil health. For example, hemp is known to suppress certain nematodes due to compounds it releases into the soil, making it an effective crop for reducing nematode infestations in subsequent plantings. Furthermore, its dense canopy shades out weeds, which minimizes the seed bank of problematic weeds and reduces the reliance on herbicides in subsequent cropping cycles.

Hemp is also highly efficient at extracting nutrients from the soil, particularly nitrogen, phosphorus, and potassium. This nutrient uptake is paired with a significant return of organic matter to the soil through hemp's residues, particularly its stems and leaves, which decompose after harvest. These residues contribute to improved soil structure, increase microbial activity, and boost organic matter content, all of which enhance soil fertility. When used as a green manure crop, hemp can be plowed back into the soil after its growth cycle to further enrich soil nutrient levels, making it ideal for supporting the growth of nutrient-demanding crops like maize or wheat, although this latter practice is probably not the ideal use of hemp, which has greater value as a source of other products (textiles, building materials, food ingredients, industrial feedstock, etc.).

In crop rotation, hemp's compatibility with a variety of crops enhances its practical use. For example, its deep rooting complements shallow-rooted crops like soybeans, which can benefit from the improved soil structure and nutrient availability following hemp cultivation. Similarly, hemp rotation with cereals or legumes can optimize nutrient cycling. In particular, legumes, which fix nitrogen in the soil, can pair well with hemp to create a synergistic effect, improving soil fertility while disrupting pest cycles.

In summary, hemp's inclusion in crop rotation systems offers multiple agronomic benefits. It breaks pest and disease cycles, enhances soil structure and fertility, suppresses weeds, and reduces reliance on chemical inputs. By contributing to long-term soil health, hemp supports sustainable and regenerative farming practices and boosts the productivity of the entire agricultural system. Its flexibility and compatibility with various crops make it a valuable component of integrated farming approaches aimed at environmental stewardship and economic viability.

6 Trends and innovations in outdoor hemp cultivation

6.1 Advances in hemp genetics and breeding

Hemp genetics and breeding are rapidly advancing, offering significant potential for improving outdoor cultivation. Breeding programs are focusing on improving key traits such as pest resistance, drought tolerance, higher yields, and cannabinoid production (Flajšman and Ačko 2022). These advancements allow for the development of hemp varieties specifically tailored to different environmental conditions, leading to improved performance in diverse climates.

Genetic Improvements: The use of advanced genetic tools, such as CRISPR-Cas9, while not yet deployed for hemp, promises to enable breeders to develop hemp strains with particularly valuable properties, such as enhanced disease resistance and optimized growth cycles, crucial for regions with short growing seasons or unpredictable weather patterns (Shiels et al. 2022). At present, traditional breeding techniques are being used to improve hemp plants so that they, e.g. produce higher levels of CBD and CBG, while also ensuring that the THC content remains within legal limits (Table 2).

Examples: In Canada, genetic advancements have led to the development of regionally adapted hemp cultivars that can tolerate colder climates and the day length/light quality of very northern latitudes, allowing farmers in regions like Manitoba and Saskatchewan to extend the growing season (Ameis and

Table 2 Summary of market demand trends for hemp

Category	Current demand (2023)	Projected demand (2028)	Key drivers	Applications
Hemp fiber	$5.5 billion globally	$11.5 billion globally	Demand for sustainable materials in textiles and construction	Textiles, ropes, paper, bioplastics, hempcrete
Hemp seeds	$1.1 billion globally	$2.2 billion globally	Growth in superfood trends and nutritional products	Hemp oil, protein powders, health foods
CBD and cannabinoids	$6.8 billion (US)	$16 billion (US)	Demand for wellness products, legalizations	CBD oils, tinctures, topicals, nutraceuticals
Hemp textiles	$1.3 billion globally	$3.4 billion globally	Eco-friendly fashion demand	Clothing, upholstery, biodegradable fabrics
Hemp-based plastics	$250 million globally	$1 billion globally	Regulations pushing for biodegradable packaging	Biodegradable packaging, automotive parts
Hempcrete (construction)	$75 million globally	$300 million globally	Growth in green building practices	Eco-friendly buildings, insulation

Sommersell 2020). Breeding efforts like these need to be implemented on a regional basis due to the high genetic variation within the species and the highly variable conditions across regions. What is bred for a particular purpose in one location may not perform well in another. As a further example, a grain hemp variety, X-59, is used extensively in the upper great plains region of North America (in Saskatchewan and Manitoba as well as right across the border in Minnesota, North Dakota, and Montana), where it appears to perform very well. However, this variety has been grown repeatedly by us in variety trials (every year since 2021), and it has always performed very poorly, where it is extremely variable in growth habit from plant to plant within the stand. Varieties need to be bred not only for specific end-use purposes but also for suitability in specific growing regions.

6.2 Technological innovations in hemp farming

Advancements in farming technologies are transforming how hemp is cultivated outdoors. These innovations improve efficiency, reduce costs, and increase yields. The integration of smart farming technologies, such as sensors, drones, and autonomous equipment, is becoming more prevalent in hemp farming.

Precision Agriculture: Precision farming uses technology to monitor and manage field variability. By employing soil moisture sensors, climate prediction tools, and GPS technology, farmers can optimize water usage, nutrient application, and planting strategies. In regions with variable soil quality, precision agriculture techniques allow for targeted interventions that improve overall crop health and productivity.

Automation and Robotics: Robotics and automation also show promise in hemp farming. From automated seed planting and harvesting to weed control using AI-powered machines, these innovations promise to reduce labor costs and improve operational efficiency. Autonomous harvesters designed specifically for hemp, which can operate in large outdoor fields, will improve harvesting speed and reduce crop loss.

Example: The use of drones to monitor hemp crops has enabled farmers to detect disease outbreaks early, allowing for targeted pesticide application and reducing chemical use (Passley 2025).

6.3 Policy and regulatory factors impacting outdoor cultivation

As the hemp industry continues to grow, policy and regulatory frameworks will play a crucial role in shaping the future of outdoor hemp cultivation. Regulations around THC content, environmental impact, and market access are currently evolving, and they will have a significant influence on cultivation practices and industry growth.

6.3.1 THC regulations

Hemp cultivation is tightly regulated, particularly regarding the allowable THC content, which must remain below 0.3% in many countries (U.S. Department of Agriculture 2020). As research into cannabinoids advances, there may be shifts in THC limits, impacting the genetic selection and breeding of hemp varieties. Understanding and complying with these regulations will be crucial for outdoor hemp cultivators.

6.3.2 Sustainability and environmental regulations

As outdoor hemp farming expands, there is growing scrutiny on the environmental impacts of cultivation. In particular, water usage, pesticide application, and the sustainability of farming practices are under review. Some regions are already adopting stricter environmental regulations to ensure that hemp farming does not contribute to water scarcity or soil degradation.

Example: In the EU, the European Commission's 'Green Deal' encourages farmers to adopt more sustainable practices, which could drive the hemp industry to adopt eco-friendly techniques like organic farming and reduced pesticide use (Schmid 2022).

6.3.3 Market access and international trade

International trade policies are also key to the growth of the hemp industry. While hemp cultivation has increased in many countries, the lack of standardized regulations and certification processes can create barriers to exporting hemp products. Future trade agreements may help facilitate smoother access to international markets, enabling hemp cultivators to access higher-value markets for both CBD and fiber (Grand View Research 2024).

7 Mistakes and failures

Mistakes can make learning interesting, but they need not be repeated, if they are known. Failure in hemp cultivation can ruin a trial or a crop, cause financial loss, or worse by far, destroy a farm's future. As hemp resurges and acceptance grows, failures can and will be reduced. This section contains examples of what can go wrong and suggestions (even if just implied) on how to avoid these problems.

7.1 Environmental issues/challenges

Examples of the environmental effects are outlined in the figures (Figs. 3, 4, 5) presented in this section.

7.2 Top ten ways to avoid failure in outdoor hemp cultivation

1. Follow in a field that had a suitable cover crop to reduce weed and disease issues;
2. Choose the right compliant seed for the ultimate end-use purpose;
3. Test irrigation equipment prior to planting;
4. Get soil and water tests from each field and follow a tailored nutrient program;
5. Get seeds early and plant them early to get a canopy in place before weeds can take over;
6. Do not let seeds dry out while germinating;
7. Plant seeds shallow and with direct, firm soil contact;
8. Do not over water;
9. Schedule compliance tests early to meet regulations; and
10. Slow down while harvesting so equipment does not break.

Examples of what happens if you do everything right as opposed to many things wrong are shown in Figure 6.

Figure 3 An autoflower variety was planted on 5 acres in central Washington in 2020, and then devastating wildfires that summer blanked the region in smoke for much of July and August (note the color of the sky). The light quality was drastically altered compared to what most plants are adapted to, and these autoflower plants flowered very early, were stunted, and drastically underperformed compared to what the grower hoped to see.

Figure 4 Plots showing the variable impact of a major stress on different varieties of hemp. These results are from the 2021 grain and fiber trials at Pear Acres, near Prosser, WA. The plots were planted about a month later than planned due to delays in getting seed. Planting date was on the 3rd of June 2021. Soon after, just 2 weeks after planning and right after emergence of the plants, a record heat wave hit (48°C high temp and 35°C low temp each day for 2 weeks straight starting on June 17th). Keeping the soil moisture up in the sandy soil was a major challenge in the heat. There were dramatic differences in response of different varieties to this combined heat/drought stress. Top, henola did poorly but was not completely killed off – note the variation in seed head height and size and the scarcity of the hemp plants compared to the weeds, but some hemp seed heads are visible. Other hemp seed varieties, like X-59, commonly grown in the United States, did not survive the heat stress at all. Bottom, in contrast, Kompolti, a fiber variety from Hungary, survived the stress relatively well, with plants reaching expected height between 3 and 4 m, and with density, although less than ideal, being high enough to outcomplete the weeds.

Figure 5 The four rows of plants most visible in this picture are from an experimental cross that did not tolerate the heat stress of 2021 (see caption for Fig. 3 for details on the heat wave stress). The plants surrounding it (to the right and behind) tolerated the heat well and performed better than expected. This experimental line, however, did not do well, and it appeared that the roots were particularly affected by the heat stress, leading to the plants collapsing at the end of the season. Size and yield were very poor. A field at large scale with such a crop type would be a disastrous failure if exposed to this stress. Selection of the right variety for the environment/local conditions is essential.

8 Future directions and conclusion

The future of outdoor hemp cultivation is filled with exciting advancements and innovations. From genetic breakthroughs that make hemp more suitable for widespread cultivation and resistance to climate change or that support more efficient variety establishment and faster to market product development to technological innovations that improve farm efficiency, hemp farming is evolving rapidly. However, the regulatory environment will be a critical factor in determining how these innovations are implemented and whether they can be scaled globally. By staying ahead of these trends, hemp growers can maximize their productivity, sustainability, and profitability in a growing market.

As discussed in Section 5, hemp holds a significant promise in advancing sustainable and regenerative agricultural practices due to its impact on the soil, potential to generate durable stored carbon, other environmental benefits, and economic potential. As a crop that requires minimal chemical inputs and thrives in diverse climates, hemp can serve as a key component of sustainable and regenerative farming systems, particularly when integrated into existing crop rotation systems that could benefit from its soil health-building properties. Below are key takeaways regarding the role of hemp in promoting sustainability in agriculture.

Figure 6 Examples of what happens if you do everything right (top) vs. many things wrong (bottom).

8.1 Environmental benefits

Hemp is a highly efficient crop in terms of water and nutrient use. It requires less water compared to many traditional crops, particularly other fiber and protein crops, making it particularly valuable in water-scarce regions. Furthermore, hemp has a deep root system that improves soil structure, enhances water retention, and prevents erosion, promoting healthier soils. Its ability to absorb carbon dioxide from the atmosphere and convert it into durable products also positions it as a promising crop for carbon sequestration efforts, contributing to climate change mitigation.

8.2 Crop rotation and soil health

Hemp can play a crucial role in crop rotation systems, helping to break pest cycles, manage weeds, and reduce soil degradation. Its deep roots can bring nutrients from deeper soil layers to the surface, benefiting subsequent crops in the rotation. Additionally, hemp's reduced need for synthetic fertilizers and pesticides compared to other fiber crops reduces the environmental footprint of farming, also aligning it with the principles of integrated pest management (IPM) and organic farming.

8.3 Reduction of synthetic inputs

Hemp cultivation requires fewer synthetic chemicals such as pesticides and herbicides compared to conventional crops. This characteristic helps reduce chemical runoff into local ecosystems, thus preventing soil and water pollution. The natural resilience of hemp, along with sustainable pest and disease management strategies, makes it an ideal candidate for organic farming and low-input agricultural systems.

8.4 Economic viability and market opportunities

Hemp offers diverse economic opportunities, ranging from fiber production for textiles, building materials, and biocomposites to its use in food products, cosmetics, and medicinal applications. This versatility provides farmers with multiple potential revenue streams, making hemp an attractive option for diversifying agricultural enterprises. Furthermore, the global market for hemp-based products is growing, with increasing demand for sustainable and eco-friendly materials. The limiting factor so far is the need to develop processing facilities that are within range of growers that support the sustainable/regenerative goals of the crop. Such facilities were non-existent in the United States and other parts of the world 10 years ago, but that is changing. Continued investment into processing facilities for all types of hemp utilization streams will support this crop's integration into broader cropping systems and make it economically viable for the average grower.

8.5 Challenges and considerations

Despite its potential, hemp cultivation faces several challenges, including regulatory hurdles, lack of infrastructure for processing (mentioned just above), and market volatility. Farmers may also face a steep learning curve in terms of cultivation practices, as hemp is a relatively new crop in many regions. Additionally, the crop's association with cannabis has led to varying regulations across different jurisdictions, which can create uncertainty for producers. Overcoming these barriers will be critical for realizing hemp's full potential in sustainable and regenerative agriculture.

8.6 Technological Innovations

Advances in agricultural technologies, including precision farming, automated systems for cultivation, harvesting and processing, and improved breeding techniques, will continue to enhance the efficiency of hemp cultivation. Research into climate-resilient hemp varieties and innovative harvesting methods could help optimize yields and quality, particularly in regions with challenging climates. For example, identification of the shattering gene would transform how hemp seed is harvested, greatly simplifying the process and leading to significant cost savings in seed processing and storage.

8.7 Conclusion

Outdoor grown hemp is well-positioned to play a leading role in sustainable and regenerative agriculture, offering environmental benefits, economic opportunities, and agricultural resilience. As farming practices continue to evolve, hemp's low environmental impact, adaptability, and diverse uses make it an ideal crop for future agricultural systems that prioritize sustainability and regenerative practices. By addressing the challenges related to cultivation, processing, and regulation, hemp can contribute to a more sustainable, resilient, and profitable agricultural sector globally. There are 80 years of trials to catch up on. The next century is sure to transform how hemp is and perceived to benefit society.

9 Acknowledgments

We thank Drs Jessica Tonani and Jay Stratton Noller for their editing suggestions. This work was supported in part by the U.S. Department of Agriculture's (USDA) National Institute of Food and Agriculture (NIFA) Hatch project 227700, NIFA Hatch Multistate Project S-1084, and the NIFA Agriculture and Food Research Initiative (AFRI), Sustainable Agricultural Systems (SAS), Grant no. 13333755/ Project accession no. 1027531.

10 Where to look for further information

Readers seeking a broader introduction to the agronomy, biology, and historical development of hemp cultivation may consult several comprehensive references. The book *The Cultivation of Hemp* by Ivan Bócsaand Michael Karus (Bócsa & Karus, 1998) remains one of the most widely cited foundational texts on industrial hemp agronomy, including cultivation practices, varieties, and harvesting techniques. Another valuable resource is *Cannabis Evolution* and *Ethnobotany* by Robert C. Clarke and Mark Merlin (Clarke &Merlin, 2016), which provides an extensive historical and biological perspective on Cannabis cultivation and use. For up-to-date scientific developments in hemp agronomy, genetics, and industrial applications, readers may consult journals such as *Journal of Cannabis Research, Industrial Crops and Products, Agronomy, Frontiers in Plant Science, Plants, and GCB Bioenergy*, which regularly publish research on hemp production systems and crop improvement. Several professional organizations and information networks also track developments in hemp research and cultivation. The International Hemp Association and the Phytochemical Society of North America provide forums for scientific exchange on plant chemistry and industrial crops. In addition, government and extension programs such as the United States Department of Agriculture hemp program and agricultural extension resources from institutions such as Oregon State University and Cornell University maintain updated technical guidance for growers. Finally, several other research centers worldwide conduct major work on hemp genetics, breeding, and agronomy. Notable programs include those at Washington State University, University of Manitoba, Wageningen University & Research, and Yunnan Academy of Agricultural Sciences. Monitoring work from these groups and related international collaborations can help researchers and growers stay informed about rapidly evolving developments in hemp cultivation. Bócsa, I. and Karus, M., 1998. *The cultivation of hemp: botany, varieties, cultivation and harvesting* (pp. vi+-184). Clarke, R. and Merlin, M., 2016. *Cannabis: evolution and ethnobotany*. Univ of California Press.

11 References

Abdollahi, M., et al. (2020) 'Impact of four hemp (Cannabis sativa L.) varieties and stage of plant growth on yield and composition of essential oils', *Industrial Crops and Products*, 155, 112793.

Abdollahi, M., et al. (2021) 'Assessment of the cannabinoid content from different varieties of Cannabis sativa L. during the growth stages in three regions', *Chemistry & Biodiversity*, 18(12), e2100247.

Adesina, I., et al. (2020) 'A review on the current state of knowledge of growing conditions, agronomic soil health practices and utilities of hemp in the United States', *Agriculture*, 10(4), 129.

Amaducci, S., et al. (2015) 'Key cultivation techniques for hemp in Europe and China', *Industrial Crops and Products*, 68, 2–16.

Ameis, D. and Sommersell, N. (2020) *Impact report: research and innovation in Manitoba grows the hemp industry*, Research Manitoba.

Angulu, M. and Gusovius, H.-J. (2024) 'Retting of bast fiber crops like hemp and flax—a review for classification of procedures', *Fibers*, 12(3), 28.

Bajić, I., et al. (2022) 'The effects of irrigation, topping, and interrow spacing on the yield and quality of hemp (Cannabis sativa L.) fibers in temperate climatic conditions', *Agriculture*, 12(11), 1923.

Bakowska-Barczak, A., et al. (2022) 'Industrial hemp-based dietary supplements and cosmetic products' in: Pojic, M. and Tiwari, B. K. (eds.) *Industrial hemp*, Elsevier, 247–299.

Bergstrom, G., et al. (2020) 'Diseases affecting hemp in New York (2020 version)', [online], available: https://hemp.cals.cornell.edu/resources/reports-factsheets/2020-factsheets/diseases-affecting-hemp-in-new-york-2020-version/ [Accessed 7 April 2025].

Bocsa, I. and Karus, M. (1998) *The cultivation of hemp: botany, varieties, cultivation and harvesting*, HEMPTECH.

Bouloc, P., et al. (2013) *Hemp: industrial production and uses*, CABI.

Buirs, L. and Punja, Z. K. (2024) 'Integrated management of pathogens and microbes in Cannabis sativa L.(Cannabis) under greenhouse conditions', *Plants*, 13(6), 786.

Callaway, J. (2004) 'Hempseed as a nutritional resource: an overview', *Euphytica*, 140, 65–72.

Chacon, F. T., et al. (2024) 'Effect of hemp extraction procedures on cannabinoid and terpenoid composition', *Plants*, 13(16), 2222.

Cherney, J. H. and Small, E. (2016) 'Industrial hemp in North America: production, politics and potential', *Agronomy*, 6(4), 58.

Cho, H.-N., et al. (2023) 'Impact of biochar and compost amendment on corn yield and greenhouse gas emissions under waterlogged conditions', *Applied Biological Chemistry*, 66(1), 87.

Citti, C., et al. (2018) 'Analysis of cannabinoids in commercial hemp seed oil and decarboxylation kinetics studies of cannabidiolic acid (CBDA)', *Journal of Pharmaceutical and Biomedical Analysis*, 149, 532–540.

Clarke, R. and Merlin, M. (2016) *Cannabis: evolution and ethnobotany*, University of California Press.

Das, P., et al. (2022) 'Postharvest operations of cannabis and their effect on cannabinoid content: a review', *Bioengineering*, 9, 364.

Deabold, K. A., et al. (2019) 'Single-dose pharmacokinetics and preliminary safety assessment with use of CBD-rich hemp nutraceutical in healthy dogs and cats', *Animals*, 9(10), 832.

Deng, G., et al. (2019) 'Planting density and fertilization evidently influence the fiber yield of hemp (Cannabis sativa L.)', *Agronomy*, 9(7), 368.

Fike, J. (2016) 'Industrial hemp: renewed opportunities for an ancient crop', *Critical Reviews in Plant Sciences*, 35(5–6), 406–424.

Finnan, J. and Burke, B. (2013) 'Nitrogen fertilization to optimize the greenhouse gas balance of hemp crops grown for biomass', *GCB Bioenergy*, 5(6), 701–712.

Fiorentino, N., et al. (2024) 'Environmental and agronomic factors affecting the chemical composition and biological activity of cannabis extracts', *Frontiers in Plant Science*, 15, 1407262.

Flajšman, M. and Ačko, D. K. (2022) 'Industrial hemp breeding and genetics' in: Pojic, M. and Tiwari, B. K. (eds.) *Industrial Hemp*, Elsevier, 37–57.

Garcia Tejero, I., et al. (2014) 'Impact of plant density and irrigation on yield of hemp (Cannabis sativa L.) in a mediterranean semi-arid environment', *Journal of Agricultural Science and Technology*, 16(4), 887–895.

Goloborod'ko, P. (1995) 'Hemp research and growing in Ukraine', [online], available: http://www.internationalhempassociation.org/jiha/iha02113.html [Accessed 3 April 2025].

Grand View Research (2024) 'Industrial hemp market size, share & trends analysis report by application (Animal care, textiles, food & beverages), by product (Seeds, fiber, shivs), by region (North America, Asia Pacific), and segment forecasts, 2024–2030', [online], available: https://www.grandviewresearch.com/industry-analysis/industrial-hemp-market [Accessed 4 April 2025].

Gwinn, K. D., et al. (2023) 'Fungal and mycotoxin contaminants in cannabis and hemp flowers: implications for consumer health and directions for further research', *Frontiers in Microbiology*, 14, 1278189.

Health Canada (2021) 'Industrial hemp regulations', [online], available: http://laws-lois.justice.gc.ca/eng/regulations/SOR-2018-145/FullText.html [Accessed 4 April 2025].

Hossain, M., et al. (2022) 'Synthesis method of hemp-derived graphene', *Journal of Scientific Research*, 14(2), 601–606.

Islam, M. M., et al. (2021) 'Industrial hemp (Cannabis sativa L.) varieties and seed pre-treatments affect seed germination and early growth of seedlings', *Agronomy*, 12(1), 6.

Jarugula, S., et al. (2023) 'First reports of beet curly top virus, citrus yellow vein-associated virus, and hop latent viroid in industrial hemp (Cannabis sativa) in Washington State', *Plant Disease*, 107(9). 2897

Jovičić, D., et al. (2019) 'Comparison of methods for germination testing of Cannabis sativa seed', *Ratarstvo i povrtarstvo/Field and Vegetable Crops Research*, 56(3), 71–75.

Kaminski, K. P., et al. (2024) 'Opportunities, challenges, and scientific progress in hemp crops', *Molecules*, 29(10), 2397.

Karus, M. and Kaup, M. (2002) 'Natural fibres in the European automotive industry', *Journal of Industrial Hemp*, 7(1), 119–131.

Laprairie, R., et al. (2015) 'Cannabidiol is a negative allosteric modulator of the cannabinoid CB1 receptor', *British Journal of Pharmacology*, 172(20), 4790–4805.

Mariz, J., et al. (2024) 'Hemp: from field to fiber—a review', *Textiles*, 4(2), 165–182.

Mark, T., et al. (2020) *Economic viability of industrial hemp in the United States: a review of state pilot programs*, Research Serice United States Department of Agriculture.

New York State Department of Agriculture and Markets (2025) 'Hemp licensing program', [online], available: https://agriculture.ny.gov/hemp [Accessed 4 April 2025].

Mazzara, E. (2023) *Industrial hemp (Cannabis sativa L.) waste products as a source of biopesticides and bioactive compounds for pharmaceutical, nutraceutical and cosmeceutical purposes*, University of Camerino.

McPartland, J. M., et al. (2000) *Hemp diseases and pests: management and biological control: an advanced treatise*, CABI.

Mead, A. (2017) 'The legal status of cannabis (marijuana) and cannabidiol (CBD) under US law', *Epilepsy & Behavior*, 70, 288–291.

Meijer, E. (1995) 'Fibre hemp cultivars: a survey of origin, ancestry, availability and brief agronomic characters', *Journal of the International Hemp Association Archives*, 2, 66–73.

Passley, S. (2025) '"High" tech: farming drones A boon to organic/earth-friendly cannabis', *Head Magazine* [online], available: https://headmagazine.com/high-tech-farming-drones-a-boon-to-organic-earth-friendly-cannabis/ [Accessed 7 April 2025].

Pieracci, Y., et al. (2021) 'Essential oil of Cannabis sativa L: comparison of yield and chemical composition of 11 hemp genotypes', *Molecules*, 26(13), 4080.

Prade, T., et al. (2011) 'Biomass and energy yield of industrial hemp grown for biogas and solid fuel', *Biomass and Bioenergy*, 35(7), 3040–3049.

Punja, Z. K. (2021) 'Emerging diseases of Cannabis sativa and sustainable management', *Pest Management Science*, 77(9), 3857–3870.

Punja, Z. K., et al. (2019) 'Pathogens and molds affecting production and quality of Cannabis sativa L', *Frontiers in Plant Science*, 10, 1120.

Punja, Z. K., et al. (2023) 'Challenges to Cannabis sativa production from pathogens and microbes–the role of molecular diagnostics and bioinformatics', *International Journal of Molecular Sciences*, 25(1), 14.

Ranalli, P. (1999) *Advances in hemp research*, CRC Press.

Rizzo, G., et al. (2023) 'The role of hemp (Cannabis sativa L.) as a functional food in vegetarian nutrition', *Foods*, 12(18), 3505.

Rock, E. M., et al. (2014) 'Effect of phytocannabinoids on nausea and vomiting' in: Pertwee, R. G. (ed.) *Handbook of cannabis*, Oxford University Press, 435–454.

Roseberg, R. (2024) 'Optimum irrigation for CBD-type hemp plants in the field', [online], available: https://extension.oregonstate.edu/crop-production/field-crops/optimum-irrigation-cbd-type-hemp-plants-field [Accessed 7 April 2025].

Russo, E. B. (2011) 'Taming THC: potential cannabis synergy and phytocannabinoid-terpenoid entourage effects', *British Journal of Pharmacology*, 163(7), 1344–1364.

Salentijn, E. M., et al. (2015) 'New developments in fiber hemp (Cannabis sativa L.) breeding', *Industrial Crops and Products*, 68, 32–41.

Schmid, L. (2022) *How hemp contributes to EU green deal objectives*, Signature Products.

Schwartz, J. and Smart, C. D. (2024) 'Abiotic vs. biotic issues in hemp fields', [online], available: https://hemp.cals.cornell.edu/resources/reports-factsheets/2024-factsheets/abiotic-vs-biotic-issues-in-hemp-fields/ [Accessed 7 April 2025].

Serpell, M., et al. (2014) 'A double-blind, randomized, placebo-controlled, parallel group study of THC/CBD spray in peripheral neuropathic pain treatment', *European Journal of Pain*, 18(7), 999–1012.

Shiels, D., et al. (2022) 'Hemp genome editing–Challenges and opportunities', *Frontiers in Genome Editing*, 4, 823486.

Small, E. and Marcus, D. (2019) 'Hemp: a new crop with new uses for North America' in: Janick, J. and Whipkey, A. (eds.) *Trends in new crops and new uses*, ASHS Press.

Stella, A., et al. (2019) 'A therapeutic effect of cbd-enriched ointment in inflammatory skin diseases and cutaneous scars', *La Clinica Terapeutica*, 170(2), e93–e99.

Stevulova, N., et al. (2013) 'Lightweight composites containing hemp hurds', *Procedia Engineering*, 65, 69–74.

Stramkale, V., et al. (2024) 'Industrial hemp variety performance in Latvia under Baltic Sea climate', *Agronomy*, 14(12), 2750.

Struik, P., et al. (2000) 'Agronomy of fibre hemp (Cannabis sativa L.) in Europe', *Industrial Crops and Products*, 11(2–3), 107–118.

Sunoj Valiaparambil Sebastian, J., et al. (2023) 'Hemp agronomy: current advances, questions, challenges, and opportunities', *Agronomy*, 13(2), 475.

Tanney, C. A., et al. (2021) 'Cannabis glandular trichomes: a cellular metabolite factory', *Frontiers in Plant Science*, 12, 721986.

Teirumnieka, Ē., et al. (2021). Product-oriented production of industrial hemp according to climatic conditions. *Agronomy Research* 19(4), 2026–2036, https://doi.org/10.15159/AR.21.123.

Thweatt, I. (2023) *Exploring IPM Strategies in Southeastern Hemp (Cannabis sativa L.)*, Unpublished thesis, Auburn University.

U.S. Department of Agriculture. (2020) 'Hemp production', [online], available: https://www.ams.usda.gov/rules-regulations/hemp [Accessed 4 April 2025].

Valizadehderakhshan, M., et al. (2021) 'Extraction of cannabinoids from Cannabis sativa L. (Hemp)', *Agriculture*, 11(5), 384.

Verma, R. S., et al. (2014) 'The essential oil of'bhang'(Cannabis sativa L.) for non-narcotic applications', *Current Science*, 645–650.

Wang, S. (2021) *Diagnosing hemp and cannabis crop diseases*, CABI.

Weston-Green, K., et al. (2021) 'A review of the potential use of pinene and linalool as terpene-based medicines for brain health: discovering novel therapeutics in the flavours and fragrances of cannabis', *Frontiers in Psychiatry*, 12, 583211.

Wylie, S. E., et al. (2021) 'Fertility management for industrial hemp production: current knowledge and future research needs', *GCB Bioenergy*, 13(4), 517–524.

Part 2

Applications, sustainability and nutraceutical properties of hemp

Chapter 5

Advances in the use of hemp fibre in the construction industry

Chun Hua Julia Liu and Bernardino D'Amico, Edinburgh Napier University, UK; and Francesco Pomponi, University of Cambridge and University of York, UK

1 Introduction

Climate change caused by anthropogenic greenhouse gas emissions has brought disastrous impact to planet Earth [1]. Earth's temperature has risen by an average of about 0.06°C per decade since 1850. The speed of warming since 1982 is more than three times fast per decade. The ten warmest years in the historical record have occurred in the past decade including 2023 being the warmest year since global records began in 1850 [2]. Additionally, the frequency and severity of extreme natural disasters have escalated. Potential consequences of climate change include heightened health risks, food shortages, increased poverty, and widespread displacement [3]. To combat climate change, the Paris Agreement has committed to limit the temperature increase to 1.5°C above pre-industrial levels and reduce emissions by 45% by 2030 and reach net zero by 2050 [4].

The building sector is the third greatest greenhouse gas emitter [5] responsible for 37% of global energy-related carbon emissions, half from operational emissions and half from materials and construction [6]. Carbon

http://dx.doi.org/10.19103/AS.2025.0161.08

emissions are generated at every stage of a building's life cycle [7]. Those emissions are typically distinguished between embodied carbon emissions and operational emissions. Embodied carbon pertains to the emissions generated from the production of building materials, the construction process, material placement, and the disposal of materials at the end of the building's life. According to current estimates, more than 50% of the embodied carbon in construction projects is attributable to the use of building materials [8].

In contrast, operational carbon is associated with the energy consumed during the building's operational phase, such as the energy required for lighting, power, heating, cooling, ventilation, and water services [9]. Among the various aspects of operational carbon, energy consumption for heating and cooling is particularly significant, accounting for approximately 40% of global energy consumption. Therefore, implementing strategies to conserve energy and reduce energy losses of the building fabric is critical for effectively reducing the carbon emissions associated with building operations [10].

Thermal insulation are materials specifically designed to mitigate the undesired transfer of heat, be that by conduction, convection, or radiation, and thus is one of the most valuable tools in achieving energy conservation in buildings [11]. It not only provides resistance to heat flow and lowers energy required for heating and cooling but also improves comfort [12]. Traditional insulation materials, such as mineral wool and plastic forms, contain high quantities of embodied carbon and energy due to an impactful material extraction and intensive material production process [13].

Natural, or bio-based, building materials are derived from renewable biomass sources, both animal and plant origins [14]. These materials characteristically have low quantities of embodied carbon. Additionally, when derived from plants, bio-based insulation materials enable the sequestered carbon to be stored over their lifespan within the building fabric, therefore contributing positively to the overall carbon balance of the environment [15]. Recent advancements have led to the development of natural plant-based aggregates and bio-composite panels specifically designed for sustainable construction, offering low thermal conductivity and superior hygrothermal efficiency.

Notably, cellulosic materials such as hemp, flax, rape shiv, straw, corn cob residue, and fibres from hemp and flax have been shown to achieve excellent microporosity and breathability [16]. When these materials are combined with natural matrix components, they yield insulation with exceptionally low thermal conductivity and enhanced hygrothermal performance [16]. Natural insulation materials represent a competitive alternative to conventional insulation not only for their comparable thermal performance but also due to their significantly lower global warming potential (GWP) and extended useful lifespan [17]. These attributes make bio-based insulation materials increasingly attractive in the context of sustainable and environmentally responsible building practices.

Hemp, for instance, is widely recognized as one of the most used plant-based insulation materials in building and construction [18]. Hemp stands out from other plant-based building materials due to its great ability to sequestrate carbon dioxide through rapid growth cycles under a wide range of climatic and soil conditions, and it has excellent hygrothermal and acoustic properties [19]. However, despite its potential, the advances in the use of hemp fibre in the construction and building industry have not been thoroughly explored.

Current hempcrete research focuses on the mechanical properties and environmental impact assessment. Compared with the other materials, there is a lack of scholarship specifically on the use of hemp fibre in construction. Therefore, the aim of this chapter is to study the advances in the use of hemp fibre in the construction through the evolution of hemp construction and hemp-based building materials, advances in construction methods, and development of standards and building codes. The objectives include a brief introduction of hemp, types of hemp-based building materials, and rationale to use hemp; an overview of notable hemp construction from ancient time to future, different methods to build with hemp, and the development of hemp standards and building codes. To achieve the goal, a combination of methodologies is applied, including the following:

1 Literature review. Review academic articles, industry reports, case studies, regulations and standards to synthesize and analyze existing research on hemp fibre's utilization in construction for the purpose to identify the knowledge gaps, emerging trends, and potential direction for future research.
2 Case studies. Explore real-world applications of hemp fibre in construction to highlight the challenge, benefits, and potential of hemp construction.
3 Comparative analysis. Compare hemp fibre with traditional building materials to determine the unique advantages or limitations of hemp as a construction material.
4 Regulatory and policy analysis. Examine regulatory frameworks and policies surround the use of hemp in construction to identify policy barriers or opportunities for promoting hemp-based materials, as well as recommendations for policymakers.

The novelty of this chapter is to apply an interdisciplinary approach, covering material science, sustainability studies, and policies analysis to provide a holistic view of advances of hemp fibre's application in construction. It is hoped that findings from this study will provide a summary of the potential of hemp fibre in transforming the construction industry and serve as a valuable source for guiding the scope of further research on this very topic.

The definition of common terms used in this chapter are listed below:

- **Hempcrete** – a natural building material made from hemp hurds (the woody cores of the hemp plant), lime (a natural carbonate binder that hardens over time), and water (used to mix and suspend the hemp hurds and lime).
- **Hemp building material** – building materials derived from the hemp plant, such as hempcrete, hemp fibreboard, or hemp insulation, used in sustainable construction.
- **Hemp building and construction** – the practice of utilizing hemp-based materials in architectural and construction projects, promoting sustainability and energy efficiency.
- **Hemp industry standards** – guidelines and benchmarks that define the quality, performance, and environmental impact of hemp products used in construction.
- **Hemp building code** – the set of rules and regulations that govern the use of hemp-based materials in construction. These codes outline permitted uses, material specifications, construction methods, fire safety, and structural integrity.
- **Hemp sustainability practices** – methods and principles that prioritize eco-friendly production and usage of hemp materials to reduce environmental impact.
- **Hemp thermal insulation** – the property of hemp-based materials like hempcrete that enhances energy efficiency by maintaining consistent indoor temperatures.
- **Hemp fire resistance** – the ability of hempcrete and other hemp-based materials to resist combustion, contributing to fire safety in buildings.
- **Hemp moisture regulation** – the capability of hemp materials to absorb and release moisture, preventing mould growth and maintaining indoor air quality.
- **Hemp carbon sequestration** – the process by which hemp plants and hempcrete lock away atmospheric carbon dioxide, hence withdrawing greenhouse gases out of the atmosphere.
- **Hemp bio-composites** – advanced materials created by combining hemp fibres with other natural or synthetic substances for enhanced strength and versatility in construction.

2 Current research trends

The literature review for this research centers around three major subjects. First, hemp-based building materials, involving a brief introduction to hemp, hemp building material types, mechanical properties, sustainability, and comparison with traditional building materials. Second, the utilization of hemp fibre in construction, involving the historic context and traditional hemp fibre uses in construction, modern construction techniques, and emerging technologies and innovations. Third, regulation and standardization, the development of hemp-related policies and industrial standards, as well as the drivers and barriers of hemp fibre's utilization in construction (Fig. 1).

The literature search process for this review was methodically structured around the subjects. The first step entailed a preliminary research and selection of relevant keywords for the search. This initial step involved screening articles based on their titles. Subsequently, abstracts were reviewed to further refine the selection, and finally, the full texts of potentially promising articles were examined to determine their suitability for inclusion in the list of relevant studies.

ScienceDirect (https://www.sciencedirect.com/) and Google scholar (https://scholar.google.com/) served as the primary search engines, complemented by supplementary website searches. The initial search was conducted using keyword 'hempcrete' in ScienceDirect, resulting in identification of 234 articles published since 2006. Notably, the number of annual publications has surged dramatically, increasing from less than 10 before 2016 to around 30 in recent years, following the annual growth in science publications of about 10% and 61 in 2024 (Fig. 2). The significant increase in 2024 with 61 new publications, a 1-year doubling, highlights a growing research interest and investment in the field of hemp-based building and construction. This jump corresponds to policy changes including the passage of 2014 and

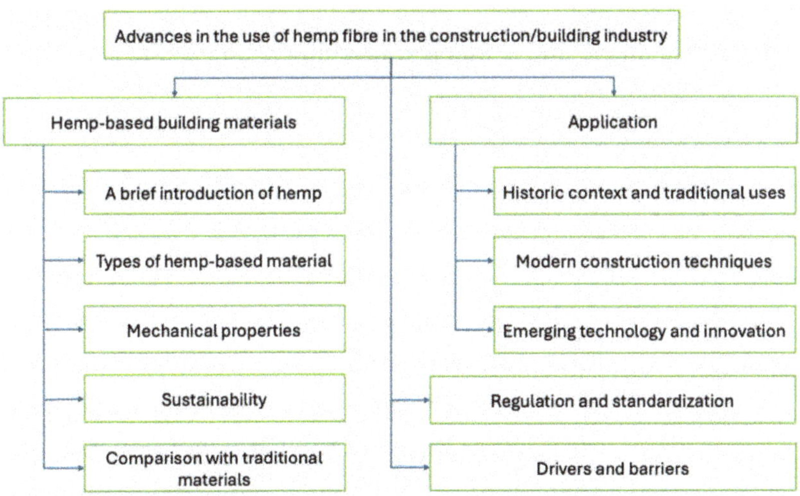

Figure 1 Literature review for this chapter followed this subject tree in approach.

Published by Burleigh Dodds Science Publishing Limited, 2026.

Figure 2 The annual growth rate follows the global rate for scientific publications until 2023–2024.

2018 U.S. Farm Bill to legalize hemp cultivation and expand provisions of hemp cultivation for commercial use, which led to the surge of hemp research in the United States [20].

As a result of this initial screening process, 137 articles that featured relevant keywords in their titles were selected for abstract evaluation. This selection compromised articles of varying focus and scope (Fig. 3), including 45 in regard to physical-mechanical properties, 34 with strong focus on sustainability aspects, 14 with reference to the utilization of hemp fibre in construction and building, 10 relating to the manufacturing process, 9 with regard to technology advancement, 8 about comparison of hempcrete with

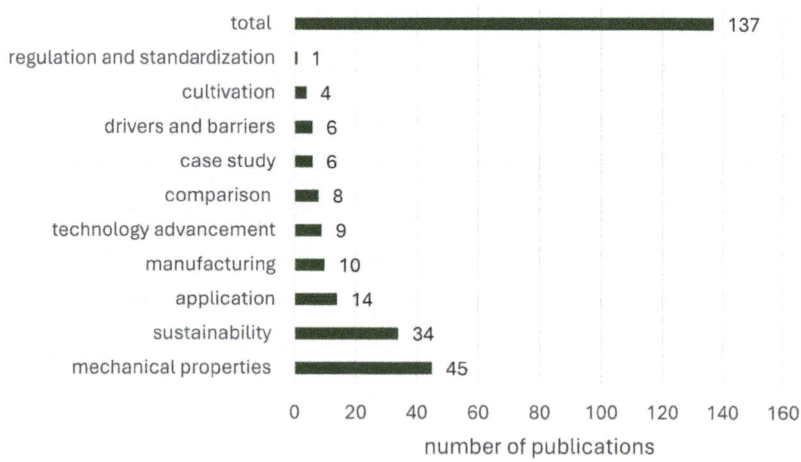

Figure 3 Population of focus and scope selection of articles selected for review.

traditional and other bio-based materials, 6 as to case studies, 6 on the subject of drivers and barriers, 4 in respect of cultivation, and 1 relating to regulation and standardization.

During the abstract review phase with Google Scholar, 88 publications aligned with the classes in Fig. 3 were included in the full text evaluation process. Consequently, 88 of which were found relevant of inclusion in the literature review presented therein. Additionally, 39 articles sourced from website (cited where relevant) have also been included in here to provide supplementary supporting information on policies, regulations, statistics, and manufacturing details.

3 Types of hemp-based construction materials

3.1 A brief introduction to hemp as a material

Hemp, or industrial hemp (*Cannabis Sativa* L.), is an annual C3 (referring to C3 pathway of photosynthesis, the most common form of photosynthesis in plants) herbaceous plant in the Cannabis genus, a species of the Cannabaceae family [21]. Hemp is considered one of the most ancient crops originated from central Asia about 10 000 years ago. China has kept a continuous record of hemp cultivation dating back from 6000 years ago to present [22]. Hemp migrated to Europe between 1000 and 2000 BC, to South America in the fifteenth century, and to North America in the sixteenth century.

Hemp is one of the fastest growing plants, typically planted in May and harvested in August and adapted to most environments from the equator to about 60° latitude in both hemispheres [18]. It is considered a successful rotation crop because of drought resistance with a deep tap root, which can remediate and replenish the soil by killing and displacing of competing plants and weeds, and absorbing heavy metals in contaminated soil through phytoremediation [23].

From an historical perspective, the cultivation and global distribution of hemp have been fundamentally shaped by humanity's wide utilization of the species, particularly to produce fibre, as well as for the provision of seeds used in food, medicinal preparation, and psychotropic substances for therapeutic and religious purposes [24]. Hemp held strategic importance during the seventeenth and eighteenth centuries as its fibres were essential in manufacturing riggings and sails, which were critical components in the advancement of both commercial and naval fleets.

However, the advent of commercial cotton production and weaving machines and steam engines in the nineteenth century brought about profound changes in global shipping and industry, leading to a notable decline to the role hemp had in the wider economy [25]. Plantation, production, and processing of cotton buds have started to replace hemp over time, thus becoming the dominant raw materials in the textile industry by the end of the nineteenth century [26]. Concomitantly, the gradual replacement of large sailing vessels with steam-powered ships marked a significant shift in maritime technology, which coupled with increasing public concerns regarding the pharmacological properties of hemp, eventually led to its prohibition to cultivation in most Western countries. Consequently, hemp cultivation experienced a substantial decline across most regions throughout the nineteenth century [27].

A renewed interest in hemp emerged in Europe during the late 1990s, driven by the search for innovative land-use strategies. Hemp started to be seen favourably due to its adaptability to a wide range of agro-ecological conditions. Its higher yield compared to many other non-food crops and its potential to be processed into a multiple of renewable products are some of the main reasons partially explaining a renewed interest to this ancient crop [28]. Inspired by the success of Europe's revival of hemp cultivation, Canada and the United States also initiated a legalization process for hemp cultivation in 1994 and 2018, respectively. As a result, hemp has emerged as a modern commodity crop in North America [29]. It is safe to say that as a re-emerging high-value multi-purpose crop, hemp is currently undergoing a renaissance around the world [30].

3.2 Hemp stalk

Hemp stalk, also known as hemp straw, provides two co-products, bast fibre and woody core, for use in construction and building [21]. Hemp stalk consists of multiple layers: epidermis layer, phloem layer, cambium layer, xylem layer, pitch layer, and hollow core (Fig. 4) [31].

The most used parts are the epidermis and phloem layers, where the primary fibres originate. The layers adjacent to the cambium layer are secondary fibres. The primary and secondary fibres are removed and bundled as technical fibres. Hemp fibre is one of the strongest and stiffest natural fibres, composed of bundles of fibres, each consisting of 10–40 individual cells or elementary fibres. These fibres are characterized by their layered structure and a central lumen within each cell. A single hemp stem contains between 20 and 50 such fibre bundles (Fig. 5) [32].

Hemp fibre constitutes approximately 20–40% of the total mass of the hemp stalk [33]. The primary component of hemp fibre is cellulose, which constitutes 70–74% of its composition. The remainder comprises various chemical compounds, including hemicellulose (15–20%), lignin (3.5–5.7%), and pectin (0.8%), as well as fats and waxes (1.2–1.6%) [32]. The typical physical and mechanical properties of hemp fibre is shown in Table 1 [34].

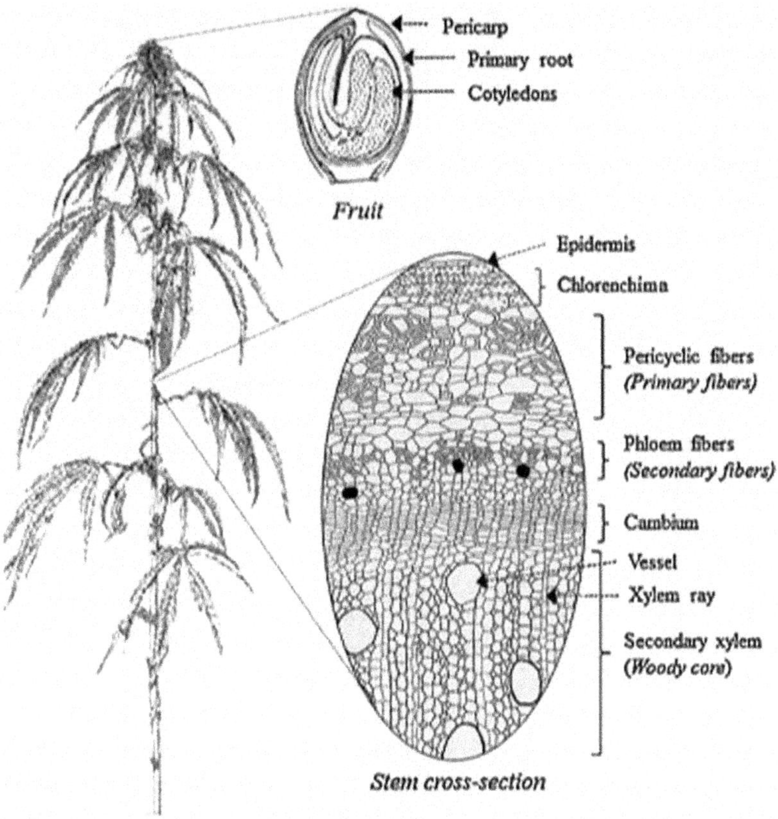

Figure 4 Hemp stem cross-section [31].

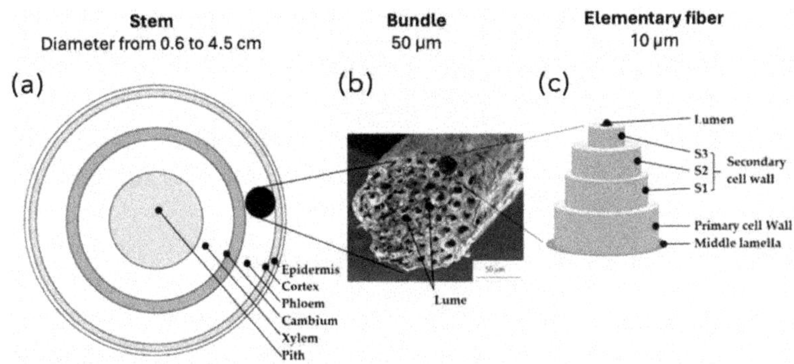

Figure 5 Structure of hemp fibre: (a) transverse hemp stem section, (b) cross-section morphology of the hemp fibre bundle, and (c) schematic depiction of hemp elementary fibre [32].

Table 1 Mechanical properties of hemp fibre [34]

Properties	Values
Length (ultimate) (mm)	8.3–1.4
Diameter (ultimate) (μm)	17–23
Aspect ratio (length/diameter)	549
Specific apparent density (gravity)	1500
Microfibril angle (θ)	6.2
Moisture content (%)	12
Cellulose content (%)	90
Tensile strength (Mpa)	310–750
Specific tensile strength (MPa)	210–510
Young's modulus (GPa)	30–60
Specific Young's modulus (GPa)	20–41
Failure strain (%)	2–4

The porous structure of the bast fibres makes hemp fibre suitable for thermal insulation. It has been used to replace synthetic fibres to produce insulation for ecological building developments, when recyclable and renewable raw materials and low-resource production techniques are desired [35]. Currently, about 20% of hemp fibre are utilized in the construction sector to replace synthetic fibres in insulation [36].

Hemp hurds, the woody core of hemp, are primarily derived from the dead cells located beneath the cambium layer of the hemp stalk [37]. The composition of hemp hurds can vary based on the growing conditions and the nutrients available to the plant. Typically, hemp hurds consist of 34–64% cellulose, 15–37% hemicellulose, 19–28% lignin, 4% pectin, 1% wax, 3% protein, and 1–5% as content [38]. Hurds are considered as agricultural by-products. About 30% of hurds are used in the construction sector. The hurds are used primarily to produce light and super light concretes and mortars, as well as infills for cavities and under floors [36]. Thanks to its highly porous structure, hemp hurds are light and can trap large amounts of air in and between hurd particles to ensure a low thermal conductivity and good sound absorption [21].

3.3 Survey of hemp-based construction materials

Typical commercial hemp-based construction products include hemp insulation panel and flooring underlay made of hemp fibre, hemp plaster, and spun made of hemp fibre and hurds, hempcrete made of hemp hurds, and innovative hemp fibre-reinforced composite, hemp fibre-reinforced concrete, and hemp fibre-reinforced thermoplastic composite filaments. Nowadays, over 100 hemp-based construction products are available on the market globally [21].

3.3.1 Hemp fibre insulation

Hemp wool is a natural fibre thermal insulation made of hemp fibre and polymer fibre [47]. Typical hemp wool blends are made by melting polyester to fuse them together with hemp fibres. As synthetic polyester is not environmentally friendly, the development of bio-based binding agent to replace polyester to reduce environmental impact is emerging [21]. The first hemp wool products were produced in 1998 in France. Since then, loose-fill hemp fibre insulation and different types of hemp-fibre mats have been developed to commercial production for modern houses. Hemp-fibre insulation can be used for floor, wall, and roof. However, it is not suitable for load-bearing due to the porosity of hemp fibre [21]. The thermal conductivity of hemp wool is compatible with conventional insulations [39]. Attribute to the CO_2 uptake associated with hemp plant photosynthesis and fibre biodegradable, hemp wool enables comparable thermal insulation performance with much higher levels of environmental compatibility. Hemp wool is particularly recommended for old timber houses due to its superior ability of regaining water than those of inorganic materials [40].

3.3.2 Hempcrete

It was an innovation to use hemp hurds as an aggregate to produce light cement in 1986 by Charles Rasetti in collaboration with *Chanvrière de l'Aube* for the renovation of the *Maison de la Turque at Nogent sur Seine*, France [21]. Hempcrete, also known as hemp concrete or lime hemp concrete, is a bio-composite mix made of hemp hurds, water, and lime [41]. A lime binder is chosen because of its abundance and low emission from its manufacture. The type of the binder and the amount used play a significant role in the mechanical, thermal, hygrothermal, and acoustic qualities of hempcrete. Hydrated lime, $Ca(OH)_2$, is the most common binding agent for hempcrete. New introduced magnesium-based binders are feasible alternatives due to their good strength, fire resistance, and compatibility with organic aggregates [38].

Thanks to their highly porous structure, hurds are light and can trap large amounts of air in and between hurd particles to ensure a low thermal conductivity and good sound absorption [21] and can passively regulate the internal humidity level [42]. The thermal properties and behaviour of a hempcrete wall were found experimentally to be identical to those of a commonly used lightweight concrete insulation material with a similar density [43]. However, hempcrete cannot be used as a direct load-bearing material due to its low compressive strength and modulus of elasticity [42]. Hempcrete can be used for roofs, walls, and floors with adjusted composition through casting, spraying, or pre-casting [44].

3.3.3 *Hemp plaster*

Hemp plaster is a mixture of hemp hurds or bast fibres, lime, and sand. Thanks to the porous structure of hemp fibres, it is breathable and able to regulate air humidity by absorbing large amounts of water and releasing it gradually [45]. It is generally used as indoor insulating plasters of building envelopes to moderate ambient relative humidity variation [46]. Hemp-based plasters have good insulation properties, allowing thermal dispersion to reduce in the winter and improve the performance of the walls in the summer by approximately 20% [47]. Thanks to the strength and stiffness of hemp fibres, the strength of lime plaster with the addition of hemp fibres can be increased more than three times [48].

3.3.4 *Hemp-reinforced composite*

The invention of polymers in early twentieth century led to a revolution of materials. The use of hemp fibre as reinforcement in composite materials has started in 1940s, when hemp and flax fibres were used in resin matrix for the bodywork of a Henry Ford car. Over the past decade, polymers reinforced with natural fibres are increasingly popular due to their low cost, good mechanical properties, low density, and biodegradable. Thanks to its super strength and stiffness, hemp fibre-reinforced thermoplastic, thermoset, and biodegradable polymers, hemp hybrid composites, and hemp-fibre surface modification are prevailing [34]. Their thermal conductivity is superior to that of traditional insulation materials such as mineral wool with enhanced mechanical strength. In most cases, they are limited to low and medium load-bearing application [49].

3D printing is a novel technology of constructing objectives by gradually adding materials layer by layer, starting from bottom. It allows the creation of complicated geometric structural accurately with minimum material waste. The technology started in the later 1980s and has become widely adopted by different industries including the building sector [50]. Hemp fibre can be blended with the thermoplastic matrix material, a carbon-chained polymer that is flexible enough to be moulded into any shape without breaking [23], to create a hemp fibre-reinforced thermoplastic composite filaments suitable for fused-deposition modelling (FDM) 3D printing, a commonly used additive manufacturing process. Compared to the plain thermoplastic, hemp fibre-reinforced thermoplastic has enhanced strength and insulation properties and less environmental impact [50]. But the disadvantage is that their processing temperature must be controlled under 230°C to avoid thermal degradation of natural fibres [34].

Published by Burleigh Dodds Science Publishing Limited, 2026.

3.3.5 Hemp fibre-reinforced concrete

Fibre-reinforced concrete, consisting of a hydraulic cement matrix reinforced with discontinuous discrete fibres, dates to 1960s [18]. Hemp fibre is a good reinforcement material due to its high-tensile strength and strong tolerance for an alkali environment [51]. Hemp fibre-reinforced concretes, under optimum conditions compared with unreinforced concrete, can increase compressive strength by 4%, flexural strength by 9%, and flexural toughness by 144% [52]. Hemp fibres can be used with concrete in two different ways: as a partial aggregate replacement or as steel reinforcement replacement in cementitious composites. Although hemp fibre-reinforced concrete may satisfy the requirements of high-performance concrete in strength, ductility, and durability, it is difficult to obtain hemp fibre with consistent quality because of the current variation in hemp growing conditions and the methods of processing [53].

3.3.6 Hemp phase-change materials

A phase-change material (PCM) refers to a substance that can store or release the thermal energy under latent form when it changes from one physical state to another without significant temperature change [54]. It has been proven that latent heat thermal energy storage incorporating PCMs is an effective technique to improve the thermal capacity, energy efficiency, and comfort of buildings [55]. Hemp hurds can be used as supporter for PCM-modified hemp/gypsum composites for construction applications like insulation, wall panes, and roofing systems. It provides a cooler environment in hot weather, shows the characteristic of leak-proof and improved thermoregulation; however, the thermal conductivity and compressive strength are reduced [56].

Hempcrete incorporating micro-capsulated PCM combines the hygric regulation capacity of hempcrete with the thermal regulation performance of PCM. The laboratory studies show that hempcrete with micro-capsulated PCM has considerably increased the heat capacity and thermal inertia, while the thermal conductivity and the moisture buffering capacity remain excellent [57]. Hemp stems can be converted into biochar with anisotropic pores and used as a support for polyethylene glycol (PEG6000) to create shape-stable composite PCMs (ss-CPCM) for energy storage. Hemp-stem-biochar ss-CPCM shows high PEG loading rate, large latent heat, high solar-thermal conversion efficiency, and good thermal conductivity and durability [58].

3.4 Rationale of using hemp-based materials

3.4.1 Thermal and hygrothermal properties

The thermal performance of hemp-based materials is a fundamental consideration for building design [59]. Thermal conductivity, which quantifies the rate of heat flow through a given material, serve as the primary metric for evaluating the thermal efficiency of insulation materials [60]. The porous structure of hemp fibres facilitates water vapour permeability and allows the material to effectively absorb moisture from the air [59]. This exceptional moisture-buffering capacity mitigates building condensation, consequently reducing the risk of indoor fungal growth. Research on the use of hempcrete in humid climates demonstrates that it can reduce fungal infection by up to 94% [61]. Additionally, the porous structure of hemp fibres, when combined with lime, enhances the antibacterial properties of hempcrete [62].

3.4.2 Acoustic properties

Noise prevention is another important function of insulation materials. Acoustic properties, usually described by the sound absorption coefficient and sound insulation, are principally dominated by the scale of interparticle pores [63]. The porosity of hempcrete is remarkably high at 70–80%. Pores of different scales exist including macropores or inter-particle pores between the particles of hemp hurds. Extensive research has demonstrated the acoustic advantages enabled by the porous nature of hemp composites [64].

3.4.3 Reaction to fire

Fire is always the critical design factor. Insulation materials used as external and internal wall coverings, accounting for a substantial proportion of potential fire load within a building. Their performance under fire is a critical factor [65]. The European standard EN13501-1 has developed a rating system to classify reaction to fire based on the parameters of temperature increase, mass loss, time to sustained flaming of the specimen, gross calorific potential, fire grown rate index, total heat release, lateral flame spread, smoke growth rate index, flame spread, and total smoke production. Hemp mats are classified as E class, low reaction to fire [60]. Flammability is another important parameter [66] and flame tests have also been proving that hempcrete is a non-flammable material [67].

3.4.4 Durability

Hempcrete is durable. The estimated life span of the hempcrete wall system in Europe is 600 – 800 years [68]. Hempcrete is resistant to freeze-thaw salt exposure, mould, and insects and other animals [69]. Those are desirable properties needed for freezing chilly winter and more frequently floodings caused by climate change [70]. Furthermore, the heat resistant of hempcrete is better than that of conventional insulation materials due to its permeability and

Published by Burleigh Dodds Science Publishing Limited, 2026.

non-flammability [67]. These characteristics are extremely beneficial under the background that wildfires are happening around world with more frequency and intensity. The spread of extreme forest fires caused by climate change has been increasing since 1980 and will continue to worsen through 2090 [71].

3.4.5 Sustainability

The sustainability assessment of hemp-based insulation materials is grounded in the three pillars of sustainability: economy, environment, and society [72].

From an economic perspective, the challenge of housing affordability is intensifying globally with urbanization. Efficient utilization of low-cost, locally sourced renewable materials can reduce construction costs by up to 60% [73]. However, the economic competitiveness of hemp-based construction materials can vary by region globally. For instance, in Italy, hempcrete costs half as much as other insulation materials, whereas in France, it is more expensive than conventional options. This economic variability is influenced by the cost-effective procurement of hemp fibre and the associated manufacturing [74].

From an environmental perspective however, hemp cultivation offers several benefits, including the sequestration of approximately 2.5 tons of CO2 per hectare, suppression of weed growth, and improvement of soil quality through the phytoremediation of heavy metals [75]. Because of the CO_2 uptake associated with hemp plant photosynthesis, hempcrete is often considered to be carbon negative [76]. However, it should be noted that such a CO_2 uptake followed by storage in the finished product is only temporary in nature.

From a social perspective, the use of agricultural by-products, such as hemp hurds, in building materials can reduce construction costs, thereby contributing to the provision of more affordable housing for low-income populations. Utilizing locally sourced materials also stimulates job creation and economic growth, which can have a positive impact on poverty reduction. Moreover, homes built with hemp-based materials offer improved indoor air quality, enhanced thermal comfort, better acoustic control, mould resistance, and fire safety, all of which contribute to the well-being of occupants. The large-scale development of hemp-based materials can also foster interdisciplinary collaboration and strengthens global partnerships for sustainable development [73].

3.5 Comparison of hempcrete with mineral/synthetic insulation materials

Globally, the most widely utilized insulation materials are composed of mineral wool (MW), which includes glass wool, stone wool, and fibreglass, and plastic forms (PFs) such as expanded polystyrene (EPS), extruded polystyrene (XPS), polyisocyanurate (PIR), and polyurethane [39]. MW is derived from inorganic raw materials whereas PFs are processed starting from fossilized organic compounds via petrochemical refining processes. As such, both insulation materials tend to have a significant level of embodied carbon and energy, attributable to the energy-intensive nature of their production processes and the substantial environmental impact associated with the extraction of materials [13].

A comparative analysis of hemp-based insulation mats with commonly used insulation materials reveals that hemp mat, which possesses a density comparable to that of other materials, exhibits superior specific heat capacity, vapour diffusion resistance, and sound absorption coefficients. In terms of fire resistance, hemp performs comparably to commonly used insulation materials, and its market cost is situated within the mid-range of the cost spectrum relative to the other materials [77]. The detail of the comparison is shown in Table 2.

4 Evolution of hemp construction

4.1 Ancient hemp construction

The use of hemp in construction and building materials has a historic precedent dating back to 1500 years ago. Notably, the ancient Merovingian stone bridge (Fig. 6a) was constructed using mortar that incorporated more than 10% of hemp. The inclusion of hemp in the mortar significantly contributed to the bridge's durability, with the material's inherent permeability playing a crucial role in preserving the structure's integrity over the centuries [78].

Another remarkable evidence of hemp's application in construction is the decorative lime plasters of Ellora caves, a World Heritage Site in India dating back to the sixth- to eleventh-century AD. Unlike traditional lime plasters, the decorative lime plasters of Ellora caves were composed of mixture of feebly dolomitic lime binder mixed with fine-to-medium grained siliceous aggregates with addition of hemp as an organic filler. They were applied in 2–3 layers above the host rock (Fig. 6b). The multilayer system, a traditional plastering technique featured with different size and numbers of pores in each layer, can significantly improve the resistance of external layer to rainwater penetration and prevent accumulation of moisture within the plaster masonry. The combination of hemp fibres in the decorative lime plasters further enhanced its permeability. It was also proved that ancient Indian builders had mastered the knowledge of hempcrete in the sixth century [79].

In the Great Seljuq, Anatolian Seljuk, and Ottoman period, hemp motif was a popular decoration and the mixture of hemp and linen fibres with porous and soft Horasan mortar was one of the three popular mortars used for mosques during Ottoman period. It was applied to the dome of Sultan Ahmet Mosque,

Table 2 Comparison of hemp with commonly used insulation materials

Insulation type	Density *r* (kg/m³)	Thermal conductivity *K* (mW/mK)	Specific heat capacity C_p (J/g°C)	Vapour diffusion resistant factor *m* (-)	Sound absorption coefficient *a* (-)	Reaction to fire *f* (-)	Cost C_i (US$/m³)	Embodied energy *EE* (MJ/kg)	Embodied carbon *EC* ($kgCO_2eq/kg$)
Fibre glass	10–100	30–50	0.8–1.0	1.0-1.3	0.45–0.8	A1	9.3–14.7	14–30.8	1.24
Rock wool	40–200	33–44	0.8–1	1.0-1.3	0.29–0.9	A1-A2	12-20	16.8	1.05
Slag wool	50	40	0.7	0.5	n/a	n/a	n/a	n/a	n/a
cellulose	30–80	37–42	1.3–1.6	1.7-3.0	0.53–0.9	B-C-E	24.6	3.3–10.5	0.31–1.83
Spray foam	16–55	25–48	n/a	n/a	n/a	n/a	n/a	n/a	n/a
EPS	18–50	29–41	1.25	20–100	0.22–3.65	E	8.6–1.7	80.8–127	6.3–7.3
XPSS	32–40	32–37	1.45–1.7	80–170	0.2–0.65	E	18–23	72.8–105	7.55
PIR	30–45	18–28	1.4–1.5	55–150	n/a	B	20–24	69.8	5.5
PUR	30–160	22–35	1.3–1.45	50–100	0.67 or 0.8	D-F	24.91	74–140.4	5.9
HEMP	25–100	39–123	1.7–1.8	1.0–10	0.52-0.6	E	15–19.4	18.71	0.14

Source: Data adapted from [77].

(a) Merovingian Stone Bridge in France

(c) Sultanahmet Mosque in Turkey (b) The decorative lime plaster in Ellora cave in India

Figure 6 Historic hemp construction and buildings. (a) and (c) [78], (b) [79].

also known as Blue Mosques, built in 1603–1617 in Istanbul, one of the most magnificent structures of Ottoman Empire (Fig. 6c). The mixture was a good sound absorber. The absorption of sound can be controlled by adjusting the hardness of the mixture [78].

Another noteworthy historical example is Mr. Nakamura's hemp house, constructed in 1698 in Miasa Mura, Japan. The structure, which now functions as a hemp museum, has been designated as a Japanese heritage site. The stone house is notable for its hemp stalk roof, which exemplifies the material's application in traditional architecture. The beams and roof of the house are securely fastened with hemp ropes, while hemp stalks were also employed in the construction of the interior walls. This example highlights the integral role hemp has played in traditional Japanese-building practices and its enduring cultural significance (Fig. 7) [80].

4.2 Modern hemp construction and building

The modern application of hemp mixed with lime (so-called hempcrete) as an alternative to replace deteriorated wattle and daub (a woven lattice of wooden strips called 'wattle' is 'daubed' with a sticky material usually made of some combination of wet soil, clay, sand, animal dung and straw) infills in medieval timber stud walls began in France during the 1980s. The innovative use of hempcrete was first implemented by Charles Rasetti in the renovation of the *Maison de la Turquie* in France (Fig. 8), marking a significant milestone in construction history. Rasetti subsequently patented hempcrete in 1986, solidifying its place as a viable building material [82].

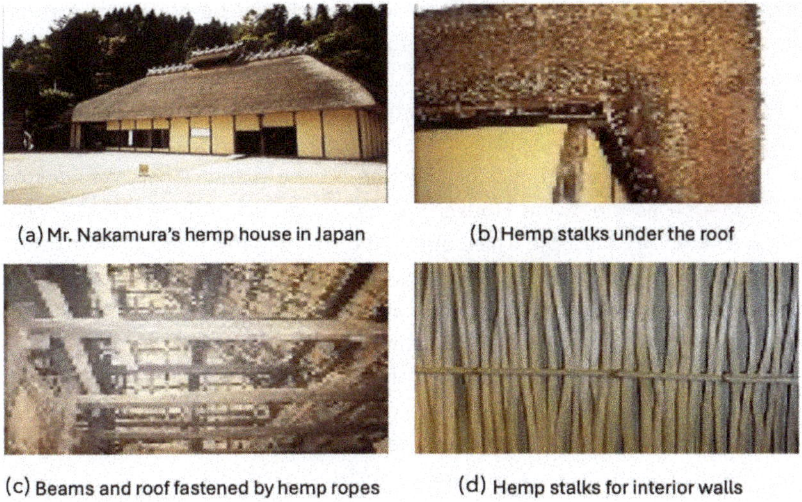

Figure 7 Mr. Nakamura's hemp house. (a) Mr. Nakamura's hemp house in Japan. (b) Hemp stalks under the roof. (c) beams and roof fastened by hemp ropes. (d) Hemp stalks for interior walls [81].

Figure 8 *Maison de la Turquie* in France [82].

Today, hempcrete is widely recognized as a multifunctional, ecological building material and is commonly used in countries such as France, England, Germany, Ireland, Belgium, Luxembourg, and Switzerland. Progress in building technology, coupled with the increased availability of hemp-based materials and allowance by building regulations, has facilitated the construction of thousands of hemp buildings across Europe, serving residential, commercial, and public purposes [83]. Its adoption has also extended to regions including North America, South Africa, Israel, Australia, New Zealand, and India [75].

Hempcrete is also an excellent choice for the extraordinary architectural projects, such as the Twine House (Fig. 9), visualized by British architectural designer Anthony Gibbon. The combination of organic design with flexible mould technology and precision engineered hempcrete bricks to form the structure of hempcrete coiled walls are an innovative response to the challenge of mathematics, architecture, and sustainability [68].

The research in the application of hemp fibres as a reinforcement for asphalt mixture for road pavement has developed in recent years. Hemp fibre can increase the tensile strength of asphalt mixtures, reducing rutting and permanent deformations, and increase resistance of fatigue cracking. Hemp fibre-reinforced pavement is competitive in the material and construction cost compared to synthetic polymer fibre-modified pavement [85].

The application of hemp fibre in structural loading is very rare. One example is the pedestrian bridge built by Eindhoven University of Technology in 2016. The structural part of the bridge is made of bio-composite of hemp and flax fibres attached to a polylactic acid (PLA)-based foam core with a 56% fibre volume fraction (Fig. 10) [86]. As a composite bridge, it requires almost no maintenance. It not only reduces environmental impact but also more economic compared with traditional bridges [87].

Figure 9 The Twine House created by Antony Gibbon [84].

Figure 10 Eindhoven bio-composite footbridge over the river Dommel, Eindhoven, The Netherlands [86].

5 Advances in construction methods

Hempcrete can be manufactured either on-site or off-site. Traditional methods include cast-in-situ and spray with a projection process for on-site application and pre-cast into blocks or framed panels off-site [48].

5.1 On-site application

Cast-in-situ hempcrete means wet-mixing of chopped hemp hurds with lime on site and casting it into moulds constructed from shuttering for the target position within a building, which could be walls, floors, or roofs (Fig. 11a, b) [48]. Cast-in-situ by hand placing is most common due to its simplicity and the result of a monolithic insulation layer with increased air-tightness [88]. However, because of the porous structure of hemp hurds and strong capillary effects inside the tubes, it constitutes of significant excess water is added during the mixing and casting process. Consequently, the setting and drying of hempcrete takes months, which is not practical at industrial scale [37].

Spray hempcrete to the target site using projection process (Fig. 11c, d) was developed to regulate the water content for mixing hemp and lime. The process consists of two parts: a lime and hemp hopper and a water pump. A pre-mix of dry hemp and lime is conducted by air through a hose, and the quantity of water necessary to slake the lime was added right before the hose outlet to prevent hemp absorbing excess water [89]. As a result, the setting and drying time was reduced compared with cast-in-situ.

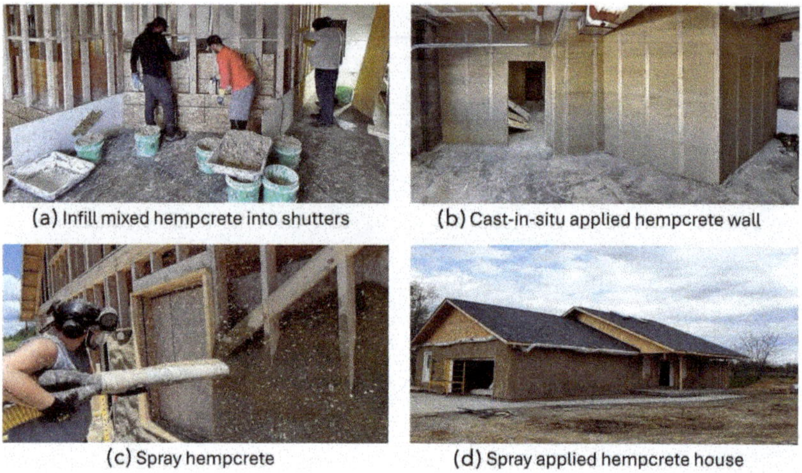

Figure 11 Hempcrete on-site manufacture. (a) Cast-in-situ. Infill mixed hempcrete into shutters. (b) Cast-in-situ applied hempcrete wall. (c) Spray hempcrete. (d) Spray applied hempcrete house [90].

5.2 Off-site application

As an alternative to in-situ casting, hempcrete can be pre-cast into hemp blocks. This approach allows massive production of hempcrete at industrial scale with consistent qualities. However, the disadvantages of pre-cast are less sustainable due to more complicated production process, which caused extra emissions, compared to cast-in-situ, and potential impact of the thermal performance due to the 'cold bridge' between blocks [48].

Most commercially available hempcrete blocks are non-load bearing, one exception is the structural hemp block developed by Canadian manufacturer Just BioFibre (JBF) [91]. JBF invented an internal structural frame system made of advanced composites for their hemp blocks, which can provide strong vertical load bearing compression strength. Furthermore, the interlocking design enables overlapping and turning 90 degrees for corners and intersecting walls. The frames and the interlocking system enable the blocks stacked like Lego. The Harmless Home, built in 2019 in Victoria, BC, Canada, was the first project built with JBF structural hempcrete blocks (Fig. 12) [92]. The project aiming to reduce environmental impact through a holistic approach by selecting sustainable materials, water, electric, and septic systems. The post occupancy assessment shows that the application of hempcrete blocks as insulation together with solar panels for electricity has significantly reduced annual greenhouse gas emission intensity factors for electricity use by 90% [93].

Modular construction is one of the modern methods of construction. It refers to the production of standardized or prefabricated components

Figure 12 The Harmless Home built with JBF structural hempcrete blocks [92].

of a structure in an off-site factory and transported to a construction site for installation [94]. Off-site methods involve manufacturing building components in a controlled environment before transporting them to the construction site, which can reduce on-site construction time, labour costs, and exposure to weather-related delays. It also facilitates a more sustainable approach to building design and construction [95].

Hemp-based construction materials can be effectively integrated into off-site methods of construction, such as prefabrication and modular building techniques. By utilizing hemp in these offsite processes, construction efficiency and quality control are significantly enhanced. Hemp-based materials, like hempcrete panels or prefabricated hemp insulation, can be produced with high precision, ensuring consistent quality and performance. Additionally, integrating hemp-based materials into offsite construction aligns with sustainability goals, as it optimizes the use of renewable resources and minimizes waste. This approach not only accelerates the construction process but also leverages the environmental benefits of hemp, such as its low embodied carbon and excellent thermal properties [95].

The technology of prefabricated hempcrete homes enables the curing of hempcrete under the controlled environment consistently. This can address the issue of extended drying times of cast hempcrete walls, which typically requires about 2–8 weeks or even longer time when hemp homes are constructed in open air conditions [96]. A house constructed of prefabricated hempcrete panels might take 1–2 weeks to be assembled on site and ready for finishing (Fig. 13) [97]. The assembly of prefabricated hempcrete home on site involves four steps. (1) Prepare foundation and timber structure in place. (2) Install

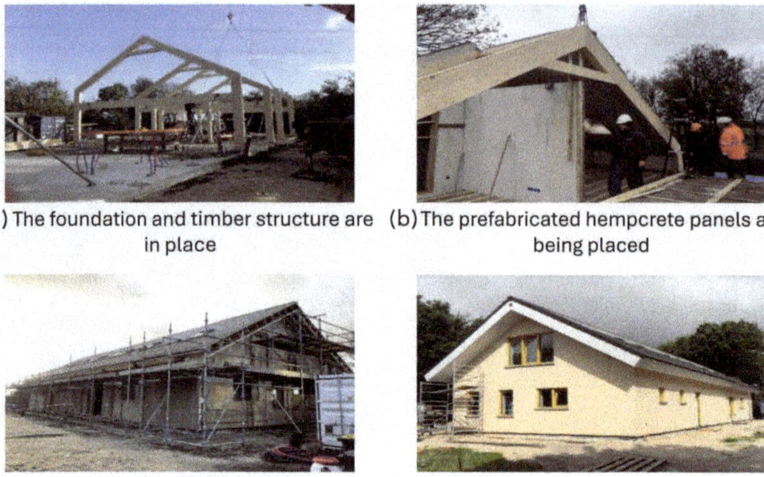

(a) The foundation and timber structure are in place

(b) The prefabricated hempcrete panels are being placed

(c) The house is wind and watertight, ready for finishing

(d) Finished exterior prefabricated hempcrete walls

Figure 13 The assembly of prefabricated hempcrete home on site. (a) The foundation and timber structure are in place. (b) The prefabricated hempcrete panels are being placed. (c) The house is wind and watertight, ready for finishing. (d) Finished exterior of prefabricated hempcrete homes [98].

prefabricated hempcrete panels. (3) Seal hempcrete walls/roofs wind and watertight. (4) Finish the exterior of buildings [98].

5.3 3D printing

The integration of hempcrete with advanced 3D-printing technology presents a promising avenue for enhancing sustainable construction practices. 3D printing has gained recognition for its exceptional precision and adaptability, pushing the boundaries of architectural design, and advancing low-carbon architecture [98]. Among the various methods of 3D printing, material extrusion and powder-based printing are the most employed for large-scale projects. While cement-based materials dominate the 3D printing market, comprising over 90% of market share, and polymers follow with 4.26%, the use of natural earth-based materials, including hemp, is gaining attention due to their sustainability [99].

Hempcrete, known for its cost-effectiveness, natural insulation properties, and environmental benefits, aligns well with 3D printing techniques. It is particularly advantageous in structures requiring compressive load resistance. When combined with subsoil and lime, hempcrete can also serve as a load-bearing material [100]. For instance, an Italian firm, WASP, in collaboration with Mario Cucinella Architects, has demonstrated the potential of 3D printing with

Figure 14 Double dome earthen 3D printed house [101].

natural, recyclable, and carbon-neutral materials by constructing double dome-shaped houses using locally sourced clay (Fig. 14) [101]. It is envisaged that the same technique lends itself well to alternative 3D printing mixes involving a hemp-based matrix.

It should be noted, however, that despite the considerable potential for 3D printing hemp-based homes to contribute to sustainable building practices, several challenges remain. These include difficulties in material formulation, issues with cracking and shrinkage, reinforcement integration, the need for skilled labour, a lack of standardized codes, and scalability concerns. Addressing these challenges will require interdisciplinary collaboration and innovation [99].

6 Regulatory framework

6.1 Europe

Since 1980s, hempcrete has been widely used in European countries including France, UK, Germany, Ireland, Belgium, Luxemburg, and Switzerland [78]. However, there are no specific guidelines to regulate bio-based materials, including hempcrete, harmonized by European nations. France, as the pioneer of using hemp in construction, has developed more specific regulations and standards in efforts to harmonize with European standards to regulate hempcrete and other hemp-based products since 1990s. In 2010, hempcrete was officially incorporated into French Building Code. While the other countries have guidelines and on-going research regarding hempcrete but do not yet fully develop national standards as France does [102].

The two key organizations who have worked collectively to incorporate hemp-based materials into French building codes are Rèseau des Professionnels du Chanvre (The Hemp Professionals Network), often referred to as France Chanvre, the key organization in France responsible for establishing hemp building codes and promoting the use of hemp-based construction materials [103], and Centre Scientifique et Technique du Bâtiment (CSTB), a government body responsible for providing technical research, certification, and standards for building materials and construction practices in France. In 2010, CSTB officially recognized hempcrete as a valid construction material under the French technical approval system. This ensures that hempcrete could be used for residential, commercial, and public buildings across France if it adhered to the necessary technical and performance standards [104].

Countries which have not yet developed their specific standards for hemp-based materials will follow European (EN) standards [102], which are organized by the European Committee for Standardization (CEN). EN standards, including construction, are developed by the national CEN members, the CEN National Standardization Bodies (NSBs), and followed by NSBs, while making necessary adoptions where required for their local conditions, as a national standard [105]. CEN has been active in developing standards for bio-based materials. While there is no single explicit EN standard dedicated especially to hempcrete, several EN standards have relation to hempcrete in the areas of thermal insulation, fire safety, and material performance to building practices [106]. Major EN standards related, or applicable, to hempcrete are

EN 15804:2012+A2:2019 provides core product category rules (PCR) for Type III Environmental Product Declarations (EPDs) for any construction product (including hemp-based building materials) and construction service. The core PCR defines how data for EPDs is collected, reported, verified, and presented, as well as the rules for the Life Cycle Inventory and the Life Cycle of Impact Assessment [107].

EN 1745:2012 sets out requirements for lime-based products in construction, which is highly relevant to hempcrete because it uses lime as binder. While not specific to hempcrete, EN1745 covers various lime-based construction materials and their performance, including aspects such as compressive strength, durability, and mortar application. This guidelines provide a foundation for the proper use of lime in hempcrete and other lime-based materials in construction [108].

EN 13162 [109], **EN 12664:2001**[110], and **EN1602** [111] are related to thermal insulation products for buildings, although they not hempcrete specific, they provide test methods for measuring the thermal conductivity and density of insulation materials, which would apply to hempcrete used as insulation.

EN 13369 sets common rules for precast concrete products, while hempcrete is not a concrete product, the production process for precast hempcrete blocks or panels may follow similar rules and guidelines for dimensional accuracy, quality control, and testing of properties [112].

EN 13501-1 establishes the fire classification for building materials, including the reaction to fire of construction products. Although hempcrete itself is typically non-combustible, it may need to undergo fire resistance testing to meet national or regional fire safety standards. Hempcrete's lime binder contributes to its fire-resistant properties, but it would still need to meet relevant performance standards under EN 13501-1 to assess its fire classification [113].

6.2 North America

Hemp is considered a modern agricultural commodity in North America. Hemp fibre's application in North America is still in its infancy [114]. There are no specific standards to govern hemp-based materials in North America.

Hempcrete was approved for the model U.S. residential building code in 2022, as an appendix for the 2024 International Residential Code (IRC) [115]. IRC is developed by the International Code Council (ICC), one of the most influential organizations for building codes and standards in the United States and around the world. It is responsible for the International Building Code (IBC) and other construction related regulations, such as IRC, International plumbing Code (IPC), International Mechanical Code (IMC), and International Fire Code (IFC) [116]. Established specifications for hemp-lime wall infill materials include technical data, ASTM standards, performance features and product attributes [103].

ASTM International, formerly known as American Society for Testing and Materials, founded in 1902, is one of the largest international standard organizations, particularly influential in North America. ASTM creates standards for materials, products, systems, and services, many of which are adopted by governments and industries. ASTM subcommittee (D37.07) is responsible developing standards for industrial hemp. There are ongoing efforts to establish standards for hemp-based materials. Two new standards have proposed addressing hempcrete and its fire-resistance properties. The goal is to establish a set of performance standards enabling any building inspector to use it to approve a hempcrete structure. Because hempcrete is an insulative, rather than structural, material, the Subcommittee decided to work on performance rather than material specifications. This will remove the bias from manufacturing processes and allow a hempcrete manufacturer to produce hempcrete in the way suitable for their circumstances and make it easier for hempcrete buildings to pass inspection [117].

Canadian Standards Association (CSA) is Canada's primary standards body. In 2022, CSA issued specifications of hemp fibre and board based on the evaluation made by Canadian Construction Materials Centre (CCMC), the organization under National Research Council Canada (NRCC) responsible for assessing compliance with Canadian building, energy and safety codes [118]. Hemp has not been integrated into Canada National Building Code yet.

6.3 International

International Organization for Standardization (ISO), founded in 1947, is an international body that creates and publishes standards across industries. It works with national standard organizations to ensure global consistency and compliance. ISO standards are especially important for materials used in multiple countries and for promoting international trade [119]. ISO14040 describes the principles and framework for life cycle assessment (LCA) [120], which is adopted by hempcrete and hemp-based materials. ISO 24260:2022 specifies the requirements for factory-made products of hemp fibre mat and board, which are used for the thermal insulation of buildings [121]. EN ISO 10140-3 provides standard method for measuring sound insulation in buildings and building elements, it is applicable for testing the acoustic properties of hempcrete when used in walls, floors, or ceilings in buildings [122].

6.4 The hemp industry and advocacy groups

The hemp industry and hemp-advocacy organizations have been collectively pushing for the inclusion of hemp-based materials in national and international building codes and standards. Their work involves research of the environmental, structural, and health benefits of using hemp as a building material; work with the local and national governments to create building relations that allow for the safe and standardized use of hemp-based materials; offer courses, certifications, and hands-on experience to industry professionals to ensure high standards in the production and application of hemp-based materials; and collaborate with manufacturers and government agencies to develop standards for hempcrete, hemp insulation, and other hemp-based building products, such as developing safety protocols, testing procedures, and quality assurance. Major organizations include the following:

- European Industrial Hemp Association (EIHA) advocates for the use of hemp-based materials and plays a key role in pushing for standardization at the European level. They work with other stakeholders, including CEN and government bodies, to facilitate the development of hemp-related building codes [123].
- International Hemp Building Association (IHBA) promotes the use of hemp in construction and has supported the development of guidelines and best practice for hempcrete and other hemp-based materials. They also

collaborate with various standards organizations to facilitate the adoption of hempcrete worldwide [124].

- U.S. Hemp Building Association (USHBA) is committed to the promotion and advancement of hemp-based building materials in the United States. This includes pushing for the development of national standards for hemp-based materials, conducting research on the durability and performance of hempcrete, and raising awareness about the environmental and health benefits of using hemp in building construction [125].
- Canadian Hemp Trade Alliance (CHTA) primarily focuses on hemp cultivation and commercialization, it also supports the use of hemp in construction. Canada has been a leader in hemp agriculture, and CHTA advocates for the use of hemp fibres and other parts of the plant in building materials such as hempcrete for insulation [126].

6.5 National and local governments

Many governments, especially in countries where hemp cultivation is being reintroduced, are exploring the role of hempcrete in sustainable housing and infrastructure. These bodies may develop national codes and regulations that incorporate hempcrete, either directly or through amendments to existing building codes. In some cases, local jurisdictions or regional authorities are also playing a key role by allowing pilot projects or encouraging research into sustainable building materials, helping to lay the groundwork for future standardization.

6.6 Research and educational institutions

Universities and research organizations worldwide are conducting studies on hempcrete's physical properties, environmental impact, and potential applications in the construction industry. The data from these studies will be essential in informing the development of codes and standards. The leading universities who have conducted research on hempcrete and may play an important role in shaping technical guidelines for its use include University of Bath (UK) [127], ETH Zurich (Switzerland) [128], and Université de Limoges (France) [129]. Oregon State University (OSU) is the leader of hemp research and education. OSU's Global Hemp Innovation Center (GHIC) is home of world-class interdisciplinary hemp experts with focus on advancing hemp utilization across all the segments of hemp supply chain [130].

Overall, no single organization is currently responsible for setting comprehensive global standards for hemp building materials, a combination of national standards bodies (like ASTM, BSI, and CSA) and international bodies (like ISO and CEN) are playing a key role in the development of standards for hemp-based materials, including hempcrete. In addition, advocacy groups, research institutions, and industry organizations are pushing for the integration of hempcrete into building codes and standards worldwide. The process is ongoing. As hempcrete continuous to gain acceptance in the construction industry, it is likely that formalized standards will emerge in the coming years.

7 Drivers and barriers

7.1 Drivers

Drivers of the application of hemp-based construction materials are many. First, the environmental benefit. Hemp-based construction materials can sequester carbon and reduce environmental impact and waste. Furthermore, they are biodegradable. Second, the superior hygrothermal and acoustic properties, which contribute to healthy indoor air quality and comfort. Third, long-term saving due to energy savings [37]. Fourth, incentives from government or industry organizations to support sustainable building practice. Fifth, growing demand for sustainable construction from consumers. Sixth, innovation and technological advancement will provide more option of hemp products with enhanced material performance [131].

7.2 Barriers

However, there are multiple barriers in advancing hemp-based materials. First, the lack of industry standards and building codes. Without official recognition and standards, architects, builders, and contractors maybe hesitant to choose hemp-based materials due to the concerns about compliance with local regulations [132] and coverage by insurance companies. Second, limited availability due to supply-chain issues and geographic limitations. Because hemp cultivation for building materials is still in its early stage, the short of consistent high-quality hemp hurds is persistent. Moreover, the regulation of hemp cultivation varies in different countries. It could be significantly challenged to source raw materials in places where hemp cultivation is not widely permitted. Third, the high initial cost. Although hempcrete can save money in long term, its initial cost is higher due to more expensive materials and labour cost. Fourth, perception and familiarity. Hemp-based materials are innovative and still lacking awareness. Therefore, the construction industry is reluctant to adopt these new materials. Fifth, performance and durability concerns. While hempcrete is durable under the right conditions, there is still relatively limited long-term performance data compared to traditional materials [102].

8 Conclusion and future trends

Hemp stalk provides two principal co-products for construction materials: fibres and hurds. Typical commercial hemp-based construction products include hemp insulation panel and flooring underlay made of hemp fibre, hemp plaster, and spun made of hemp fibre and hurds, hempcrete made of hemp hurds. With technology advancement, innovative hemp-fibre-reinforced composite, hemp-fibre-reinforced concrete, and hemp-fibre-reinforced thermoplastic composite filaments are being developed.

The porous structural of hemp fibre/hurds enables hemp-based construction materials to have superior hygrothermal and acoustic properties yet are not strong enough for load bearing. Attributed to the CO_2 uptake associated with hemp plant photosynthesis and fibre biodegradability, hemp-based insulation materials have comparable thermal performance to conventional insulation materials with much higher level of environmental compatibility. Furthermore, hempcrete is very durable, permeable, non-flammable, resistant to freeze-thaw salt exposure, mould, and insects and other animals. These qualities are very desirable with the increases in the frequency and severity of extreme weather events, flooding, and wildfire of climate change.

The utilization of hemp fibre/hurds in construction can be evidenced from the existing infrastructure and buildings built about 1500 years ago, where hemp plasters made of hemp fibre/hurds were applied to protect the structural integrity and regulate sound reflections. The stone house built with hemp stalk roof and walls in Japan has lasted for centuries also. However, hemp construction was terminated in the last century because of the ban of hemp cultivation around the world. The renaissance of hemp construction started in 1980s following the legalization of hemp cultivation in Europe. Today, hempcrete is widely recognized as a multiple functional, ecological building material and is commonly used in European countries, which is extending to regions including North America, South Africa, Isreal, Australia, and New Zealand.

Cast-in-situ is the most common way to manufacture hempcrete because of its simplicity and the result of a monolithic insulation layer with increased airtightness. However, its long drying time (about three months) is not practical for industrial scale. Alternatively, spray with a projection process was invented to shorten the drying process by reducing the water needed to slake the lime. However, this is less popular due to the complexity to operate the machine. The development of prefabrication technology enables massive production of pre-cast hemp blocks and panels off-site with consistent quality. The recent development of integration of hempcrete with 3D printing technology is another promising avenue for hemp building in the future.

Although hempcrete has been commonly used in European countries and emerging in North America and other regions, there is no incorporated hemp standards or building codes applied in EU or globally. France is the only country who has established specific technical guidelines for hemp-based materials and successfully integrated them into French building code in 2010. While the other European countries harmonize relevant EN standards and adopt their local requirements to regulate hemp construction. Recently, hempcrete was approved for use in residential construction in the United States in 2022. Canada Standard Association issued specifications of hemp fibre board in 2022 also.

The drivers of hemp fibre's application in construction includes the growing demand for sustainable construction from consumers, environmental sustainability, healthy indoor comfort because of superior hygrothermal and acoustic properties, innovation and technology advancements, and incentives. The barriers are high initial costs, lack of standardization and certification, limited awareness and knowledge, supply chain and scaling challenges, public perception and legal issues, and limited market availability.

Further research regarding the establishment of explicit hemp technical guidelines and building codes at national and international level, and effective local hemp building materials supply chain will enable the scalability of hemp fibre's application in construction.

9 Where to look for further information

9.1 Further reading

- A good book to review all aspects of hemp is 'HEMP Industrial Production and Use' Edited by P Bouloc, first printed in 2013, reprinted in 2019, 2020, transferred to digital print on demand in 2021.
- A good book to review the status of industry hemp in the United States is 'Industrial Hemp as a modern Commodity Crop' Edited by DW Williams, 2019.
- A good book to review designing and building with hempcrete is 'The Hempcrete Book: Designing and Building with hemp-lime' by W Stanwix and A Sparrow, 2014.
- A comprehensive review of the properties of hempcrete and its application in buildings is 'A review of the properties of hemp concrete for green building applications' by T Jami et al., 2019.
- A comprehensive review of the energy and environmental performance of hempcrete is 'Energy and environmental assessment of industrial hemp for building applications' by C Ingrao et al., 2015.

Published by Burleigh Dodds Science Publishing Limited, 2026.

9.2 Key journals/conferences

- *Construction and Building Materials*, *Journal of Building Engineering*, and *Journal of Cleaner Production* are top three journals most relevant to hemp construction/building.
- International Hemp Building Symposium (IHBS) is an annual event. IHBS focus specifically on the application of hemp in building and construction with an emphasis on the performance, benefits, and challenges of using hemp as a construction material.
- Hemp Industry Association (HIA) Conference is one of the largest annual hemp events in North America. While it covers the full spectrum of hemp applications, the conference regularly features sessions on hemp-based building materials, hempcrete, and future of hemp in construction. It is an essential event for those interested in learning about the latest trends in hemp construction in the United States.
- European Industrial Hemp Association (EIHA) Conference is an annual event covering agricultural, industrial, and commercial uses. It frequently includes sessions on hemp-based construction materials, innovation in building techniques, and environmental benefits of using hemp in construction.

10 References

1. R. Tuckett, "Greenhouse Gases," in *Reference Module in Chemistry, Molecular Sciences and Chemical Engineering*, Elsevier, 2018. https://doi.org/10.1016/B978-0-12-409547-2.14031-4.
2. R. Lindsey and L. Dahlman, "Climate change: global temperature, 2023 global summary," [Online]. Available: https://www.ncei.noaa.gov/access/monitoring/monthly-report/global/202313.
3. Intergovernmental Panel on Climate Change, *Climate Change 2021 – The Physical Science Basis*. Cambridge University Press, 2023. https://doi.org/10.1017/9781009157896.
4. "The Paris Agreement," *UNCC*, 2015, [Online]. Available: https://unfccc.int/process-and-meetings/the-paris-agreement.
5. W. F. Lamb et al., "A review of trends and drivers of greenhouse gas emissions by sector from 1990 to 2018," *Environ. Res. Lett.*, vol. 16, no. 7, pp. 073005–073005, Jul. 2021. https://doi.org/10.1088/1748-9326/abee4e.
6. "8dff87ea3e1e4e7ba7d349a83ed04cbd.pdf." Accessed: Oct. 27, 2024. [Online]. Available: https://unfccc.int/ttclear/misc_/StaticFiles/gnwoerk_static/tn_meetings/00cf22a4049c4ece9f414e190def4202/8dff87ea3e1e4e7ba7d349a83ed04cbd.pdf.
7. E. Marsh, et al., "Tackling uncertainty in life cycle assessments for the built environment: a review," *Build. Environ.*, vol. 231, pp. 109941–109941, Mar. 2023. https://doi.org/10.1016/j.buildenv.2022.109941.
8. J. Giesekam, et al., "The greenhouse gas emissions and mitigation options for materials used in UK construction," *Energy Build.*, vol. 78, pp. 202–214, Aug. 2014. https://doi.org/10.1016/j.enbuild.2014.04.035.

9. A. Akbarnezhad and J. Xiao, "Estimation and minimization of embodied carbon of buildings: a review," *Buildings*, vol. 7, no. 1, pp. 5–5, Jan. 2017. https://doi.org/10.3390/buildings7010005.
10. B. Abu-Jdayil, et al., "Traditional, state-of-the-art and renewable thermal building insulation materials: an overview," *Constr. Build. Mater.*, vol. 214, pp. 709–735, Jul. 2019. https://doi.org/10.1016/j.conbuildmat.2019.04.102.
11. L. Christie, "Reducing the whole life carbon impact of buildings," 2021. [Online]. Available: post.parliament.uk/research-briefings/post-pb-0044/.
12. F. Pomponi and A. Moncaster, "Scrutinising embodied carbon in buildings: the next performance gap made manifest," *Renew. Sustain. Energy Rev.*, vol. 81, pp. 2431–2442, Jan. 2018. https://doi.org/10.1016/j.rser.2017.06.049.
13. B. P. Jelle, "Traditional, state-of-the-art and future thermal building insulation materials and solutions–Properties, requirements and possibilities," *Energy Build.*, vol. 43, no. 10, pp. 2549–2563, 2011.
14. M. Viel, et al., "Development and characterization of thermal insulation materials from renewable resources," *Constr. Build. Mater.*, vol. 214, pp. 685–697, Jul. 2019. https://doi.org/10.1016/j.conbuildmat.2019.04.139.
15. G. Grazieschi, et al., "Embodied energy and carbon of building insulating materials: a critical review," *Clean. Environ. Syst.*, vol. 2, pp. 100032–100032, Jun. 2021. https://doi.org/10.1016/j.cesys.2021.100032.
16. M. P. Ansell et al., "10 - Natural plant-based aggregates and bio-composite panels with low thermal conductivity and high hygrothermal efficiency for applications in construction," in *Nonconventional and Vernacular Construction Materials (Second Edition)*, K. A. Harries and B. Sharma, Eds., in Woodhead Publishing Series in Civil and Structural Engineering. Woodhead Publishing, 2020, pp. 217–245. https://doi.org/10.1016/B978-0-08-102704-2.00010-X.
17. "Comparative performance evaluation of conventional and renewable thermal insulation materials used in building envelops," *Teh. Vjesn. - Tech. Gaz.*, vol. 27, no. 1, Feb. 2020. https://doi.org/10.17559/TV-20171228212943.
18. C. Ingrao et al., "Energy and environmental assessment of industrial hemp for building applications: a review," *Renew. Sustain. Energy Rev.*, vol. 51, pp. 29–42, 2015.
19. O. Gencel et al., "Enhancing sustainability with waste hemp-shive and phase change material: novel gypsum-based composites with advanced thermal energy storage properties," *J. Clean. Prod.*, vol. 451, pp. 142000–142000, Apr. 2024. https://doi.org/10.1016/j.jclepro.2024.142000.
20. D. R. Enarevba and K. R. Haapala, "The emerging hemp industry: a review of industrial hemp materials and product manufacturing," *AgriEngineering*, vol. 6, no. 3, pp. 2891–2925, Aug. 2024. https://doi.org/10.3390/agriengineering6030167.
21. P. Bouloc, *Hemp: Industrial Production and Uses*. CABI, 2013. [Online]. Available: cabidigitallibrary.org/doi/abs/10.1079/9781845937935.0000.
22. H.-L. Li, "An archaeological and historical account of cannabis in China," *Econ. Bot.*, vol. 28, no. 4, pp. 437–448, 1974.
23. A. T. M. F. Ahmed, et al., "Hemp as a potential raw material toward a sustainable world: a review," *Heliyon*, vol. 8, no. 1, pp. e08753–e08753, Jan. 2022. https://doi.org/10.1016/j.heliyon.2022.e08753.
24. X. Lu and R. C. Clarke, "The cultivation and use of hemp (Cannabis sativa L.) in ancient China," *J. Int. Hemp Assoc.*, vol. 2, no. 1, pp. 26–30, 1995.

25. D. W. Williams, *Industrial Hemp as a Modern Commodity Crop, 2019*. John Wiley & Sons, 2020.
26. G. Crini, et al., "Applications of hemp in textiles, paper industry, insulation and building materials, horticulture, animal nutrition, food and beverages, nutraceuticals, cosmetics and hygiene, medicine, agrochemistry, energy production and environment: a review," *Environ. Chem. Lett.*, vol. 18, no. 5, pp. 1451–1476, 2020.
27. G. Richter, et al., "Cannabis sativa: an overview," in *Nutraceuticals*, G. Richter, et al., Eds., Elsevier, 2021, pp. 603–624. https://doi.org/10.1016/B978-0-12-821038-3.00038-0.
28. P. C. Struik, et al., "Agronomy of fibre hemp (Cannabis sativa L.) in Europe," *Ind. Crops Prod.*, vol. 11, no. 2–3, pp. 107–118, Mar. 2000. https://doi.org/10.1016/S0926-6690(99)00048-5.
29. J. H. Cherney and E. Small, "Industrial hemp in North America: production, politics and potential," *Agronomy*, vol. 6, no. 4, pp. 58–58, 2016.
30. I. Adesina, A. Bhowmik, H. Sharma, and A. Shahbazi, "A Review on the Current State of Knowledge of Growing Conditions, Agronomic Soil Health Practices and Utilities of Hemp in the United States," *Agriculture*, vol. 10, no. 4, pp. 129–129, Apr. 2020. https://doi.org/10.3390/agriculture10040129.
31. S. Amaducci et al., "Key cultivation techniques for hemp in Europe and China," *Ind. Crops Prod.*, vol. 68, pp. 2–16, 2015.
32. J. P. Manaia, et al., "Industrial hemp fibers: an overview," *Fibers*, vol. 7, no. 12, pp. 106–106, Dec. 2019. https://doi.org/10.3390/fib7120106.
33. "Hemp: from field to fiber: a review." Accessed: Nov. 06, 2024. [Online]. Available: https://www.mdpi.com/2673-7248/4/2/11.
34. A. Shahzad, "Hemp fiber and its composites – a review," *J. Compos. Mater.*, vol. 46, no. 8, pp. 973–986, Apr. 2012. https://doi.org/10.1177/0021998311413623.
35. H.-R. Kymäläinen and A.-M. Sjöberg, "Flax and hemp fibres as raw materials for thermal insulations," *Build. Environ.*, vol. 43, no. 7, pp. 1261–1269, 2008.
36. F. Scrucca et al., "Energy and carbon footprint assessment of production of hemp hurds for application in buildings," *Environ. Impact Assess. Rev.*, vol. 84, pp. 106417–106417, Sep. 2020. https://doi.org/10.1016/j.eiar.2020.106417.
37. T. Jami, et al., "A review of the properties of hemp concrete for green building applications," *J. Clean. Prod.*, vol. 239, pp. 117852–117852, 2019.
38. R. Walker and S. Pavía, "Moisture transfer and thermal properties of hemp–lime concretes," *Constr. Build. Mater.*, vol. 64, pp. 270–276, Aug. 2014. https://doi.org/10.1016/j.conbuildmat.2014.04.081.
39. L. Aditya et al., "A review on insulation materials for energy conservation in buildings," *Renew. Sustain. Energy Rev.*, vol. 73, pp. 1352–1365, Jun. 2017. https://doi.org/10.1016/j.rser.2017.02.034.
40. A. Korjenic, et al., "Development and performance evaluation of natural thermal-insulation materials composed of renewable resources," *Energy Build.*, vol. 43, no. 9, pp. 2518–2523, 2011.
41. F. Collet and S. Pretot, "Experimental highlight of hygrothermal phenomena in hemp concrete wall," *Build. Environ.*, vol. 82, pp. 459–466, 2014.
42. A. Mukherjee and C. MacDougall, "Structural benefits of hempcrete infill in timber stud walls," *Int. J. Sustain. Build. Technol. Urban Dev.*, vol. 4, no. 4, pp. 295–305, 2013.

Published by Burleigh Dodds Science Publishing Limited, 2026.

43. Y. Florentin, D. Pearlmutter, B. Givoni, and E. Gal, "A life-cycle energy and carbon analysis of hemp-lime bio-composite building materials," *Energy Build.*, vol. 156, pp. 293–305, 2017.
44. E. Alexander and J. Hirst, "Characterisation of hemp-lime as a composite building material," 2013, [Online]. Available: https://purehost.bath.ac.uk/ws/portalfiles/portal/187944938.
45. P. Nováková, "Use of technical hemp in the construction industry," *MATEC Web Conf.*, vol. 146, p. 03011, 2018. https://doi.org/10.1051/matecconf/201814603011.
46. B. Mazhoud, et al., "Hygric and thermal properties of hemp-lime plasters," *Build. Environ.*, vol. 96, pp. 206–216, Feb. 2016. https://doi.org/10.1016/j.buildenv.2015.11.013.
47. R. Agliata, et al., "Historic building energy audit and retrofit simulation with hemp-lime plaster–a case study," *Sustainability*, vol. 12, no. 11, Art. no. 11, Jan. 2020. https://doi.org/10.3390/su12114620.
48. W. Stanwix and A. Sparrow, *The Hempcrete Book: Designing and Building with Hemp-Lime*. Bloomsbury Publishing, 2014.
49. S. Sair, et al., "Mechanical and thermal conductivity properties of hemp fiber reinforced polyurethane composites," *Case Stud. Constr. Mater.*, vol. 8, pp. 203–212, Jun. 2018. https://doi.org/10.1016/j.cscm.2018.02.001.
50. K. Karthik, et al., "3D printing with natural fiber composites: a review for advanced manufacturing applications," *Interactions*, vol. 245, no. 1, pp. 231–231, Aug. 2024. https://doi.org/10.1007/s10751-024-02098-3.
51. Z. Li, et al., "Properties of hemp fibre reinforced concrete composites," *Compos. Part Appl. Sci. Manuf.*, vol. 37, no. 3, pp. 497–505, Mar. 2006. https://doi.org/10.1016/j.compositesa.2005.01.032.
52. Z. Li, et al., "Compressive and flexural properties of hemp fiber reinforced concrete," *Fibers Polym.*, vol. 5, no. 3, pp. 187–197, Sep. 2004. https://doi.org/10.1007/BF02902998.
53. L. Di Sarno, et al., "Exploring innovative resilient and sustainable bio-materials for structural applications: hemp-fibre concrete," *Structures*, vol. 68, pp. 107096–107096, Oct. 2024. https://doi.org/10.1016/j.istruc.2024.107096.
54. "Phase change material - an overview | ScienceDirect topics." Accessed: Dec. 2, 2024. [Online]. Available: https://www.sciencedirect.com/topics/engineering/phase-change-material.
55. M. Sawadogo, et al., "Investigation of a novel bio-based phase change material hemp concrete for passive energy storage in buildings," *Appl. Therm. Eng.*, vol. 212, p. 118620, Jul. 2022. https://doi.org/10.1016/j.applthermaleng.2022.118620.
56. O. Gencel et al., "Enhancing sustainability with waste hemp-shive and phase change material: novel gypsum-based composites with advanced thermal energy storage properties," *J. Clean. Prod.*, vol. 451, p. 142000, Apr. 2024. https://doi.org/10.1016/j.jclepro.2024.142000.
57. A. M. O. Ada et al., "Hygrothermal performance of hemp lime concrete embedded with phase change materials for buildings," *J. Phys. Conf. Ser.*, vol. 2069, no. 1, p. 012005, Nov. 2021. https://doi.org/10.1088/1742-6596/2069/1/012005.
58. R. Yang et al., "Anisotropic hemp-stem-derived biochar supported phase change materials with efficient solar-thermal energy conversion and storage," *Biochar*, vol. 4, no. 1, p. 38, Jul. 2022. https://doi.org/10.1007/s42773-022-00162-1.

59. S. Barbhuiya and B. Bhusan Das, "A comprehensive review on the use of hemp in concrete," *Constr. Build. Mater.*, vol. 341, pp. 127857–127857, Jul. 2022. https://doi.org/10.1016/j.conbuildmat.2022.127857.
60. S. Schiavoni, et al., "Insulation materials for the building sector: a review and comparative analysis," *Renew. Sustain. Energy Rev.*, vol. 62, pp. 988–1011, Sep. 2016. https://doi.org/10.1016/j.rser.2016.05.045.
61. K. Chau, et al., "Hempcrete as a substrate for fungal growth under high humidity and variable temperature conditions," *Constr. Build. Mater.*, vol. 398, pp. 132373–132373, Sep. 2023. https://doi.org/10.1016/j.conbuildmat.2023.132373.
62. E. Horszczaruk, et al., "Investigation of durability properties for lightweight structural concrete with hemp shives instead of aggregate," *Appl. Sci.*, vol. 13, no. 14, pp. 8447–8447, Jul. 2023. https://doi.org/10.3390/app13148447.
63. M. Degrave-Lemeurs, et al., "Acoustical properties of hemp concretes for buildings thermal insulation: Application to clay and lime binders," *Constr. Build. Mater.*, vol. 160, pp. 462–474, Jan. 2018. https://doi.org/10.1016/j.conbuildmat.2017.11.064.
64. O. Kinnane, et al., "Acoustic absorption of hemp-lime construction," *Constr. Build. Mater.*, vol. 122, pp. 674–682, Sep. 2016. https://doi.org/10.1016/j.conbuildmat.2016.06.106.
65. M. Fan, et al., "Fire performance of natural fibre composites in construction," in *Advanced High Strength Natural Fibre Composites in Construction*, Elsevier, 2017, pp. 375–404. https://doi.org/10.1016/B978-0-08-100411-1.00014-5.
66. R. Kozlowski, et al., "Bast fibres (flax, hemp, jute, ramie, kenaf, abaca)," in *Biodegradable and Sustainable Fibres*, Elsevier, 2005, pp. 36–88. https://doi.org/10.1533/9781845690991.36.
67. S. Pochwała, et al., "The heat conductivity properties of hemp–lime composite material used in single-family buildings," *Materials*, vol. 13, no. 4, pp. 1011–1011, Feb. 2020. https://doi.org/10.3390/ma13041011.
68. M. Yadav and A. Saini, "Opportunities & challenges of hempcrete as a building material for construction: an overview," *Mater. Today Proc.*, vol. 65, pp. 2021–2028, 2022. https://doi.org/10.1016/j.matpr.2022.05.576.
69. R. Walker, et al., "Mechanical properties and durability of hemp-lime concretes," *Constr. Build. Mater.*, vol. 61, pp. 340–348, Jun. 2014. https://doi.org/10.1016/j.conbuildmat.2014.02.065.
70. A. Gaur, et al., "Flooding related consequences of climate change on Canadian cities and flow regulation infrastructure," *Water*, vol. 11, no. 1, pp. 63–63, Jan. 2019. https://doi.org/10.3390/w11010063.
71. X. Wang, et al., "Increasing frequency of extreme fire weather in Canada with climate change," *Clim. Change*, vol. 130, no. 4, pp. 573–586, Jun. 2015. https://doi.org/10.1007/s10584-015-1375-5.
72. G. Kaur and R. Kander, "The sustainability of industrial hemp: a literature review of its economic, environmental, and social sustainability," *Sustainability*, vol. 15, no. 8, pp. 6457–6457, Apr. 2023. https://doi.org/10.3390/su15086457.
73. N. Aghamohammadi and M. Shahmohammadi, "Towards sustainable development goals and role of bio-based building materials," in *Bio-Based Materials and Waste for Energy Generation and Resource Management*, R. N. Bharagava, Eds., Elsevier, 2023, pp. 243–279. https://doi.org/10.1016/B978-0-323-91149-8.00004-1.

74. R. Budhathoki, et al., "Enviro-economic and feasibility analysis of industrial hemp value chain: a systematic literature review," *GCB Bioenergy*, vol. 16, no. 6, Jun. 2024. https://doi.org/10.1111/gcbb.13141.
75. G. Crini and E. Lichtfouse, Eds., *Sustainable Agriculture Reviews 42*, vol. 42. Springer International Publishing, 2020. https://doi.org/10.1007/978-3-030-41384-2.
76. A. Arrigoni, et al., "Life cycle assessment of natural building materials: the role of carbonation, mixture components and transport in the environmental impacts of hempcrete blocks," *J. Clean. Prod.*, vol. 149, pp. 1051–1061, 2017.
77. D. Kumar, et al., "Comparative analysis of building insulation material properties and performance," *Renew. Sustain. Energy Rev.*, vol. 131, pp. 110038–110038, Oct. 2020. https://doi.org/10.1016/j.rser.2020.110038.
78. E. G. Özdamar, "Hemp as a potential material in architecture: is it possible in Turkey?," *ICONARP Int. J. Archit. Plan.*, vol. 9, no. 1, pp. 131–154, 2021.
79. M. Singh, et al., "Characterization of 6–11th century A.D decorative lime plasters of rock cut caves of Ellora," *Constr. Build. Mater.*, vol. 98, pp. 156–170, Nov. 2015. https://doi.org/10.1016/j.conbuildmat.2015.08.039.
80. R. C. Clark, "Hemp museum tour of nagano prefecture, Japan," *J. Ind. Hemp*, vol. 11, pp. 95–101, 2006.
81. "Asa no e: Miasa village." Accessed: Jan. 25, 2025. [Online]. Available: http://www.japanhemp.org/en/miasahouse.htm.
82. C. H. J. Liu, et al., "The extent to which hemp insulation materials can be used in Canadian residential buildings," *Sustainability*, vol. 15, no. 19, Art. no. 19, Jan. 2023. https://doi.org/10.3390/su151914471.
83. T. Woolley, "Hemp lime construction: a guide to building with hemp lime composites," 2008. http://dx.doi.org/10.4995/vitruvio-ijats.2019.11771
84. "Twine: Series One | antonygibbondesigns.com." Accessed: Jan. 25, 2025. [Online]. Available: https://antonygibbondesigns.com/twine-series-one/.
85. I. B. Muhit, et al., "A holistic sustainability overview of hemp as building and highway construction materials," *Build. Environ.*, vol. 256, pp. 111470–111470, May 2024. https://doi.org/10.1016/j.buildenv.2024.111470.
86. J. Qureshi, "A review of fibre reinforced polymer bridges," *Fibers*, vol. 11, no. 5, Art. no. 5, May 2023. https://doi.org/10.3390/fib11050040.
87. J. Smits and R. Gkaidatzis, "A bio-based fibre-reinforced plastic pedestrian bridge for Schiphol," presented at the IABSE Conference, Nara 2015: Elegance in structures, Nara, Japan, 2015, pp. 124–125. https://doi.org/10.2749/222137815815774232.
88. U. Dhakal, et al., "Hygrothermal performance of hempcrete for Ontario (Canada) buildings," *J. Clean. Prod.*, vol. 142, pp. 3655–3664, 2017.
89. S. Elfordy, et al., "Mechanical and thermal properties of lime and hemp concrete ('hempcrete') manufactured by a projection process," *Constr. Build. Mater.*, vol. 22, no. 10, pp. 2116–2123, Oct. 2008. https://doi.org/10.1016/j.conbuildmat.2007.07.016.
90. "New Build Projects," Americhanvre. Accessed: Jan. 26, 2025. [Online]. Available: https://americhanvre.com/new-build-projects/.
91. "Info – Hempcrete Walls: Building Journal." Accessed: Nov. 14, 2024. [Online]. Available: https://hempcretewalls.com/info/.
92. "Just BioFiber | The Future of Building." Accessed: Nov. 17, 2024. [Online]. Available: https://justbiofiber.com/.

93. R. W. C. Sousa and T. Froese, "Performance assessment of the harmless home," in *Proceedings of the Canadian Society of Civil Engineering Annual Conference 2022*, R. Gupta, et al., Eds., Springer Nature Switzerland, 2024, pp. 531–543. https://doi.org/10.1007/978-3-031-35471-7_39.
94. Y.-W. Lim, et al., "Planning and coordination of modular construction," *Autom. Constr.*, vol. 141, p. 104455, Sep. 2022. https://doi.org/10.1016/j.autcon.2022.104455.
95. M. Sutkowska, et al., "Recent advances in prefabrication techniques for biobased materials towards a low-carbon future: from modules to sustainability," *J. Build. Eng.*, vol. 91, pp. 109558–109558, Aug. 2024. https://doi.org/10.1016/j.jobe.2024.109558.
96. "Hempcrete – Natural Building Alliance." Accessed: Jan. 26, 2025. [Online]. Available: https://natural-building-alliance.org/hempcrete/.
97. "Hemp Construction," Dunagro Hemp Group. Accessed: Jan. 25, 2025. [Online]. Available: https://dunagrohempgroup.com/hemp-construction/.
98. Y. Lu, et al., "3D printing recycled concrete incorporating plant fibres: a comprehensive review," *Constr. Build. Mater.*, vol. 425, pp. 135951–135951, Apr. 2024. https://doi.org/10.1016/j.conbuildmat.2024.135951.
99. H. Hassan, et al., "Towards innovative and sustainable buildings: a comprehensive review of 3D printing in construction," *Autom. Constr.*, vol. 163, pp. 105417–105417, Jul. 2024. https://doi.org/10.1016/j.autcon.2024.105417.
100. E. Binega Yemesegen and A. M. Memari, "A review of experimental studies on Cob, Hempcrete, and bamboo components and the call for transition towards sustainable home building with 3D printing," *Constr. Build. Mater.*, vol. 399, pp. 132603–132603, Oct. 2023. https://doi.org/10.1016/j.conbuildmat.2023.132603.
101. "3d Printing Architecture," [Online]. Available: https://www.3dwasp.com/en/3d-printing-architecture/.
102. "Industry Experts’ Perspectives on the difficulties and opportunities of the integration of bio-based insulation materials in the European construction sector." Accessed: Nov. 22, 2024. [Online]. Available: https://www.mdpi.com/2071-1050/16/17/7314.
103. "InterChanvre - Cultivation." Accessed: Nov. 22, 2024. [Online]. Available: https://www.interchanvre.org/la_culture.
104. "Rénovation, fiabilisation de l'acte de construire, innovation." Accessed: Nov. 22, 2024. [Online]. Available: https://www.cstb.fr/recherche-expertise/approche/renovation-fiabilisation-construire-innovation.
105. "European Standards," CEN-CENELEC. Accessed: Nov. 20, 2024. [Online]. Available: https://www.cencenelec.eu/european-standardization/european-standards/.
106. "About CEN," CEN-CENELEC. Accessed: Nov. 18, 2024. [Online]. Available: https://www.cencenelec.eu/about-cen/.
107. "EN 15804:2012+A2:2019 - Sustainability of construction works - Environmental product declarations - Core rules for the product category of construction products," iTeh Standards. Accessed: Jan. 24, 2025. [Online]. Available: https://standards.iteh.ai/catalog/standards/cen/c98127b4-8dc2-48a4-9338-3e1366b16669/en-15804-2012a2-2019.
108. "BS-EN-1745-2012.pdf." Accessed: Nov. 20, 2024. [Online]. Available: https://www.modiranahan.com/uploads/pdf/BS-EN-1745-2012.pdf.

109. "EN 13162: Thermal insulation products for buildings - Factory made mineral wool (MW) products - Specification." Accessed: Nov. 21, 2024. [Online]. Available: https://www.intertek.com/building/standards/en-13162/.
110. E. Standards, "BS EN 12664:2001 Thermal performance of building materials and products. Determination of thermal resistance by means of guarded hot plate and heat flow meter methods. Dry and moist products of medium and low thermal resistance," https://www.en-standard.eu. Accessed: Nov. 21, 2024. [Online]. Available: https://www.en-standard.eu/bs-en-12664-2001-thermal-performance-of-building-materials-and-products-determination-of-thermal-resistance-by-means-of-guarded-hot-plate-and-heat-flow-meter-methods-dry-and-moist-products-of-medium-and-low-thermal-resistance/.
111. E. Standards, "BS EN 1602:2013 thermal insulating products for building applications. Determination of the apparent density," https://www.en-standard.eu. Accessed: Nov. 21, 2024. [Online]. Available: https://www.en-standard.eu/bs-en-1602-2013-thermal-insulating-products-for-building-applications-determination-of-the-apparent-density/.
112. E. Standards, "BS EN 13369:2023 Common rules for precast concrete products," https://www.en-standard.eu. Accessed: Nov. 21, 2024. [Online]. Available: https://www.en-standard.eu/bs-en-13369-2023-common-rules-for-precast-concrete-products/.
113. K. K. Measurlabs, "EN 13501-1 Fire Classification | Performance Classes & Criteria | Measurlabs." Accessed: Nov. 21, 2024. [Online]. Available: https://measurlabs.com/blog/en-13501-1-fire-classification-performance-classes-and-criteria/.
114. J. Fike, "Industrial hemp: renewed opportunities for an ancient crop," *Crit. Rev. Plant Sci.*, vol. 35, no. 5–6, pp. 406–424, 2016.
115. "RB316-22," [Online]. Available: https://ushba.org/wp-content/uploads/proposal_8646-Final.pdf.
116. "2024 International Building Code (IBC)." Accessed: Nov. 17, 2024. [Online]. Available: https://codes.iccsafe.org/content/IBC2024P1.
117. "Home | ASTM Standardization News." Accessed: Nov. 18, 2024. [Online]. Available: https://sn.astm.org/.
118. G. of C. N. R. C. Canada, "[CCMC 14272-R] CCMC Canadian code compliance evaluation." Accessed: Nov. 19, 2024. [Online]. Available: https://nrc.canada.ca/en/certifications-evaluations-standards/canadian-construction-materials-centre/ccmc-publications/document.html?id=14272-R&type=cert.
119. "ISO - International Organization for Standardization," ISO. Accessed: Nov. 19, 2024. [Online]. Available: https://www.iso.org/home.html.
120. ISO14040:2006(en), "Environmental management - Life cycle assessment - Principles and framework", [Online]. Available: https://www.iso.org/obp/ui#iso:std:iso:14040:ed-2:v1:en.
121. "ISO 24260:2022," ISO. Accessed: Nov. 19, 2024. [Online]. Available: https://www.iso.org/standard/78231.html.
122. "ISO 10140-3:2021(en), Acoustics – Laboratory measurement of sound insulation of building elements – Part 3: Measurement of impact sound insulation." Accessed: Nov. 21, 2024. [Online]. Available: https://www.iso.org/obp/ui/en/#iso:std:iso:10140:-3:ed-2:v1:en.
123. "EIHA," EIHA. Accessed: Nov. 22, 2024. [Online]. Available: https://eiha.org/.

124. "International Hemp Building Association - IHBA," International Hemp Building Association. Accessed: Nov. 22, 2024. [Online]. Available: https://internationalhempbuilding.org/.
125. "Home," USHBA. Accessed: Nov. 22, 2024. [Online]. Available: https://www.ushba.org.
126. "Canada Hemp Trade Alliance," [Online]. Available: https://www.hemptrade.ca/.
127. "Hemp-lime construction panels on test at University's new Building Research Park." Accessed: Nov. 22, 2024. [Online]. Available: https://www.bath.ac.uk/announcements/hemp-lime-construction-panels-on-test-at-universitys-new-building-research-park/.
128. "Climate-friendly renovations using straw and hemp," ETH Zurich. Accessed: Nov. 22, 2024. [Online]. Available: https://ethz.ch/en/news-and-events/eth-news/news/2024/03/climate-friendly-renovations-using-straw-and-hemp.html.
129. "Heterogeneous Materials Research Group (GEMH)," Université de Limoges. Accessed: Nov. 22, 2024. [Online]. Available: https://www.unilim.fr/heterogeneous-materials-research-group-gemh/?lang=en.
130. "Global Hemp Innovation Center." Accessed: Jan. 24, 2025. [Online]. Available: https://agsci.oregonstate.edu/hemp.
131. S. B. et al., "Development of sustainable alternative materials for the construction of green buildings using agricultural residues: a review," *Constr. Build. Mater.*, vol. 368, pp. 130457–130457, Mar. 2023. https://doi.org/10.1016/j.conbuildmat.2023.130457.
132. D. L. Le, et al., "Status and barriers to circular bio-based building material adoption in developed economies: The case of Flanders, Belgium," *J. Environ. Manage.*, vol. 367, p. 121965, Sep. 2024. https://doi.org/10.1016/j.jenvman.2024.121965.

Chapter 6

Advances in the use of hemp in livestock feed

N. B. Parker, A. Irawan, S. Ates and M. Bionaz, Oregon State University, USA

1 Introduction

In recent years, hemp (*Cannabis sativa* L.) and its derivative products, including cannabidiol (CBD) oil, seeds, and fiber, have gained recognition as versatile agricultural commodities with the potential to benefit multiple industries worldwide. In the United States, hemp production, primarily for CBD use, which accounts for approximately 62% of cultivated hemp, has experienced a resurgence following its removal from the Comprehensive Drug Abuse Prevention and Control Act with the passing of the 2018 Farm Bill (USDA, 2020). Despite this shift, hemp and its by-products remain ineligible for commercial use in animal agriculture in the United States, as they require approval from the Food and Drug Administration's Center for Veterinary Medicine (FDA-CVM) before they can be used in animal feed.

Studies in Canada and European Union (EU) countries have long examined hemp by-products, particularly hempseed cake and meal, as feed ingredients in livestock diets, demonstrating their safety and effectiveness (Mustafa et al., 1999; Silversides and LefranÇois, 2005; Gakhar et al., 2012; Mourot and Guillevic, 2015). Food products sourced from hemp-fed animals

http://dx.doi.org/10.19103/AS.2025.0161.12

improved nutritional composition in meat (Gibb et al., 2005), eggs (Jing et al., 2017; Kasula et al., 2021; Skřivan et al., 2019), and milk (Karlsson et al., 2010; Cozma et al., 2015). Given these positive impacts of hemp by-products used in international animal feeding systems and the new legal status, US researchers have begun to assess hemp feeding in domestic animals (Altman et al., 2023; Irawan et al., 2024a; Kleinhenz et al., 2020; Parker et al., 2022).

This chapter discusses the current legal status of hemp use in livestock and poultry diets. It reviews existing research on the inclusion of hemp and its by-products in animal feed, focusing on their effects on animal health, performance, and food quality. The chapter also explores hemp as an alternative feed ingredient in livestock and poultry production. Additionally, it summarizes studies on the impact of feeding hemp by-products to both ruminant and monogastric animals and examines future opportunities for incorporating hemp and its by-products into animal feeding systems.

2 Common terms

- Hemp: commonly referred as 'industrial hemp' (*Cannabis sativa* L.), a plant that is grown for fiber, seed, and oil production with a very low content of psychoactive compound tetrahydrocannabinol (THC).
- Marijuana is a *Cannabis sativa* L. plant grown for recreational purposes, which has a high content of THC.
- Cannabinoids: compounds that occur naturally in the *Cannabis sativa* plant.
- THC: psychoactive cannabinoid known for its multiple adverse effects on the neural system.
- Cannabidiol (CBD): a non-psychoactive cannabinoid that is often associated with health benefits.
- Acid detergent fiber (ADF; cellulose and lignin) and neutral detergent fiber (NDF; cellulose, hemicellulose, and lignin): fiber fractions of feed material/ingredients mainly consisted of cellulose, hemicellulose, and lignin. The ADF and NDF terms are used in ruminant feed, while crude fiber is commonly used in poultry feed.
- Antinutritional factors: compounds such as tannins, phytic acid, saponins, cyanogenic glycosides, and trypsin inhibitors are present in feed ingredients that can compromise nutrient digestion by animals when presented at certain levels.
- Polyunsaturated fatty acids (PUFA): long-chain fatty acids with two or more double bonds, such as 18:3ω-3, 18:2ω-6, and *cis*-9, *trans*-11 conjugated linoleic acid (CLA), with health benefits for humans.
- Dry matter intake (DMI): a term to express feed intake in livestock.

3 Legal status of hemp in livestock diets

Despite their biotypic differences, in addition to different industrial and recreational applications and uses, hemp and marijuana (a.k.a., cannabis, marihuana) have remained inextricably linked to one another throughout history. From a botanical standpoint, hemp genotypes are often referred to as 'industrial hemp,' and are primarily utilized for fiber, seed, and oil production. Conversely, marijuana cultivars, often referred to colloquially as unitalicized 'cannabis,' are almost exclusively grown and utilized for their psychotropic effects. Marijuana contains high concentrations (≥3% on a dry weight basis) of Δ^9-tetrahydrocannabinol (THC), the primary chemical compound that induces intoxication-like effects. Conversely, cultivars grown for hemp production possess low threshold values of THC, ordinarily ≤ 0.3% (dried weight basis) (Cherney and Small, 2016; Visković et al., 2023). Presently, much of the world uses the United Nations' (UN) threshold of 0.3% THC as the legal basis to distinguish hemp from marijuana (Cherney and Small, 2016; Houtzager et al., 2024).

The 1970 Comprehensive Drug Abuse Prevention and Control Act, in concert with the US Drug Enforcement Agency and the UN, treated industrial hemp and marijuana interchangeably. Consequently, policy and law enforcement were applied equally to hemp and marijuana production (Cherney and Small, 2016). Ultimately, *Cannabis* and its derivatives were codified as Schedule I drugs (Arsenault et al., 2024). Passage of the 2014 and 2018 US Farm Bills effectively removed hemp from the Controlled Substances Act's definition of 'marijuana' (FDA, 2018a). The 2018 hemp legislation granted the US Food and Drug Administration (FDA) to retain authority over hemp product approval prior to commercial sale and use pursuant to the Federal Food, Drug, and Cosmetic Act (FD&C), in addition to section 351 of the Public Health Act. Consequently, hemp products must adhere to and comply with the FDA's Generally Recognized as Safe (GRAS) approval process prior to entering commerce (FDA, 2018b). Under current FDA guidelines, only hempseed-derived products have met GRAS specifications and are approved for interstate commercial use and sale, such as hulled hemp seed (also known as 'hemp grain'), hempseed protein powder, hempseed oil, and hempseed cake or meal. FDA-CVM has currently not approved any hemp by-products as a feed ingredient for livestock and poultry (FDA, 2024). Because feeding hemp to animals is currently not permitted in the United States (and other nations), food products derived from these animals, such as meat and milk, are considered adulterated and ineligible for sale (USDA, 2023).

4 Hemp composition, nutritional value and cannabinoid content

4.1 Hemp composition

Hemp fiber is the global dominant hemp crop. It is ordinarily harvested when male species achieve 100% flower bloom and female plants possess no greater than 20% flower bloom, owing to the diminishing strength of fibers after this period of the plant growth cycle (Visković et al., 2023). The fiber is composed of long, outer phloem fibers, analogous to bark, while shorter, hardened xylem (woody) fibers formed centrally in the stem are referred to as hurds. While shorter, xylem and hurd fibers carry less value, the longer, phloem layers are highly coveted for fiber production (Cherney and Small, 2016). Compositionally, hemp fibers are composed of ~75% cellulose, ~8% hemicellulose, and ~5% lignin on a dried weight basis, respectively (Żuk-Gołaszewska and Gołaszewski, 2020).

Hemp grown for seed and oil production is traditionally harvested 3–4 months after planting when ~90% of seeds have reached maturation (Visković et al., 2023). In addition, to yield variances, the harvest timing of hemp seed may also have an impact on the quality of the seed, as the fat composition and phenolic profile are altered by the stage of seed maturation (Burton et al., 2022). Hemp seed can be further categorized into three forms often used for industrial feed applications: whole hemp seed, dehulled hemp seed, and hulls. Dehulled hemp seed is largely composed of the seed heart (embryo), containing elevated levels of fat and protein, while the seed hull is compositionally higher in fiber. Hemp seed is largely utilized for food production, with whole seeds milled into flour, or, after grinding, used to extract oil and protein concentrates. The extraction processes also yield nutrient-rich by-products, such as hempseed cake or meal (Burton et al., 2022).

Hemp seed is recognized for a decent nutritional profile. On average, the whole hempseed macronutrient profile is composed of 25–35% fat, 20–25% protein, and 20–30% carbohydrate, with insoluble fibers making up the majority of the carbohydrate fraction. Of commercial cultivars analyzed, whole hemp seed consists of 90.8–96.0% dry matter, 25.4–33.0% fat, 21.3–27.3% crude protein (CP), and 3.7–5.9% ash. Of crude fiber present in whole hemp seed, 21.8–26.1% consists of acid detergent fiber (ADF), while neutral detergent fiber (NDF) constitutes 27.8–36.2%, respectively (Burton et al., 2022). Hempseed oil is especially rich in phytonutrients, including phenolic compounds, tocopherols, carotenoids, and phytosterols, with γ-tocopherol comprising the largest concentration (Visković et al., 2023). The fat composition of hemp seed consists predominantly of polyunsaturated fatty acids (PUFA), which make up >70% of the lipid fraction (Burton et al, 2022). Of the PUFA present in hemp seed, linoleic acid (18:2ω-6) constitutes the largest percentage of total fatty acids (52–62%), followed by α-linolenic acid (18:3ω-3; 12–23%) (Rupasinghe et al., 2020). Hemp seed is also denoted by its mineral composition, containing elevated levels of P, K, Mg, Ca, Na, Fe, Mn, Zn, and Cu. Despite an excellent nutritional profile, hemp seed does contain antinutritional compounds, such as tannins, phytic acid, saponins, cyanogenic glycosides, and trypsin inhibitors (Burton et al., 2022).

4.2 Phytocannabinoids and other compounds

Hemp for the extraction of essential oil and cannabinoids, mainly CBD, is a new crop grown in the United States and elsewhere. *Cannabis sativa* L. is known to contain over 550 different bioactive compounds, including more than 144 known phytocannabinoids (Montero et al., 2022). Phytocannabinoids can act on mammalian cannabinoid receptors, initiating physiological responses (Duggan, 2021). The main phytocannabinoids present in *Cannabis* are Δ-THC, CBD, cannabigerol (CBG), cannabichromene (CBC), cannabinol (CBN), Δ^8-trans-tetrahydrocannabinol (Δ^8-THC), cannabicyclol (CBL), cannabinodiol (CBND), cannabielsoin (CBE), and cannabitriol (CBT). THC and CBD comprise the majority of cannabinoid content in *Cannabis*, with THC primarily responsible for psychotropic effects, while CBD generally lacks psychoactive properties (Montero et al., 2022). Besides the neutral derivatives, industrial hemp is characterized by a higher content of cannabidiolic acid (CBDA) and cannabichromenic acid (CBCA) (Montero et al., 2022), with CBDA being the most abundant (Duggan, 2021).

Other key secondary metabolites in hemp are phenols, terpenes, amides, amines, and phytosterols (Montero et al., 2022). Terpenes comprise a significant percentage (1.28–2.14% in inflorescence material on a dry weight basis) of bioactive compounds in hemp (Chen and Pan, 2021). Terpenes are volatile compounds principally accountable for the distinctive aroma of hemp. Sensorial characteristics and palatability of hemp, namely aroma, are largely attributed to monoterpene compounds found within the plant and can vary by cultivar and genotype (Duggan, 2021).

Hemp biomass from fiber, grain, and chemical production has experienced a wide range of uses in recent years, such as construction materials, paper, textiles, and biofuels (Ahmed et al., 2022). Though not as prolific at biomass production as other crops, such as corn and potatoes, hemp yields a relatively moderate volume of biomass, which has been estimated to be ~8 tons/acre. Hemp biomass production in North America has remained limited in recent years, owing to financial interest and investment in lumber and other crop by-products (Cherney and Small, 2016).

5 Hemp and by-products as alternative feed ingredients

Commercial livestock and poultry diets in the United States consist primarily of two commodities and their derivatives – soybeans and corn (Hoffman and Baker, 2011). Soybeans are employed in dietary rations largely to provide a source of protein, while corn supplies a high-density source of energy for the growing animal (Shen et al., 2015; Nascimento et al., 2022). The diets of livestock and poultry have a profound effect on the nutritive value of food products sourced from these animals, such as meat, milk, and eggs. Dietary feeds also play an instrumental role in other qualitative factors of animal food products, such as shelf life, processing characteristics, and overall palatability (Falowo et al., 2014; Wood and Scollan, 2022). Nutritionally, animal products provide humans with a significant source of protein, B-complex vitamins, and trace minerals (zinc, iron, and selenium). Furthermore, animal-based protein offers higher absorption and digestibility to humans compared to protein derived from plant sources (Day et al., 2022).

Since livestock feeding systems profoundly influence health, welfare, and food quality attributes in livestock, novel research has focused on how alternative feeds and feed programs may impact animals and food products sourced from them. This area of study has largely been driven by consumer demand for animal-derived food products that consider factors such as sustainability, environmental impact, food waste, animal welfare, nutritional value, and product quality (Barone et al., 2021; Hoffman et al., 2018; Renna et al., 2022).

In conventional feeding systems, feed constitutes up to 70% of production expenses (McGrath et al., 2018). While feedstocks are produced primarily for animals (e.g. hay, silage, and soybeans), many of the alternative or additive feed ingredients used in the livestock industry are by-products of agricultural commodities. Often these commodity by-products possess little economic value or viability after harvest and processing, yet account for an estimated 30% of waste biomass produced from worldwide agricultural production (Salami et al., 2019). Thus, the use of by-products as feed ingredients offers potential money savings to producers and animal feed operations alike. The utilization of these agricultural by-products in livestock feeding systems may increase their value by creating new markets through repurposing efforts in concert with enhancing livestock health and qualitative properties of food products (Sandström et al., 2022).

6 Hemp by-products in ruminant diets

By-products of the nascent hemp industry worldwide offer a new, beneficial source of animal feed. Until recently, our knowledge concerning the use of hemp by-products in livestock comes from studies in Canada and countries in the EU (Gibb et al., 2005; Silversides and LefranÇois, 2005). Research in the United States was not permitted given legal restrictions on industrial hemp cultivation until the passage of the 2014 and 2018 Farm Bills (Višković et al., 2023). However, since industrial hemp's decriminalization there has also been a resurgence in evaluating hemp and its derivatives in US livestock feeding systems. Just one decade in, significant hemp research has been conducted in the United States in particular, and elsewhere on ruminant and monogastric animals. The current status and future implications for hemp as a ruminant feed ingredient are discussed below, focusing on hemp seed, hempseed meal, and hemp biomass. These are examined along the lines of ruminant health, performance, and food product quality.

6.1 Hempseed effects on ruminant health, performance and meat quality

Improvements to carcass traits (Dayani et al., 2022) and feed performance (Antunović et al., 2021) in lambs fed dietary hemp seed were observed, in concert with significant increases in blood HDL and cholesterol. These are likely due to higher feed intake (Dayani et al., 2022) with no major impact on inflammation and antioxidant biomarkers (Antunović et al., 2021). Gibb and others (2005) fed steers full-fat hemp seed at 0%, 9%, and 14%. In that study, dietary hemp inclusion did not negatively impact dry matter intake (DMI) or average daily gain (ADG) compared to a barley-based control ration. Hemp-fed steers did not experience detriments to carcass characteristics compared to control animals, in addition to experiencing a greater proportion of 18:3ω-3, 18:2ω-6, and *cis*-9, *trans*-11-conjugated linoleic acid (CLA) in adipose and muscle tissues compared to barley-fed animals. Although hemp-fed animals in the study possessed greater levels of fatty acids that are often viewed as healthier constituents in the human diet, the same animals also showed a greater proportion of *trans* and saturated fatty acids (SFA) in adipose and muscle samples compared to control, which may be detrimental to human health (Gibb et al., 2005; Clifton and Keogh, 2017).

6.2 Hempseed cake improves ruminant health

The findings by Gibb et al. (2005) regarding the enrichment of PUFA in meat tissues are also in alignment with results from Semwogerere and others (2023), who also observed significant reductions in SFA content and increase in shelf life of meat products from goats fed hempseed cake up to 100 g/kg DM in a finishing ration. Semwogerere et al. (2022) have also reported hempseed cake/meal in goat diets led to a linear increase in intramuscular fat deposition as hemp inclusion increased, in addition to effectively replacing soybean meal altogether without negatively affecting performance and meat quality. A 2024 study by Semwogerere et al. found that increasing levels of dietary hempseed cake in goat finishing rations resulted in linear increases of magnesium in blood, liver, and urine samples, and manganese and copper in feces. Conversely, increasing levels of dietary hempseed cake led to a decrease of selenium in liver and feces, while hemp supplementation also led to a linear increase in antioxidant capacity in the liver and blood (Semwogerere et al., 2024).

6.3 Hempseed cake effects on dairy animals and milk quality

Hempseed cake has also been utilized in feeding trials evaluating dairy cattle performance and milk quality. Karlsson et al. (2010) found that increasing concentrations of hempseed cake in dairy cattle rations increased fat, CP, and NDF intake and improved energy-corrected milk, while decreasing starch intake. However, increasing hempseed cake by over 14% (dry matter basis) in the diet resulted in a decrease in milk fat and protein concentration, a decrease in the efficiency of N utilization, and an increase in milk urea (Karlsson et al., 2010). Mierliță (2016) observed hemp seed fed to lactating ewes increased fat-corrected milk, milk fat content, and fat yield compared to non-supplemented sheep, in addition to milk fatty acid alterations via a decrease in the proportion of short-chain, medium-chain, and hypercholesterolemic fatty acids. Hempseed feeding resulted in increases of trans-vaccenic acid (t11-18:1), 18:3ω-3, CLA, eicosopentaenoic acid (EPA), and docosahexaenoic acid (DHA) (Mierliță, 2016). Mierliță (2018) also evaluated dietary hemp seed and hempseed cake effects on antioxidant capacity in ewe milk, noting an increase in total antioxidant capacity in both hemp groups compared to control, observed by higher levels of α-tocopherol and lower levels of malondialdehyde. Cozma and others (2015) evaluated hempseed oil supplementation at 4.7% (dry matter basis) in goats, noting an increase in milk fat and protein production compared to goats fed a concentrate diet, while also observing a reduction in milk of the proportion of SFA and a marked increase in the proportion of *cis*-9 and *trans*-11 CLA, and *trans*-18:1 in hemp-supplemented goats. Hemp oil supplementation did not affect plasma cholesterol, triglycerides, phospholipids, or alanine aminotransferase and γ-glutamyl transferase enzymes (Cozma et al., 2015).

6.4 Spent hemp biomass improves ruminant health, performance and food product quality

The use of spent hemp biomass (SHB) in ruminant diets has also been recently evaluated by Parker et al. (2022) and Irawan et al. (2024a). SHB, a by-product of CBD extraction, comprises hemp inflorescence material found to be nutritionally comparable to alfalfa, possessing similar DM, CP, and metabolizable energy (Parker et al., 2022). Parker et al. (2022) evaluated the inclusion of SHB at 0%, 10%, and 20% in lamb-finishing diets, respectively, with either a 4-week feeding period and 4-week withdrawal prior to slaughter, or SHB fed for the entire 8-week period prior to slaughter. The authors noted that the inclusion of SHB did not have detrimental effects on lamb performance or carcass characteristics with a minimal effect on meat quality parameters. However, feed intake was lower in lambs fed 20% SHB compared to control in the first 4 weeks of the trial but was not different between the two groups during the second period after SHB withdrawal with lambs receiving 10% SHB having a higher feed intake compared to the control group. The authors also noted that SHB had a minor metabolic effect on SHB-fed lambs, with a potential decrease of liver clearance ability as suggested by the increase in bilirubin, though the levels in blood were still within normal ranges for sheep. The data also suggested a possible enhancement of the antioxidant capacity (Parker et al., 2022).

A study conducted by Irawan et al. (2024a) examined feeding pelleted SHB to lactating dairy cattle at 13% level as top fed on a total mixed ration (TMR) for 4 weeks, followed by a 4-week withdrawal period. The authors noted that SHB feed led to a decrease in DMI compared to cattle consuming pelleted alfalfa (13% added to TMR; control ration), owing to the low palatability of SHB. SHB feeding did not lead to changes in milk yield compared to the alfalfa TMR, with an increase in milk yield/DMI compared to the control ration. As for the lambs, there was a minimal effect on metabolism, with SHB-fed cattle experiencing lower cholesterol concentration and higher β-hydroxybutyric acid during the 4-week withdrawal period. However, cattle fed SHB exhibited low-grade inflammation compared to control cows fed alfalfa, as denoted by ceruloplasmin and reactive oxidative metabolites. The concentration of bilirubin also increased in SHB-fed cattle, indicating an effect on hepatic drug clearance. The study also revealed higher proportion of oleate and bacterial fatty acid in milk fat in SHB-fed compared to control cattle (Irawan et al., 2024a). Iommelli et al. (2024) also noted goats fed hemp hay experienced yield compositional changes to milk compared to control animals. Hemp-fed goats exhibited overall higher milk yield compared to control, in addition to lower 18:2ω-6 and 20:4 (arachidonic acid (AA)), and higher 20:5ω-3 (EPA) fatty acids in milk versus control goats. The nutritional profile of milk was enhanced via dietary hemp hay, with decreases in ω-6:ω-3, 18:2:18:3, and AA:EPA acid ratios (Iommelli et al., 2024).

6.5 Hemp biomass (stubble, forage) is an equivalent feed to other sources

Researchers have noted that lambs fed hemp stubble at 0%, 28%, and 56% in the diet had no decrease in feed intake, liveweight gain, or ADG, but had a positive impact on nutrient digestibility while increasing the production of ruminal volatile fatty acids (VFA) (Krebs et al., 2021). Sheep fed green hemp biomass exhibited no differences in VFA production or digestibility of DM, organic matter, and CP compared to control sheep. However, green hemp biomass inclusion resulted in lower digestibility of ADF and NDF, in addition to lower water intake, fecal water output, and nitrogen urinary output in hemp-fed sheep (Stevens et al., 2022). In goats, Ran et al. (2024) noted dietary hemp forage improved feed conversion efficiency yet resulted in lower DM intake when substituted for alfalfa. The authors did not observe any differences in final body weight (BW), ADG, carcass traits, meat quality, and most blood parameters of goats based on dietary treatment. However, linear increases in BUN and ruminal ammonia concentrations were detected in hemp-fed goats, with changes also detected in VFA production (Ran et al., 2024). Wang et al. (2023) noted that feeding dairy cattle SHB in place of alfalfa hay did not negatively impact animal health, improved milk quality, and reduced production costs. The authors also observed alterations to ruminal microflora as dietary hemp inclusion increased, suggesting a hemp effect on microbial populations and subsequent synthesis and expression of downstream metabolites from dietary hemp (Wang et al., 2023). Feeding SHB in dairy cattle has also demonstrated effects on increased nitrogen efficiency and utilization, though methane emissions were unaffected (Irawan et al., 2024a).

6.6 Effects of feeding hemp biomass with cannabinoids on ruminants

Like most animals, bovines possess an endocannabinoid system that can be activated by (exo)cannabinoids found in hemp (Kent-Dennis and Klotz, 2024). Male Holstein cattle supplied with a single oral dose (35 g) of industrial hemp exhibited detectable concentrations in blood plasma of CBDA, tetrahydrocannabinolic acid (THCA), cannabidivarinic acid (CBDVA), and CBCA. A maximal concentration was detected at 14 h after initial dosing, indicating that acidic cannabinoids are easily absorbed during ruminal digestion (Kleinhenz et al., 2020). A follow-up study performed by Kleinhenz et al. (2022) also assessed plasma and pharmacokinetics of cannabinoids during a 14-day feed trial, in which castrated Holstein cattle were fed either a daily hemp-grain total mixed ration targeting 5.5 mg/kg of CBDA or a control ration. Of the two groups, the hemp-fed cattle experienced significantly lower cortisol and prostaglandin E_2 levels compared to control, in addition to increased lying time. These results suggest feeding hemp with elevated levels of CBDA may improve animal well-being through reductions in stress and inflammatory biomarkers

(Kleinhenz et al., 2022). Irawan et al. (2024b) did not observe any behavioral changes in dairy cattle fed SHB, other than a shorter duration of lying time. Wagner et al. (2022) observed changes in behavior and, potentially, negative health effects, in dairy cows fed hemp silage.

6.7 Cannabinoid residuals in ruminant products after feeding hemp biomass

In a study in sheep fed pelleted rations containing hemp, Krebs et al. (2021) noted that plasma samples evaluated for cannabinoids detected only THCA, at low concentrations (<16 μg/L), while Δ^9-THC was not detected in any of the lambs sampled. In that study, THCA and CBD were detected in liver samples of some sheep fed 28% and 56% hemp stubble. Δ^9-THC was detected in the kidney and subcutaneous fat depots in all lambs fed dietary hemp stubble, with the highest concentration in sheep fed at 28% inclusion, and muscle tissue sourced from the *longissimus lumborum* from sheep fed dietary hemp stubble (Krebs et al., 2021). No cannabinoids were detected in muscle samples derived from the *longissimus dorsi* of goats fed hemp forage (Ran et al., 2024). Researchers have observed that hemp inclusion in ruminant diets may result in the transfer of Δ^9-THC above permissible levels in milk from dairy cattle (Wagner et al., 2022) and subcutaneous fat in sheep (Stevens et al., 2022), exceeding the acute reference dose (ARfD) in food products prescribed by some governmental agencies (Beitzke and Pate, 2021).

Studies have demonstrated hemp inclusion in ruminant diets does not result in the transfer of cannabinoids into carcass tissues (Addo et al., 2023; Ran et al., 2024). A study by Irawan et al. (2024b) found that finishing lambs fed SHB possessed residual levels of cannabinoids in carcass tissues when fed SHB at 10% and 20% in the diet for 8 weeks up to slaughter (LH2 and HH2 groups), with highest concentrations of cannabinoids detected in the liver, followed by adipose and muscle tissues, respectively. Adipose tissue contained the highest concentration of Δ^9-THC. However, the authors noted that lambs withdrawn from SHB for 4 weeks prior to slaughter (LH1 and HH2 groups) did not possess detectable levels of residual cannabinoids in muscle or fat tissues. Despite the detectable levels of cannabinoids in lamb tissues fed SHB up to slaughter, the THC levels quantified in muscle tissue were equal to or below ARfD and tolerable daily intake (TDI) levels for several population groups established by governmental agencies in some countries. This finding suggests meat from lambs fed SHB does not pose a risk to adult consumer health in these populations, though children and adolescent populations may be at risk (Irawan et al., 2024b).

7 Hemp in monogastric diets

7.1 Hemp grain has positive effects on laying hens and eggs

The majority of studies evaluating the suitability of hemp by-products inclusion in monogastric diets have explored their effects on poultry, while others have focused on swine (Mourot and Guillevic, 2015; Kim and Nyachoti, 2017; Kim et al., 2018). Three common hemp by-products that have been examined in poultry include hempseed meal/cake, hemp seed, and hemp oil. In the available studies, those three ingredients had minimal or no impact on laying performance and feed intake (Silversides and Lefrancois, 2005; Goldberg et al., 2012; Jing et al., 2017; Taaifi et al., 2023) but appeared to be promising to enhance the fatty acid profile of eggs, especially feeding hemp oil. A cumulative finding from five studies has revealed that increasing levels of hemp oil in the diets of laying hens strongly elevated omega-3 fatty acid concentrations of the eggs (Fig. 1) while decreasing the ratio of n-6/n-3 fatty acids, which are beneficial for human health (Swanson et al., 2012).

Hempseed meal has shown to be a suitable alternative for soybean meal, the major protein source for poultry. An early study by Silversides and LefranÇois (2005) assessed the effect of feeding hempseed meal to laying hens at varying levels in the diet, replacing soybean meal in a conventional poultry ration. Hempseed meal did not significantly impact egg production, feed intake and efficiency, ADG, or egg quality compared to soybean diets. Hempseed meal also positively impacted egg nutritive quality through decreased levels of palmitic acid and increased levels of 18:2ω-6 and 18:3ω-3 fatty acids (Silversides and LefranÇois, 2005). Gakhar et al. (2012) also evaluated performance and egg quality in layer hens fed varying concentrations of hemp seed, hempseed oil, or a standard corn-barley poultry diet. The authors noted that egg production was

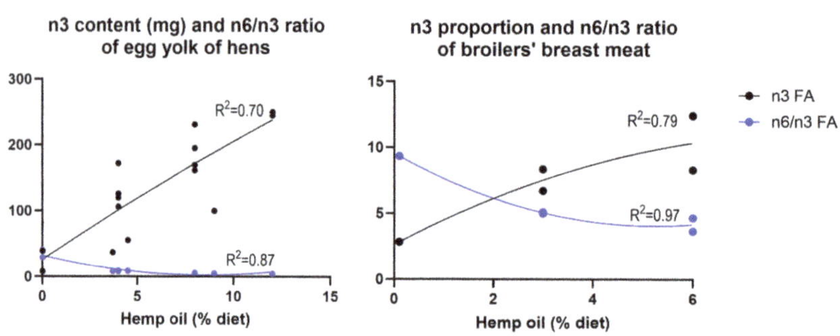

Figure 1 Relationship between increasing the inclusion of hemp oil on total n-3 fatty acids and n-6/n-3 ratio in eggs of laying hens and breast meat of broiler chickens. Data were collectively obtained from Goldberg et al. (2012), Jing et al. (2017), Gakhar et al. (2012), Neijat et al. (2016), Kanbur et al. (2023) for laying hens and Tufarelli et al. (2023) and Jing et al. (2017) for broiler chickens.

not impacted by either hemp diet, though feed intake was lower in hens fed the lowest concentration of hemp oil (4%) compared to the hens receiving higher concentrations of hemp oil (8% and 12%, respectively) and the ones receiving hemp seed and a control diet.

Not only did the hemp oil improve the fatty acid profile of eggs, but also hemp seed improved the levels of ω-3 fatty acids in egg yolks, including significantly greater levels of DHA and EPA acids compared to the control diet (Gakhar et al., 2012). Similar enhancement of egg quality was reported by Goldberg et al. (2012), where an increase in ω-3 fatty acids and color intensity in egg yolks from layers fed hemp seed and hempseed oil compared to birds fed control diets was observed, without resulting in adverse sensorial effects to cooked egg quality. The fatty acid composition of Quail eggs was improved by feeding hemp seed, with a decreased proportion of SFA and increased proportions of ω-3, while not adversely affecting the performance parameters (Cufadar et al., 2021; Razmaitė et al., 2022). DHA and EPA inclusion in the human diet is of particular interest, as these compounds in human health have anti-inflammatory properties, improve cardiovascular protection and cellular maintenance/repair, mitigate birth defects in pre- and post-natal infants, and improve cognitive health (Calder 2004; Simopoulos, 2010; Swanson et al., 2012). In addition to alterations of the fat profile, hemp feeding in layer diets has also been shown to significantly increase fat-soluble antioxidants, such as γ-tocopherol and α-tocopherol, in egg yolks. Hemp also significantly increased the concentration of calcium in tibias of laying hens, mitigating health issues common in commercial operations (Skřivan et al., 2019).

A study conducted by Kasula et al. (2021) sought to assess the effect on egg production and quality in layer hens fed 0%, 10%, 20%, or 30% hempseed cake. Similar to previously mentioned studies, hemp inclusion was denoted by significant increases in PUFA in eggs and abdominal fat in birds. In that study, THC residuals were undetected in egg, blood, breast muscle, fat, liver, kidney, and spleen.

7.2 Hemp effects on poultry health, performance and meat quality

Feeding hemp seed to broilers has a positive effect on meat quality. A meta-analysis revealed that inclusion of hemp (by)products (hemp seed, hempseed cake/meal, and hemp oil) did not impair growth performance and feed intake of broiler chickens (Fig. 2). Similar to laying hens, elevating hemp oil inclusion up to 6% in the diet strongly increased total n-3 fatty acids concentration and lowered n6/n3 ratio in breast meat of broilers (Fig. 1). Jing et al. (2017) noted that inclusion of dietary hemp oil up to 6% (by weight) resulted in significant increases in 18:3ω-3, EPA, DHA, and docosapentaenoic acid compared to control

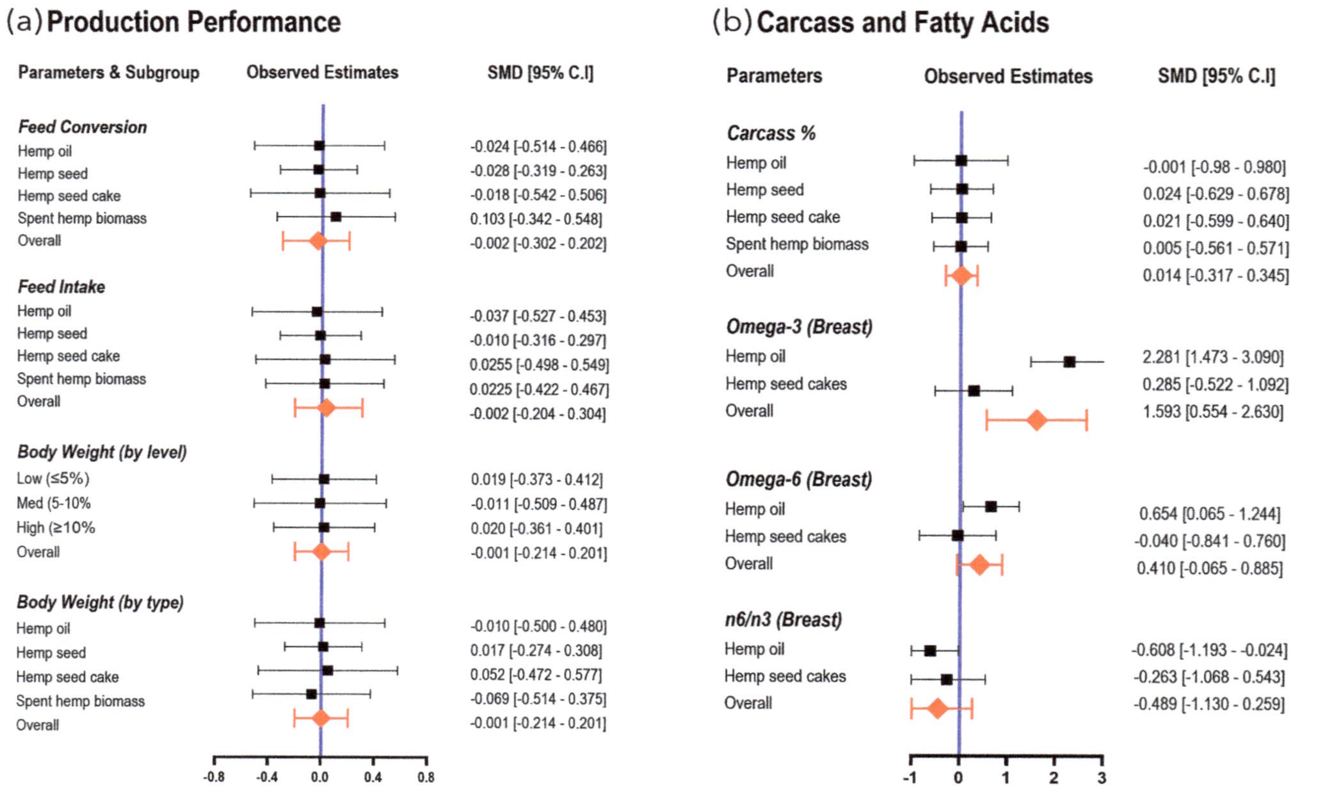

Figure 2 A summary of results of hemp by-products feeding in broiler chickens using a meta-analysis approach. The data were generated from 10 studies (Demirci et al., 2024; Eriksson and Wall, 2012; Jing et al., 2017; Khan et al., 2010; Mahmoudi et al., 2015; Sana et al., 2024; Sopian et al., 2024; Šťastník et al., 2019; Tufarelli et al., 2023; Vispute et al., 2019).

birds fed a wheat-soybean meal diet in thigh and breast tissue. Mahmoudi et al. (2012) reported broilers fed hemp seed at 7.5% in the diet experienced significantly lower blood serum cholesterol and SFA levels compared to birds fed no or lower concentrations of hemp seed. Sana et al. (2024) observed dietary hemp seed did not adversely affect various performance parameters and gut histomorphology, though white blood cell and immunoglobulin counts were significantly higher in hemp-fed birds compared to control, suggesting improvements to bird health and the immune system. Broilers supplemented with dietary hemp leaves had lower bacterial counts and enhanced immunomodulatory response through increased lymphocyte populations, improving overall health and welfare of birds (Balenović et al., 2024). Tufarelli et al. (2023) observed broilers fed hempseed cake at 5% and 10% in the diet achieved (1) significantly greater ω-3 fat deposition in thigh and breast fat, with a concomitant decrease in ω-6/ω-3 ratio; (2) decreases in malondialdehyde and lipid hydroperoxides in breast meat, indicating lower levels of oxidation in meat tissues; and (3) benefits to digestive health compared to control, denoted by increases in villus height, surface area, and villus/crypt ratio (Tufarelli et al., 2023). Other studies have also found dietary incorporation of hemp seed results in increases in BW compared to birds fed a control ration (Skřivan et al., 2019; Khan et al., 2010; Parr et al., 2020). Though these studies highlight the positive benefits of dietary hempseed incorporation into poultry diets, there are contrasting studies utilizing hemp seed (Mahmoudi et al., 2015; Stastnik et al., 2015) and hempseed oil (Kanbur, 2022) in broiler diets, which reported detriments to ADG, BW, feed conversion ratio (FCR), and overall performance.

Vispute et al. (2019) noted broilers fed hemp seed did not experience adverse performance characteristics compared to birds fed a basal diet, or a diet containing dill seed, though feed intake in hemp-fed birds was lower in the first 3 weeks of the trial. Hemp supplementation resulted in an enhanced blood serum lipid profile and gut health, via reductions in serum cholesterol and triglycerides, in concert with a reduction in bacterial coliform counts in the intestinal tract of hemp-fed birds (Vispute et al., 2019). A 2021 study by Vispute et al. also demonstrated that hemp seed fed to broilers led to an enhanced antioxidative profile in broiler meat, with reductions in lipid oxidation and increased shelf life, without altering sensory attributes (pH, color, odor) of the meat compared to control and dill seed-fed birds (Vispute et al., 2021).

Dietary hempseed oil has also been shown to not adversely affect meat quality parameters in quail when substituted for soybean oil (Göçmen and Kanbur, 2021), though Juodka et al. (2023) observed detriments to quail performance via dietary hempseed inclusion. Nevertheless, dietary hemp seed improved the meat quality of quail through the increase of total ω-3 in meat tissues compared to quails fed a control diet, though lower than birds fed camelina cake (Juodka et al., 2023). Konca and Beyzi (2012) also noted quail fed hemp seed at 10% in the diet experienced increases in carcass yield compared to birds fed 0%, 5%, or 20% hemp seed, respectively. Experiments evaluating hempseed cake in duck diets positively impacted carcass yield and growth performance without an effect on liver ω-3 concentration (Juodka et al., 2022) but did result in significant increases in 18:2ω-6 and total ω-6 in muscle samples (Juodka et al., 2018).

7.3 Spent hemp biomass in poultry diets

Researchers have investigated the effects of SHB in poultry diets and their effects on bird health, performance, and meat quality. A study conducted by Sopian et al. (2024) incorporating dietary SHB noted birds fed residue at 0.5%, 1%, and 2% in the diet (respectively) experienced lower feed intake than control birds with no differences in performance and improvements to broiler ileal villus height/crypt depth ratio (positive effects on intestinal morphology and digestive parameters) (Sopian et al., 2024). SHB fed at higher rates in broiler diets up to 5% reached a peak in positive outcomes between 5% and 10% inclusion and did not adversely impact bird performance, meat quality, or nutritional profile of breast meat. Fat metabolism and fatty acid deposition were also altered by >5% SHB inclusion in the diet, with supplementation at 5% and 10% increasing the proportion of 18:3ω-3 and total PUFA (Parker, 2024).

Surpassing an optimal peak, dietary SHB inclusion at ≥10% resulted in decreased performance, negatively impacting finishing weight, ADG, FCR, fecal score, and breast weight, while most importantly adversely affecting meat quality. SHB at 20% in the diet led to decreased intramuscular fat, in addition to detriments to the nutritional profile of fatty acids in breast tissue via increases in SFA and decreases in PUFA. The benefits of SHB inclusion in the diet of broiler liver weight are not consistent among studies, with Sopian et al. (2024) noting lower liver weights, while Parker (2024) recorded higher liver weights as dietary SHB increased. These differences are likely attributed to the higher levels of hemp biomass used by Parker (2024) (5%, 10%, and 20%) compared to the lower levels fed by Sopian et al. (2024) (0.5%, 1%, and 2%).

7.4 Hemp effects on swine health, performance and meat quality

Swine health, performance, and meat quality were improved by inclusion of hemp by-products in the diet. A study evaluating hemp fiber in weaned piglet diets demonstrated improvements in antioxidative status via phenolic compounds imparted by hemp, improving weight gain and feed conversion, while not adversely affecting hematological parameters (Świątkiewicz et al., 2022). Hempseed oil fed to primiparous sows enhanced milk quality. Hăbeanu et al. (2022) reported that sows supplemented with hempseed oil at 1.6% in the diet possessed milk with elevated levels of 18:2ω-6 and 18:3ω-3 acids compared to control sows supplemented with soybean oil, in addition to increased concentration of immunoglobulin M. Perhaps more important, sows receiving hempseed oil had less diarrhea incidence and methane production, while also exhibiting increased the proportion of lean meat (Hăbeanu et al., 2022). These results were in alignment with Vodolazska and Lauridsen (2020), who observed that sows supplemented with dietary hempseed oil exhibited greater proportion of long-chain ω-3 fatty acids in milk, which their offspring were able to in turn convert into EPA and DHA, suggesting benefits to early development in piglets. Additional studies have shown improvements to piglet performance parameters via dietary hempseed oil (Idriceanu et al., 2021) and enhanced antioxidative status in piglets through hempseed meal inclusion (Palade et al., 2019) in sow diets and their milk production.

Hemp hulls added to swine diets, in addition to dietary phytase, resulted in an increase in apparent and total tract digestibility, while decreasing fecal output of P in growing pigs (Kim et al., 2018). Madsen et al. (2022) reported that hemp hulls as a dietary fiber source in finishing pig diets resulted in reduced incidence of gastric ulcers and improvement to animal health.

Nutritional quality of pork meat benefited from supplementing hemp oil to pigs due to significantly greater deposition of 18:3ω-3 in meat tissues compared to pigs fed palm and rapeseed oil (Mourot and Guillevic, 2015). Similar results were reported by Mourot et al. (2008), who found that finishing pigs supplemented with dietary hemp oil had a greater proportion of ω-3 fats in the adipose tissue, compared to pigs fed either rapeseed or palm oil. However, no differences in ω-3 fatty acid content in meat were found in samples derived from the *longissimus dorsi* (Mourot et al., 2008). Pigs fed hemp seed at 5% exhibited increased proportion of ω-3 and PUFA in *longissimus dorsi*, brain, liver, and heart tissues, without causing detriments to animal performance compared to control animals (Hăbeanu et al., 2018).

Including hempseed meal at 15% in pigs did not adversely affect feed intake but did decrease ADG and feed:gain compared to pigs fed 0% or 10% hempseed meal (Zhang, 2024). In a follow-up study, the author reported that CP intake, initial and final BW, daily feed intake, daily gain, and gain-to-feed ratio were not adversely affected by 10% hempseed meal inclusion compared to control (Zhang, 2024).

8 Conclusion

Feeding hemp-derived constituents to livestock and poultry has shown a range of promising benefits to animal health, animal performance, the environment, and the quality of animal-sourced food products. Specifically, hemp feed has been linked to improved immunological function and overall health status in animals, likely due to its unique profile of bioactive compounds, including omega-3 and omega-6 fatty acids, antioxidants, and phytochemicals. In most studies, hemp has generally demonstrated either beneficial or neutral effects on growth and performance metrics, whereas a few studies have noted a decrease in certain performance outcomes when compared to conventional feed, underscoring the need for further optimization and research on feed formulations and dietary levels. Hemp feeding systems in livestock diets also possess the ability to positively impact environmental conditions, namely through a reduction in methane emissions.

The inclusion of hemp in animal feed has been shown to improve the nutritional profile of products such as meat, milk, and eggs. This enhancement is primarily due to the increased deposition of long-chain, omega-3 fatty acids, including 18:3ω-3, EPA, and DHA, all of which are beneficial to human health. The enrichment of these fatty acids in animal products offers a valuable pathway for meeting consumer demand for nutrient-rich foods, especially given the limited dietary sources of omega-3 fatty acids in typical Western diets.

Despite these benefits, concerns remain over the potential transfer of cannabinoids from hemp feed into animal tissues and, subsequently, into food products. This transfer raises questions about food safety and public health implications, especially as cannabinoids accumulate in the food supply chain. In the United States, these concerns currently represent a significant regulatory barrier, as legislators and health officials work to better understand the safety implications of cannabinoids in food products. Consequently, hemp is not yet widely approved as a livestock feed ingredient in many regions. However, ongoing research aims to clarify the metabolic pathways and residue levels of cannabinoids in animals, which could inform guidelines for their safe use as feed ingredients. Future studies and technological advances could guide policymakers to develop scientifically sound, risk-based legislation that allows the controlled use of hemp in animal feed. By addressing public health concerns, this legislation would enable the agricultural sector to harness hemp's nutritional benefits safely and effectively, fostering new markets for hemp products. For both hemp and livestock producers, this could offer a sustainable alternative to conventional feed, driving economic opportunities and supporting more resilient, diversified feeding systems.

9 Where to look for further information

- Irawan, A., Buffington, H., Ates, S., and Bionaz, M. 2025. Use of industrial hemp byproductsin ruminants: A review of the nutritional profile, animal response, constraints,and global regulatory environment, *Journal of Cannabis Research 7*, 25. https://doi.org/10.1186/s42238-025-00279-7.

- Muedi, H. T. H., Kujoana, T. C., Shai, K., Mabelebele, M., and Sebola, N. A. 2024. The use of industrial hemp (Cannabis sativa) on farm animal's productivity, health and reproductive performance: a review. *Animal Production Science*, *64*(2), AN23268.
- Mohamed, N., and House, J. D. 2024. Safety and efficacy of hemp-derived products in animal feeds–a narrative review. *Canadian Journal of Animal Science*, *104*(4), 390–410.

10 References

Addo, F., et al. 2023. Quality and safety of hemp meal as a protein supplement for nonlactating dairy cows. *Journal of Dairy Science*, 106(11), 7602–7612. https://doi.org/10.3168/jds.2023-23222.

Ahmed, A. T. M. F., et al. 2022. Hemp as a potential raw material toward a sustainable world: a review. *Heliyon*, 8(1), e08753. https://doi.org/10.1016/j.heliyon.2022.e08753.

Altman, A. W., et al. 2023. Utilizing industrial hemp (Cannabis sativa L.) by-products in livestock rations. *Animal Feed Science and Technology*, 307, 115850.

Antunović, Z., et al. 2021. The influence of hempseed cake on production traits, metabolic profile and antioxidant status of Merinolandschaf lambs. *Annals of animal science*, 21(3), 991–1006. https://doi.org/10.2478/aoas-2020-0107.

Arsenault, T. L., et al. 2024. Compliance testing of hemp (Cannabis sativa L.) cultivars for total delta-9 THC and total CBD using gas chromatography with flame ionization detection. *Plants*, 13(4), 519. https://doi.org/10.3390/plants13040519.

Balenović, M., et al. 2024. Immunostimulatory and antibacterial effects of Cannabis sativa L. leaves on broilers. *Animals*, 14(8), 1159. https://doi.org/10.3390/ani14081159.

Barone, A. M., et al. 2021. The usual suspect: how to co-create healthier meat products. *Food Research International*, 143, 110304. https://doi.org/10.1016/j.foodres.2021.110304.

Beitzke, B. and Pate, D. W. 2021. A broader view on deriving a reference dose for THC traces in foods. *Critical Reviews in Toxicology*, 51(8), 695–722. https://doi.org/10.1080/10408444.2021.2008867.

Burton, R. A., et al. 2022. Industrial hemp seed: from the field to value-added food ingredients. *Journal of Cannabis Research*, 4(1), 45. https://doi.org/10.1186/s42238-022-00156-7.

Calder, P. 2004. n–3 Fatty acids and cardiovascular disease: evidence explained and mechanisms explored. *Clinical Science*, 107(1), 1–11. https://doi.org/10.1042/cs20040119.

Chen, C. and Pan, Z. 2021. Cannabidiol and terpenes from hemp–ingredients for future foods and processing technologies. *Journal of Future Foods*, 1(2), 113–127. https://doi.org/10.1016/j.jfutfo.2022.01.001.

Cherney, J. H. and Small, E. 2016. Industrial hemp in North America: production, politics and potential. *Agronomy (Basel)*, 6, 58. https://doi.org/10.3390/agronomy6040058.

Clifton, P. M. and Keogh, J. B. 2017. A systematic review of the effect of dietary saturated and polyunsaturated fat on heart disease. *Nutrition, Metabolism and Cardiovascular Diseases*, 27(12), 1060–1080. https://doi.org/10.1016/j.numecd.2017.10.010.

Cozma, A., et al. 2015. Effect of hemp seed oil supplementation on plasma lipid profile, liver function, milk fatty acid, cholesterol, and vitamin A concentrations in Carpathian goats. *Czech Journal of Animal Science*, 60(7), 289–301. https://doi.org/10.17221/8275-CJAS.

Cufadar, Y., et al. 2021. Effect of hempseed meal on performance, egg quality and egg yolk fatty acids content in laying quails. *Journal of the Hellenic Veterinary Medical Society*, 72(4), 3427–3434. https://doi.org/10.12681/jhvms.29388.

Day, L., et al. 2022. Food proteins from animals and plants: differences in the nutritional and functional properties. *Trends in Food Science & Technology*, 119, 428–442. https://doi.org/10.1016/j.tifs.2021.12.020.

Dayani, O., et al. 2022. Feeding full-fat hemp seeds in low crude protein diets of fattening lambs: effects on performance, carcass characteristics and blood metabolites. *Carcass Characteristics and Blood Metabolites*. https://doi.org/10.2139/ssrn.4299990.

Demirci, M., et al. (2024). Performance and Some Blood Biochemical Parameters of Broilers Fed Diets Containing Hemp Seed Oil. *Poultry Science Journal*, 12(2).

Duggan, P. J. 2021. The chemistry of cannabis and cannabinoids. *Australian Journal of Chemistry*, 74(6), 369–387. https://doi.org/10.1071/CH21006.

Eriksson, M., and Wall, H. 2012. Hemp seed cake in organic broiler diets. *Animal feed science and technology*, 171(2–4), 205–213.

Falowo, A. B., et al. 2014. Natural antioxidants against lipid–protein oxidative deterioration in meat and meat products: a review. *Food Research International*, 64, 171–181. https://doi.org/10.1016/j.foodres.2014.06.022.

FDA. 2018a. Statement from FDA Commissioner Scott Gottlieb, M.D., on signing of the Agriculture Improvement Act and the agency's regulation of products containing cannabis and cannabis-derived compounds. [accessed August 14, 2019]. https://www.fda.gov/news-events/press-announcements/statement-fda-commissioner-scott-gottlieb-md-signing-agriculture-improvement-act-and-agencys.

FDA. 2018b. The safety and the generally recognized as safe (GRAS) status of the proposed use of hulled hemp seeds in human food (Issue 765, pp. 1–153).

FDA. 2024. FDA regulation of cannabis and cannabis-derived products, including cannabidiol (CBD). [accessed March 11, 2024]. https://www.fda.gov/news-events/public-health-focus/fda-regulation-cannabis-and-cannabis-derived-products-including-cannabidiol-cbd.

Gakhar, N., et al. 2012. Effect of feeding hemp seed and hemp seed oil on laying hen performance and egg yolk fatty acid content: evidence of their safety and efficacy for laying hen diets. *Poultry Science*, 91, 701–711. https://doi.org/10.3382/ps.2011-01825.

Gibb, D., et al. 2005. Effect of full-fat hemp seed on performance and tissue fatty acids of feedlot cattle. *Canadian Journal of Animal Science*, 85, 223–230. https://doi.org/10.4141/A04-078.

Göçmen, R. and Kanbur, G. 2021. A comparative study on the effects of use hemp seed oil substitute to soybean oil in growing quail diets. *Turkish Journal of Agriculture-Food Science and Technology*, 9(8), 1390–1394. https://doi.org/10.24925/turjaf.v9i8.1390-1394.3957.

Goldberg, E. M., et al. 2012. Fatty acid profile and sensory characteristics of table eggs from laying hens fed hempseed and hempseed oil. *Journal of food science*, 77(4), S153–S160. https://doi.org/10.1111/j.1750-3841.2012.02626.x.

Hăbeanu, M., et al. 2018. Changes in fatty acids composition of *Longissimus dorsi* muscle, brain, liver and heartas effect to hempseed addition in pigs feeding. *AgroLife Scientific Journal*, 7(2), 61–67.

Hăbeanu, M., et al. 2022. Alterations in essential fatty acids, immunoglobulins (IgA, IgG, and IgM), and enteric methane emission in primiparous sows fed hemp seed oil and their offspring response. *Veterinary Sciences*, 9(7), 352. https://doi.org/10.3390/vetsci9070352.

Hoffman, L. A. and Baker, A. J. 2011. *Estimating the substitution of distillers' grains for corn and soybean meal in the US feed complex*. Washington, DC: US Department of Agriculture.

Hoffman, E., et al. 2018. Energy use and greenhouse gas emissions in organic and conventional grain crop production: Accounting for nutrient inflows. *Agricultural Systems*, 162, 89–96. https://doi.org/10.1016/j.agsy.2018.01.021.

Houtzager, D., et al. 2024. Miniaturized paper-based analytical device for the portable analysis of phyto-cannabinoids in plant and oral fluids. *Analytical and Bioanalytical Chemistry*, 416(1), 255–264. https://doi.org/10.1007/s00216-023-05013-x.

Idriceanu, L., et al. 2021. Preliminary study of diets effects on performance, Co_2 emission and microclimate variation of primiparous sows. *Scientific Papers. Series D. Animal Science*, 64(1), 164–168.

Iommelli, P., et al. 2024. The effects of Hemp Hay (Canapa sativa L.) in the diets of grazing goats on milk production and fatty acid profile. *Animals*, 14(16), 2373. https://doi.org/10.3390/ani14162373.

Irawan, A., et al. 2024a. Feeding spent hemp biomass to lactating dairy cows: effects on performance, milk components and quality, blood parameters, and nitrogen metabolism. *Journal of Dairy Science*, 107(1), 258–277. https://doi.org/10.3168/jds.2023-23829.

Irawan, A., et al. 2024b. Cannabinoid residuals in tissues of lambs fed spent hemp biomass and consumer's exposure assessment. *Food and Chemical Toxicology*, 191, 114848. https://doi.org/10.1016/j.fct.2024.114848.

Jing, M., et al. 2017. Performance and tissue fatty acid profile of broiler chickens and laying hens fed hemp oil and HempOmegaTM. *Poultry Science*, 96(6), 1809–1819. https://doi.org/10.3382/ps/pew476.

Juodka, R., et al. 2018. The effect of feeding with hemp and Camelina cakes on the fatty acid profile of duck muscles. *Archives Animal Breeding*, 61(3), 293–303. https://doi.org/10.5194/aab-61-293-2018.

Juodka, R., et al. 2022. The effect of dietary hemp and camelina cakes on liver fatty acid profile of ducks. *Journal of Applied Animal Research*, 50(1), 152–160. https://doi.org/10.1080/09712119.2022.2044332.

Juodka, R., et al. 2023. Effects of dietary hempseed or camelina cakes on fatty acid composition of quail meat. *Life*, 14(1), 53. https://doi.org/10.3390/life14010053.

Kanbur, G. 2022. Growth-depressing effect of dietary hempseed oil on broiler performance in the starting period and alterations in meat oxidation, serum parameters and abdominal fatty acids. *Animal Science Papers and Reports*, 40(2), 203–216.

Kanbur, G., et al. 2023. A comparative study on the effects of hemp seed oil versus four different UFA-rich seed oils' dietary supplementation on egg production performance, egg quality, and yolk fatty acids in laying hens. *Tropical Animal Health and Production*, 55(1), 6.

Karlsson, L., et al. 2010. Effects of increasing amounts of hempseed cake in the diet of dairy cows on the production and composition of milk. *Animal*, 4, 1854–1860. https://doi.org/10.1017/S1751731110001254.

Kasula, R., et al. 2021. Effect of dietary hemp seed cake on the performance of commercial laying hens. *International Journal of Livestock Production*, 12, 17–27. https://doi.org/10.5897/IJLP2020.0747.

Kent-Dennis, C. and Klotz, J. L. 2024. The endocannabinoid system in bovine tissues: characterization of transcript abundance in the growing Holstein steer. *BMC Veterinary Research*, 20(1), 481. https://doi.org/10.1186/s12917-024-04319-x.

Khan, R. U., et al. 2010. Influence of feed supplementation with Cannabis sativa on quality of broilers carcass. *Pakistan Veterinary Journal*, 30(1), 34–38.

Kim, J. W. and Nyachoti, C. M. 2017. Net energy of hemp hulls and processed hemp hull products fed to growing pigs and the comparison of net energy determined via indirect calorimetry and calculated from prediction equations1. *Journal of Animal Science*, 95, 2649–2657. https://doi.org/10.2527/jas.2016.1255.

Kim, J. W., et al. 2018. Effects of extrusion and microbial phytase on the apparent and standardized total tract digestibility of phosphorus in hemp hulls fed to growing pigs. *Journal of Animal Science*, 96(5), 1838–1845. https://doi.org/10.1093/jas/sky097.

Kleinhenz, M. D., et al. 2020. Plasma concentrations of eleven cannabinoids in cattle following oral administration of industrial hemp (Cannabis sativa). *Scientific Reports*, 10(1), 12753. https://doi.org/10.1038/s41598-020-69768-4.

Kleinhenz, M. D., et al. 2022. Short term feeding of industrial hemp with a high cannabidiolic acid (CBDA) content increases lying behavior and reduces biomarkers of stress and inflammation in Holstein steers. *Scientific Reports*, 12, 3683. https://doi.org/10.1038/s41598-022-07795-z.

Konca, Y. and Beyzi, S. B. 2012. Effect of hempseed (Cannabis sativa sp.) supplementation to diet on performance, carcass and intestinal organ traits in Japanese quail (Coturnix coturnix japonica). In *XV International Feed Technology Symposium Feed-to Food/ Cost Feed for Health Joint Workshop*. Institute of Food Technology, Novi Sad, Serbia.

Krebs, G. L., et al. 2021. Intake, nutrient digestibility, rumen parameters, growth rate, carcase characteristics and cannabinoid residues of sheep fed pelleted rations containing hemp (Cannabis sativa L.) stubble. *Translational Animal Science*, 5(4), txab213. https://doi.org/10.1093/tas/txab213.

Madsen, P. A., et al. 2022. Non-targeted metabolomics of saliva to explore potential biomarkers for gastric ulceration in pigs fed hemp. *Animal*, 16(3), 100477.

Mahmoudi, M., et al. 2012. Effects of graded levels of hemp seed (Cannabis sativa L.) on performance, organ weight and serum cholesterol levels on broilers. *Journal of Medicinal Plants*, 11(42), 121–129.

Mahmoudi, M., et al. 2015. Effects of different levels of hemp seed (Cannabis sativa L.) and dextran oligosaccharide on growth performance and antibody titer response of broiler chickens. *Italian Journal of Animal Science*, 14(1), 3473. https://doi.org/10.4081/ijas.2015.3473.

McGrath, J., et al 2018. Nutritional strategies in ruminants: a lifetime approach. *Research in Veterinary Science*, 116, 28–39. https://doi.org/10.1016/j.rvsc.2017.09.011.

Mierliță, D. 2016. Fatty acid profile and health lipid indices in the raw milk of ewes grazing part-time and hemp seed supplementation of lactating ewes. *South African Journal of Animal Science*, 46, 237–246. https://doi.org/10.4314/sajas.v46i3.3.

Mierliță, D. 2018. Effects of diets containing hemp seeds or hemp cake on fatty acid composition and oxidative stability of sheep milk. *South African Journal of Animal Science*, 48, 504–5015. https://doi.org/10.4314/sajas.v48i3.11.

Montero, L., et al. 2022. Differentiation of industrial hemp strains by their cannabinoid and phenolic compounds using LC× LC-HRMS. *Analytical and Bioanalytical Chemistry*, 414(18), 5445–5459. https://doi.org/10.1007/s00216-022-03925-8.

Mourot, J. and Guillevic, M. 2015. Effect of introducing hemp oil into feed on the nutritional quality of pig meat. *OCL*, 22(6), D612. https://doi.org/10.1051/ocl/2015035.

Mourot, J., et al. 2008. Effect of the incorporation of hemp oil in the diet of growing finishing pigs on growth performance and nutritional quality of pork. In *Proceedings title* (pp. 87–88). Retrieved from CABI Digital Library.

Mustafa, A. F., et al. 1999. The nutritive value of hemp meal for ruminants. *Canadian Journal of Animal Science*, 79, 91–95. https://doi.org/10.4141/A98-031.

Nascimento, C. D. O., et al. 2022. Whole corn germ as an energy source in the feeding of feedlot lambs: metabolic and productive performance. *Animals*, 12(10), 1261. https://doi.org/10.3390%2Fani12101261.

Neijat, M., et al. 2016. Increasing levels of dietary hempseed products leads to differential responses in the fatty acid profiles of egg yolk, liver and plasma of laying hens. *Lipids*, 51(5), 615–633.

Palade, L. M., et al. 2019. Effect of dietary hemp seed on oxidative status in sows during late gestation and lactation and their offspring. *Animals*, 9(4), 194.

Parker, N. B. 2024. *Spent hemp biomass in livestock and poultry diets: effects on animal health, performance and meat quality attributes* [Doctoral dissertation, Oregon State University]. Graduate Theses and Dissertations (GTD). https://ir.library.oregonstate.edu/concern/graduate_thesis_or_dissertations/m039kd512.

Parker, N. B., et al. 2022. Assessment of spent hemp biomass as a potential ingredient in ruminant diet: nutritional quality and effect on performance, meat and carcass quality, and hematological parameters in finishing lambs. *Journal of Animal Science*, 100(10), skac263. https://doi.org/10.1093/jas/skac263.

Parr, B., et al. 2020. Effects of using hemp hearts as a feed additive in broiler chicks. *Journal of Agricultural Hemp Research*, 2(1), 1. https://doi.org/10.61611/2688-5182.1009.

Ran, T., et al. 2024. Partially substituting alfalfa hay with hemp forage in the diet of goats improved feed efficiency, ruminal fermentation pattern and microbial profiles. *Animal Nutrition*, 17, 49–60. https://doi.org/10.1016/j.aninu.2024.01.003.

Razmaitė, V., et al. 2022. Effects of dietary rapeseed (Brassica napus), Hemp (Cannabis sativa) and Camelina (Camelina sativa) seed cakes supplementation on Yolk and Albumen colour and nutritional value of Yolk Lipids in Estonian Quail Eggs. *Animals*, 12(22), 3110. https://doi.org/10.3390/ani12223110.

Renna, M., et al. 2022. Full-fat insect meals in ruminant nutrition: in vitro rumen fermentation characteristics and lipid biohydrogenation. *Journal of Animal Science and Biotechnology*, 13(1), 138. https://doi.org/10.1186/s40104-022-00792-2.

Rupasinghe, H. V., et al. 2020. Industrial hemp (Cannabis sativa subsp. sativa) as an emerging source for value-added functional food ingredients and nutraceuticals. *Molecules*, 25(18), 4078. https://doi.org/10.3390/molecules25184078.

Salami, S. A., et al. 2019. Sustainability of feeding plant by-products: a review of the implications for ruminant meat production. *Animal Feed Science and Technology*, 251, 37–55. https://doi.org/10.1016/j.anifeedsci.2019.02.006.

Sana, B., et al. 2024. Effects of hempseed (Cannabis sativa L.) in diet on growth, gut health, and immunity in broilers. *Veterinary Quarterly*, 44(1), 1–10. https://doi.org/10.1080/01652176.2024.2367214.

Sandström, V., et al. 2022. Food system by-products upcycled in livestock and aquaculture feeds can increase global food supply. *Nature Food*, 3(9), 729–740. https://doi.org/10.1038/s43016-022-00589-6.

Semwogerere, F., et al. 2022. Chevon production and quality of Kalahari Red goats fed increasing levels of hempseed cake substituted for soybean meal. *Meat Science*, 187, 108749. https://doi.org/10.1016/j.meatsci.2022.108749.

Semwogerere, F., et al. 2023. Health value and keeping quality of chevon from goats fed finisher diets containing hemp (Cannabis sativa L.) seed cake. *Meat Science*, 198, 109114. https://doi.org/10.1016/j.meatsci.2023.109114.

Semwogerere, F., et al. 2024. Nutrient intake, digestibility, and utilization in goats fed graded levels of hempseed cake finisher diets. *Tropical Animal Health and Production*, 56(1), 21. https://doi.org/10.1007/s11250-023-03864-1.

Shen, J. S., et al. 2015. Effects of corn and soybean meal types on rumen fermentation, nitrogen metabolism and productivity in dairy cows. *Asian-Australasian Journal of Animal Sciences*, 28(3), 351. https://doi.org/10.5713%2Fajas.14.0504.

Silversides, F. G. and LefranÇois, M. R. 2005. The effect of feeding hemp seed meal to laying hens. *British Poultry Science*, 46, 231–235. https://doi.org/10.1080/00071660500066183.

Simopoulos, A. P. 2010. The omega-6/omega-3 fatty acid ratio: health implications. *Oléagineux, Corps Gras, Lipides*, 17(5), 267–275. https://doi.org/10.1051/ocl.2010.0325.

Skřivan, M., et al. 2019. Hempseed increases gamma-tocopherol in egg yolks and the breaking strength of tibias in laying hens. *PLoS One*, 14(5), e0217509. https://doi.org/10.1371/journal.pone.0217509.

Sopian, Y., et al. 2024. Dietary supplementation of Cannabis sativa residues in broiler chickens affects performance, carcass characteristics, intestinal morphology, blood biochemistry profile and oxidative stability. Poultry Science, 104117. https://doi.org/10.1016/j.psj.2024.104117.

Stastnik, O., et al. 2015. The effect of hempseed cakes on broiler chickens performance parameters. In *Proceedings of the 22nd International PhD Students Conference Location, Mendel University, Brno, Czech Republic* (pp. 11–12).

Šťastník, O., et al. 2019. The effect of hempseed expellers on selected quality indicators of broiler chicken's meat. *Acta Veterinaria Brno*, 88(1), 121–128.

Stevens, S. A., et al. 2022. Nutrient digestibility, rumen parameters, and (cannabinoid) residues in sheep fed a pelleted diet containing green hemp (Cannabis sativa L.) biomass. *Translational Animal Science*, 6(4), txac141. https://doi.org/10.1093/tas/txac141.

Swanson, D., et al. 2012. Omega-3 fatty acids EPA and DHA: health benefits throughout life. *Advances in nutrition*, *3*(1), 1–7. https://doi.org/10.3945/an.111.000893.

Świątkiewicz, M., et al. 2022. Assessment of flax and hemp fibres in terms of their impact on the growth performance and health status of weaned piglets. *Animal*, 16(12), 100677. https://doi.org/10.1016/j.animal.2022.100677.

Taaifi, Y., et al. 2023. Impact of Cannabis seed incorporation in layer diet on productive performance and egg quality traits. *Scientifica*, 1, 5565825.

Tufarelli, V., et al. 2023. Hemp seed (Cannabis sativa L.) cake as sustainable dietary additive in slow-growing broilers: effects on performance, meat quality, oxidative stability and gut health. *Veterinary Quarterly*, 43(1), 1–12. https://doi.org/10.1080/01652176.2023.2260448.

USDA. 2020. FSA crop acreage data reported to FSA: 2020. https://www.fsa.usda.gov/news-room/efoia/electronic-reading-room/frequently-requested-information/crop-acreage-data/index.

USDA. 2023. askFSIS Public Q&A: Hemp as an Ingredient in Meat, Poultry, and Egg Product. [accessed February 5, 2024]. https://ask.usda.gov/s/article/askFSIS-Public-Q-A-Hemp-as-an-Ingredient-in-Meat-Poultry-and-Egg-Product.

Visković, J., et al. 2023. Industrial hemp (Cannabis sativa L.) agronomy and utilization: a review. *Agronomy*, 13(3), 931. https://doi.org/10.3390/agronomy13030931.

Vispute, M. M., et al. 2019. Effect of dietary supplementation of hemp (Cannabis sativa) and dill seed (Anethum graveolens) on performance, serum biochemicals and gut health of broiler chickens. *Journal of Animal Physiology and Animal Nutrition*, 103(2), 525–533. https://doi.org/10.1111/jpn.13052.

Vispute, M. M., et al. 2021. Dietary hemp (Cannabis sativa L.) and dill seed (Anethum graveolens) improve physicochemical properties, oxidative stability, and sensory attributes of broiler meat. *ACS Food Science & Technology*, 1(3), 453–461. https://doi.org/10.1021/acsfoodscitech.0c00049.

Vodolazska, D., and Lauridsen, C. 2020. Effects of dietary hemp seed oil to sows on fatty acid profiles, nutritional and immune status of piglets. *Journal of Animal Science and Biotechnology*, 11(1), 28.

Wagner, B., et al. 2022. Transfer of cannabinoids into the milk of dairy cows fed with industrial hemp could lead to Δ9-THC exposure that exceeds acute reference dose. *Nature Food*, 3(11), 921–932. https://doi.org/10.1038/s43016-022-00623-7.

Wang, Y., et al. 2023. Replacing alfalfa hay with industrial hemp ethanol extraction byproduct and Chinese wildrye hay: Effects on lactation performance, plasma metabolites, and bacterial communities in Holstein cows. *Frontiers in Veterinary Science*, 10, 1061219. https://doi.org/10.3389/fvets.2023.1061219.

Wood, J. D. and Scollan, N. 2022. Fatty acids in meat: effects on nutritional value and meat quality. In: Purslow, P. P. (ed.) *New aspects of meat quality* (pp. 609–647). Cambridge: Woodhead Publishing. https://doi.org/10.1016/B978-0-323-85879-3.00008-8.

Zhang, Y. 2024. Hemp seed meal as an alternative protein source in growing pigs diet: a pilot study. Master's thesis. https://doi.org/10.2478/aoas-2025-0085.

Żuk-Gołaszewska, K. and Gołaszewski, J. 2020. Hemp Production. In: Crini, G. and Lichtfouse, E. (eds) *Sustainable agriculture reviews 42. Sustainable agriculture reviews*, vol 42. Cham: Springer. https://doi.org/10.1007/978-3-030-41384-2_1.

Chapter 7

Advances in the use of hemp as feed for aquaculture

Femi John Fawole, Craig Schluttenhofer and Brandy E. Phipps, Central State University, USA

1 Introduction

Aquaculture has experienced unprecedented growth over the past few decades, and as of 2022, it accounted for 51% of aquatic animal production and 57% of fish used for food (FAO, 2024). This growth has been accelerated by changes in food choices among consumers, caused by increasing purchasing power and growing awareness of the health benefits of seafood. In addition, the surge in feed-based aquaculture production has also contributed to the industry's growth. While the recent aquaculture production output is noteworthy and expected to increase over the next decades, the production of equivalent quantities of feed and the reduction of fishmeal (FM) in feed are vital for the sustainability of the industry (Fawole et al., 2023). Feed accounts for 50–70% of the production cost, depending on the species and farming system, with FM constituting the major expense, especially for carnivorous species. Although the use of FM in aquafeed offers a benefit of supplying adequate levels of

http://dx.doi.org/10.19103/AS.2025.0161.13

nutrients necessary for optimal growth of fish and shellfish, the rising cost of this commodity and other associated limitations are posing a serious threat to sustainable growth of the industry. Thus, more effort must be made to increase the alternative ingredient portfolio for aquaculture.

Over the past decades, studies have been carried out to assess the suitability of several alternative protein sources in fish feed, including plant materials (Dossou et al., 2018; Fawole et al., 2017; Wang et al., 2020; Mounes et al., 2024), animal by-products (Dawson et al., 2018; Gaudioso et al., 2021; Ibrahim et al., 2022), insects (Kroeckel et al., 2012; Zarantoniello et al., 2019; Fawole et al., 2020; Adeoye et al., 2020), algae (Xuan et al., 2019; Nagappan et al., 2021), single-cell proteins (Hardy et al., 2018; Vasilaki et al., 2023), etc., with varying degrees of success. For instance, the salmon industry has reduced the amount of FM to an average inclusion level of approximately 15–18% (Sarker et al., 2013) compared to the 20–25% level used in 2005 (Tacon, 2005). Although studies on alternative animal-based ingredients have been done, the issue of their production in sustainable quantities remains a major concern. This has led to an increase in the use of plant-based protein in the manufacturing of commercial aquafeed.

Hemp (*Cannabis sativa* L.) is an ancient annual crop belonging to the family Cannabaceae. It is commonly grown for its fiber, seeds, and metabolites, which are used in industrial, food or feed, and medicinal products, respectively (Adams and Martin, 1996; Bonini et al., 2018; Saoud et al., 2018; Phipps and Schluttenhofer, 2022). The plant is tolerant to a wide range of environmental variables and has the capacity to survive at low levels of soil water (Eriksson and Wall, 2012; Gill et al., 2022). Hemp is grown for a wide range of purposes due to the multiple applications of its end products, and under optimal growing conditions it can yield up to 2000 kg of seed per hectare (Callaway, 2004). The analysis of the nutritional composition of the hempseed reveals that it has high levels of protein (21.3–28%), lipids (23.2–37.5%), carbohydrates (20–30%), and micronutrients (5.5–8.8%), all of which are important for fish and shellfish (House et al., 2010; Hofmanová et al., 2014; Banskota et al., 2022; Parker et al., 2022).

Despite the huge potential and nutritional attributes of hemp-based products (seeds and oil), little attention has been paid to their use in aquaculture feed production, thus creating a gap in knowledge on their utilization for aquatic animal feeding. Nevertheless, since the passage of the Agriculture Improvement Act in 2018 by the US Congress, researchers are intensifying efforts to explore the possibility of replacing conventional ingredients used in animal feed production with hemp-based products (e.g. seed cake, seed meal, and oil). While the effort thus far has shown impressive results in a variety of animals, including fish, the plant's production of psychoactive Δ^9-tetrahydrocannabinol (Δ^9-THC) remains an issue. Due to its lipophilic nature,

Δ^9-THC may be absorbed along with lipids and deposited in the flesh of fish. Deposition and accumulation of Δ^9-THC remain subject of scrutiny by regulatory agencies globally and a major holdback for feed producers. This chapter aims to discuss the current state of knowledge regarding hemp-based products and their application in aquatic animal feeding, as well as the prospects of hempseeds and/or oil for aquafeed development.

2 Alternative protein sources used in aquaculture

A plethora of studies on plant-based proteins have established their suitability for fish feed manufacturing and are considered the major viable replacement for ocean-sourced proteins given their availability, reasonable price, and relatively balanced amino acid (AA) composition (Kari et al., 2023). Such benefits have led to increased inclusion of ingredients of plant origin – soybean meal (SBM), cotton seed meal, sunflower meal, lupin meal, linseed, rapeseed meal/canola meal, wheat gluten, peas, etc. – in aquafeed formulation to reduce dependency on FM. While these ingredients offer several advantages, their inclusion at a higher level raises other challenges. Plants produce defense chemicals (i.e. antinutritional factors; ANFs) to prevent consumption by other organisms. ANFs in plant-based ingredients limit fish from efficiently utilizing nutrients or impairs feed intake (for review, see Francis et al., 2001). Furthermore, the issues of high quantities of non-starch polysaccharides (NSPs), low palatability, and AA imbalances are other limiting side effects, especially for some species of carnivorous fish and shrimp (Venold et al., 2012; Jannathulla et al., 2019; Kumar et al., 2021). These were evident in several studies on SBM, where a high level of inclusion negatively impacted growth, nutrient digestion, and absorption, and resulted in damage to the intestinal wall thereby inducing enteritis and altered gut microbiome (Lin and Cheng, 2017; Kumar et al., 2021). For instance, Fuentes-Quesada et al. (2018) fed three different levels of SBM (22%, 44%, and 64%) to totoaba (*Totoaba macdonaldi*) as a replacement for FM. According to the study, the fed fish showed a progressive decrease in growth with increasing levels of SBM compared to the FM-fed group. Furthermore, it was reported that the fish showed decreased digestive enzyme activities and severe pathohistological changes in the intestinal gut, such as hyperplasia of goblets cells, reduction in supranuclear vacuoles, increase in lamina propria width, and compromised immune system. Fuentes-Quesada and colleagues thereby concluded that the use of SBM in totoaba diets should be done with caution and should not be higher than 22%. In a similar vein, Kumar et al. (2021) reported that rainbow trout (*Oncorhynchus mykiss*) fed high SBM (21%) performed poorly, had inflammation of the distal intestine and lamina propria, and had a reduction in supranuclear absorptive vacuoles. The histological evidence of inflammation was further reinforced by the upregulation of inflammation mediators, interferon regulatory factor 1 (*IRF-1*) and prostaglandin D (*ProstD*), in the intestine. In another study, the inclusion of full-fat SBM at a level greater than 41% in African catfish (*Clarias gariepinus*) diets caused growth depression, and likely due to the adverse effect of soya saponin on the proteolytic enzyme activity relative to those fed FM diets (Abdel-Warith et al., 2019). Several other studies have reported similar results in other fish species including turbot (*Scophthalmus maximus*) (Gu et al., 2016), Atlantic cod (*Gadus morhua*) (Olsen et al., 2007), Atlantic salmon (*Salmo salar*) (Krogdahl et al., 2003), pearl gentian grouper (*Epinephelus fuscoguttatus* x *E. lanceolatus*) (Pang et al., 2023), etc. On the other hand, SBM was reported to partially or completely replace FM when supplemented with methionine, with no adverse effects observed in growth, nutrient utilization, or the physiological well-being of African catfish (Ajani et al., 2016). Badillo-Zapata et al. (2021) also reported a complete replacement of FM with SBM in the Pacific fat sleeper (*Dormitator latifrons*) fish without negative effects on growth or survival while also reducing production costs. The majority of these findings noted poor feed palatability and ANFs inhibitory or damaging effects as the causes for fish inability to effectively utilize plant protein, especially in carnivorous fish. However, species differences and sensitivities to plant proteins and AA imbalances may also be a contributing factor. Nevertheless, many of these aforementioned setbacks toward effective plant protein utilization by fish have been mitigated using methods such as fermentation, functional additives, exogenous enzymes, etc., either to remove or inactivate the toxic components present in plant material. As the aquaculture industry continues to grow and FM production remains unsustainable and costly, finding new alternative ingredients will be advantageous.

3 Overview of hempseed components and nutrient compositions

Hempseeds are a plant-based material known for their high-quality and easily digestible protein, healthy fats, and appropriate levels of micronutrients. The protein in hempseeds is rich in edestin (globulin) and albumin, while the oil contains 71–90% polyunsaturated fatty acids (PUFA), depending on the variety, with linoleic acids being the major fatty acid (Parker et al., 2003; Zając et al., 2019; Gimeno-Martinez et al., 2023). In Table 1, the data of the nutrient composition reported in literature indicate whole hempseeds were constituted of 205–257 g/kg (average 243.75 ± 16.87 g/kg DM) of protein, 292–355 g/kg (average 315.88 ± 20.81 g/kg DM) lipids, 276–428 g/kg (average 351.86 ± 47.98 g/kg DM) fiber, and 41–100 g/kg (average 58.67 ± 20.97 g/kg DM) ash. Similarly, the hempseed cake, also called meal, showed higher values after the pressing process removed 70.04%, on average, of the oil, thus increasing the protein and fiber contents to about 340.25 g/kg and 431.63 g/kg, respectively.

Table 1 Chemical composition of whole hempseed and hempseed cake/meal

Nutrient component (g/kg)	Dry matter	Protein	Lipid	Fiber	Ash	References
Whole hempseed	935	248	355	276	56	Callaway et al. (2004)
	941	240	304	321	48	House et al. (2010)[a]
	929	242	304	–	–	Gakhar et al. (2012)
	938	256	292	357	55	Vonapartis et al. (2015)[b]
	910	253	339	370	41	Wang et al. (2017)
	882	257	316	334	–	Mierliță (2018)
	932	250	307	428	100	Zając et al. (2019)
	941	205	310	377	52	Alonso-Esteban et al. (2022)[c]
Average	926	243.75	315.88	351.86	58.67	
SD	20.39	16.87	20.81	47.98	20.97	
Hempseed cake/meal	892	377	96	442	65	Hessle et al. (2008)
	–	321	52	508	82	Mustafa et al. (1999)
	944	335	111	426	72	Callaway et al. (2004)
	951	407	102	305	67	House et al. (2010)[b]
	912	281	110	447	72	Halle and Schöne (2013)
	–	298	97	326	72	Stastnik et al. (2019)
	896	364	115	495	58	Gurung et al. (2022)
	909	339	74	504	85	Winders et al. (2022)
Average	917.33	340.25	94.63	431.63	71.63	
SD	24.66	41.52	21.46	78.06	8.77	

a = average of 11 varieties, b = average of 10 varieties, c = average of 8 varieties, d = average of 10 varieties.

The high fiber content in the pressed cake may significantly impact nutrient digestibility, particularly in carnivorous fish species with limited capacity to hydrolyze complex carbohydrates such as cellulose, hemicellulose, and lignin. These components were found to be present at levels of 14.1–18.1%, 6.3–9.4%, and 10.5–11.8%, respectively (Vonapartis et al., 2015), potentially influencing the efficiency of nutrient utilization. Most of the fibrous content found in seed is located within the seed coat (House et al., 2010). Removal of the seed coat via the dehulling process would be the ideal approach to reduce most of these complex carbohydrates so hemp could be efficiently used in aquatic animal feed.

Ingredient quality plays an important role in aquafeed formulation. This was emphasized by Fawole et al. (2023), who reiterated that the quality of an ingredient is determined by the composition of its AAs, ease of digestibility, and the amount of limiting AAs contained within the ingredient. AAs are fundamental components contributing to the structural and functional aspects of the fish body, playing a crucial role in essential physiological processes such as muscle growth and immune responses. Extensive literature, notably by Li et al. (2021), has provided comprehensive insights into the significance of both essential and non-essential AAs in fish. Studies on hempseeds evidenced that the seed meal has a favorable AA profile with high levels of arginine, aspartic acid/asparagine, and glutamic acid/glutamine in addition to good levels of sulfur-containing AAs (methionine and cysteine), compared to other high-quality proteins currently used in aquafeed production (Table 2). The rich level of sulfur-containing AA is due to the high level of edestin (Norajit et al., 2011). Moreover, the high level of arginine and glutamic acid is noteworthy due to their important role in fish immunity. In hybrid striped bass (*Morone chrysops* × *Morone saxatilis*), supplementation of 1% or 2% of either arginine or glutamine as well as both the AAs (1% arginine and 1% glutamine) enhanced the production of extracellular superoxide anion, lysozyme, and influenced the neutrophil respiratory burst activity and intracellular superoxide anion production by head–kidney macrophages (Cheng et al., 2012). Likewise, arginine, when supplemented in a channel catfish (*Ictalurus punctatus*) diet, significantly improved the activities of neutrophil respiratory burst and phagocyte superoxide production (Pohlenz et al., 2014). Several other studies have demonstrated the efficacy of these AAs and methionine in modulating immune cells in fish (Li et al., 2021). The abundance of these AAs and other essential AAs in adequate amounts, except lysine, in hempseed meal underscores the nutritional value of its protein, which not only has the potential to promote aquatic animal growth but also indicates its capacity to strengthen fish immunity. Furthermore, hemp protein and its derivatives are highly digestible. A study by House et al. (2010) demonstrated that hempseed meal is approximately 98% digestible, while whole hempseed and dehulled hempseed have digestibility of 86% and 92%, respectively. In

Table 2 Amino acids composition of whole hempseed, hempseed meal and other commonly used plant protein in aquafeed

Amino acid composition (g/100g)	Whole hempseed[a] (24.0% CP)	Hempseed meal[b] (39.4% CP)	Soybean meal (32% CP)	Rapeseed meal[c] (36.6% CP)	Linseed meal (20.9% CP)
Essential amino acids (EAA)					
Arginine	2.28	3.91	2.14	1.74	1.94
Histidine	0.55	0.93	0.76	0.90	ND
Isoleucine	0.8	1.45	1.62	1.29	0.79
Leucine	1.49	2.35	2.58	2.34	1.2
Lysine	0.86	1.32	1.73	1.54	0.83
Methionine (M)	0.56	0.88	0.53	0.24	0.32
Phenylalanine (P)	1.03	1.62	1.78	1.21	0.84
Threonine	1.01	1.35	1.35	1.57	0.93
Tryptophan	0.23	0.39	0.41	ND	ND
Valine	1.14	1.91	1.6	1.78	0.72
Non-essential amino acids (NEAAs)					
Alanine	0.96	1.61	1.39	1.51	0.99
Aspartic acid	2.39	3.66	3.62	2.09	1.53
Cysteine (C)	0.41	0.70	0.54	0.04	0.28
Glutamic acid	3.74	6.03	5.89	5.30	3.87
Glycine	1.06	1.66	1.29	1.10	0.75
Proline	0.9	1.59	1.65	3.46	ND
Serine	1.19	1.73	1.54	1.13	0.87
Tyrosine (T)	0.68	1.15	1.14	0.74	0.63
M+C	0.97	1.58	1.07	0.28	0.60
P+T	1.71	2.77	2.92	1.95	1.47
References	House et al. (2010)	House et al. (2010)	Callaway (2004)	Jiang et al. (2024)	Panaite et al. (2017)

[a]Average of 11 varieties.[b]Average of 9 varieties.[c]Average of 5 varieties.
ND, data not found; CP, crude protein.

addition, *in vitro* pepsin–trypsin digestibility of hemp protein isolate showed a digestibility range of 88–91%, surpassing the 71% digestibility of soybean protein isolate (Wang et al., 2008).

Hemp oil (HO) is a co-product of hemp cake/meal and is known to contain high levels of PUFAs, especially linoleic acid (18:2n-6; 55% of total FA) and α-linolenic acid (18:3n-3; 13–17.4% of total FA), with the ratio of n-6/n-3 in the range of 2.5:1 to 3:1 (Vodolazska and Lauridsen, 2020; Gimeno-Martínez et al., 2023). The unsaturated FA content of HO is about 90.21%, of which 78.3% are PUFA (Table 3). Comparatively, the levels of 18:2n-6 in HO and soy

Table 3 Fatty acid composition of the total lipid fraction (% of total fatty acids) of hempseed oil and other plant-based oil

Fatty acid	Common name	Hempseed oil[a]	Soybean oil[a]	Linseed oil[b]	Rapeseed oil[c]
C16:0	Palmitic acid	5.58	10.7	5.58	6.06
C18:0	Stearic acid	2.50	2.99	4.31	2.08
C18:1n-9	Oleic acid	10.7	23.0	20.16	55.22
C18:1n-7	Vaccenic acid	0.82	1.83	0.593	ND
C18:2n-6	Linoleic acid	56.0	53.6	13.9	24.24
C18:3n-6	γ-Linolenic acid	3.64	0.03	0.0064	ND
C18:3n-3	α-Linolenic acid	17.4	6.17	54.11	10.34
C18:4n-3	Stearidonic acid	1.26	0.01	0.1688	ND
C20:0	Arachidic acid	0.83	0.31	0.021	0.27
C20:1n-9	11-Eicosenoic acid or Gondoic acid	0.39	0.27	0.023	1.00
∑SFA		8.91	14	9.911	8.41
∑MUFA		11.91	25.1	20.776	56.22
∑PUFA		78.3	59.81	68.1852	34.58
∑USFA		90.21	84.91	88.9612	90.80
n-3		18.66	6.18	54.2788	10.34
n-6		59.64	53.63	13.9064	24.24
n6/n3 ratio		3.196	8.677	0.256	2.344
Vitamin E (mg/kg)					
α-tocopherol		38.2	48.1	8.0[d]	251.4[e]
γ-tocopherol		568	442	390[d]	364.3[e]
δ-tocopherol		22.3	256	5.0[d]	6.6[e]
Total tocopherols		628.5	746.1	403	622.3

[a]Vodolazska and Lauridsen (2020).[b]Wang et al. (2017).[c]Sagan et al. (2019). [d]Matthäus and Özcan (2015).[e]Kmiecik et al. (2019).
SFA, saturated fatty acid; MUFA, monounsaturated fatty acid; PUFA, polyunsaturated fatty acid; USFA, unsaturated fatty acid. ND, data not found.

oil (SO) were similar but higher than linseed oil (LO), while the concentration of 18:3n-3 in LO was higher than both HO and SO (Table 3). Additionally, HO has a relatively good quantity of γ-linolenic acid (GLA, 18:3n-6) and stearidonic acid (SDA, 18:4n-3), which are distinctive characteristics of hempseed. These fatty acids are notably absent or only present in negligible quantities in the majority of plant-based oils commonly utilized in aquafeed production. Although information on the role of GLA in fish is scarce, it has been reported to attenuate various inflammatory responses in humans and mice (see review by Fan and Chapkin, 1998; Chang et al., 2010). It has been established that the consumption of oil rich in GLA could result in an elevated level of dihomo-gamma-linolenic acid (DGLA; 20:3n-6) within the cell membrane in humans and fish. Upon stimulation, DGLA can be enzymatically converted to 15-(S)-hydroxy-8,11,13-eicosatrienoic acid and prostaglandin E1 (PGE1), which possess anti-inflammatory and antiproliferative properties (Fan and Chapkin, 1998). The benefits of incorporating GLA-rich oil (i.e. borage oil) into the diet of turbot (*Scophthalmus maximus*) were demonstrated in a study by Bell et al. (1995). Turbot with an initial weight of 1.4 g was fed for a period of 12 weeks with fish oil (EPA-rich) and borage oil (GLA-rich). After the feeding trial, carcass analysis revealed that, in comparison to the initial carcass composition, the fish-fed borage oil had a significant increase in the deposition of GLA (53%) and its derivative product, DGLA (16%), accompanied by a notable decrease in arachidonic acid (20:4n-6; 90% reduction). This indicates that the fatty acid composition of fish tissue reflects that of their diet and the oils used. Thus, HO, which is abundant in linoleic and α-linolenic fatty acids and contains moderate levels of GLA, can be utilized by fish for the synthesis of long-chain polyunsaturated fatty acids (LC-PUFA), especially eicosapentaenoic acid (EPA) and docosahexaenoic acid (DHA), due to their inherent biosynthetic capacity (Tocher, 2003; Fawole et al., 2021).

Furthermore, a study by Lefort et al. (2016) found that healthy human subjects who consumed diets supplemented with *Buglossoides arvensis* seed oil, which is rich in α-linolenic acid and SDA, showed a more significant increase in cellular levels of EPA compared to those who were subjected to diets supplemented with flaxseed oil that contains only α-linolenic acid. This was also evidenced in Atlantic salmon fed with *Echium* oil (EO), rich in SDA. The study investigated whether feeding Atlantic salmon with an EO-based diet in both freshwater and seawater would enhance more deposition of n-3 LC-PUFA in the liver and white muscle compared to those fed diets based on rapeseed oil (RO) and FO (Codabaccus et al., 2011). In comparison to fish that were fed RO, those fed with EO showed a greater accumulation of EPA, eicosatetraenoic acid (ETA; 20:4n-3), and docosapentaenoic acid (DPA; 22:5n-3) and n-3 FAs in the liver in both freshwater and seawater. Although DHA and total n-3 LC-PUFA did not differ significantly between the groups in freshwater, after an extended culture period in seawater, fish fed with EO showed higher levels of total n-3 LC-PUFA compared to those fed with RO. Furthermore, in the white muscle tissue, EO-fed fish showed higher levels of EPA, ETA, DPA, and DHA, resulting in a higher total n-3 LC-PUFA relative to the RO-fed group. This enhanced synthesis of LC-PUFAs may be attributed to the more efficient supply of SDA from EO. Moreover, a higher expression of the Δ6 desaturase gene was observed in both the liver and white muscle of fish fed with RO but not in those fed with EO (Codabaccus et al., 2011). This could be explained by the fact that the initial Δ6-desaturase enzyme-driven step required to bio-convert α-linolenic acid to SDA in RO fish is already bypassed in EO, and the presence of SDA in EO could have inhibited the mRNA expression of the Δ6-desaturase enzyme, which was similarly reported by Fawole et al. (2021) that high dietary concentrations of EPA and DHA in FO-based diets inhibited the mRNA expression of the Δ5- and Δ6-desaturase genes in rainbow trout. Based on the findings of Codabaccus and colleagues, it can be inferred that the presence of SDA in HO may enhance the synthesis of EPA more efficiently than other alternative plant oils, especially those lacking SDA, which are currently used in aquafeed production.

Hempseeds are a rich source of lipid-soluble vitamins, such as tocopherols and provitamin A, known for their physiological functions and efficacy in preventing oil oxidation (Xu et al., 2021; Montero et al., 2023). Tocopherols are essential for enhancing the shelf life of oils by inhibiting the formation of free radicals and preventing oxidative degradation (Berto et al., 2020). The levels of tocopherols found in HO are comparable to those in rapeseed oil but lower than the levels in SO (Table 3). Additionally, they contain significant amounts of both macro- and trace minerals. The major microminerals found in hempseeds are phosphorus (P), calcium (Ca), magnesium (Mg), sodium (Na), and potassium (K), while the microminerals include iron (Fe), manganese (Mn), zinc (Zn), and copper (Cu) (Barbara et al., 2020; Montero et al., 2023). Beyond their nutritional value, hempseeds also contain bioactive components such as polyphenols, phytosterols, lignanamides, and cannabinoids, which have diverse biological functions such immunomodulatory, anti-inflammatory, antibacterial, antiproliferative, antioxidant, and neuroprotective functions (Eriksson and Wall, 2012; Yan et al., 2015; Mattila et al., 2018; Xu et al., 2021; Semwogerere et al., 2023; Sun et al., 2024), and this may have beneficial effects on aquatic animal health. According to Wang et al. (2019), cannabisin F, a lignanamide derivate from hempseed, inhibits the nuclear factor kappa B (NF-κB) signaling pathway, suppresses the production of proinflammatory cytokines (IL-6 and TNF-α), and reduces the production of reactive oxygen species (ROS) in lipopolysaccharide-stimulated BV2 microglia cells. These exceptional nutritional and functional attributes have the potential to enhance fish and shellfish growth, modulate the immune system, and promote overall physiological well-being. Therefore, the incorporation of hempseed and its products in aquaculture feed application could represent a major advancement in the industry, as it could serve as a sustainable alternative or a complementary ingredient to some conventional ingredients that are currently used in aquafeed production.

4 Cannabinoid and antinutrient compounds in hempseed

The nutritional attributes of industrial hempseeds have led to their consideration as potential ingredients for aquatic animal feed production. However, studies have revealed a notable presence of cannabinoids on hempseeds, and these naturally occurring compounds have the capacity to interact with the endocannabinoid systems of both animals and humans (Wagner et al., 2022; Xin et al., 2022). Of all the cannabinoids isolated from the *Cannabis sativa* plant, cannabidiolic acid (CBDA), cannabigerolic acid (CBGA), Δ^9-tetrahydrocannabinolic acid (Δ^9-THCA), and cannabichromenic acid (CBCA) are the most abundant constituents (Gülck and Møller, 2020). Upon degradation, CBDA, CBGA, THCA, and CBCA become cannabidiol (CBD), cannabigerol (CBG), Δ^9-tetrahydrocannabinol (Δ^9-THC), and cannabichromene (CBC), respectively. Notably, Δ^9-THC, a psychoactive cannabinoid, has garnered significant attention due to its lipophilic nature and possible accumulation in fish or shellfish fillets when hempseed meal or its related products are used as ingredients in aquafeed formulations. Cannabinoids are biosynthesized in the glandular trichomes present in the aerial components of female flowers, resulting in these compounds being primarily present in limited quantities in hempseeds (Montero et al., 2023). The hempseed itself does not produce any cannabinoids, thus their presence on the seed is a result of contamination during processing. Consequently, various regulatory bodies have established permissible thresholds for these constituents in hulled hempseed, particularly for Δ^9-THC, which must not exceed 3 mg/kg, 4 mg/kg, or 5 mg/kg in European Union countries, North America, and Australia/New Zealand, respectively (see review by Burton et al., 2022). In recent developments, the United States has approved the incorporation of hempseed cake/meal in laying hens' feed, with stringent regulations in place. These regulations include a total Δ^9-THC limit of 2 mg/kg and a total CBD limit of 20 mg/kg, while the incorporation level of the cake/meal must not exceed 200 g/kg of the diet. Ongoing research efforts to understand the response of aquatic animals to cannabinoids, the metabolism of cannabinoids by fish and shellfish, and potential bioaccumulation or elimination will provide valuable information for researchers, regulatory bodies, and feed manufacturers.

Similar to other oilseeds and their byproducts currently utilized in aquafeed formulation, hempseeds are recognized for containing feeding deterrents that possess the ability to hinder normal physiological processes in fish and impede nutrient bioavailability. Notable ANFs present in hempseed cake/meal include condensed tannins, phytic acid, cyanogenic glycosides, saponins, and trypsin inhibitors (Russo and Reggiani 2013; Galasso et al., 2016). Table 4 illustrates that the concentration of phytic acid and condensed tannin in hempseed cake/meal ranges from 3.5 g/100 g to 7.55 g/100 g dry weight and 1.36–2.14 mg/g dry weight, respectively. In comparison to SBMs, the most commonly used plant protein in animal feed, hempseed cake/meal exhibited higher levels of phytic acid and tannin. Phytic acid (PA) is a secondary metabolite and heat-stable antinutrient commonly found in various plant-based ingredients. It serves as the primary storage form of phosphorus, constituting 65–80% of the total phosphorous content in grains (Kumar et al., 2023). The presence of PA in fish diet often reduces the bioavailability of essential minerals (i.e. zinc, iron, calcium, and magnesium) necessary for normal physiological processes in fish, primarily due to its chelating properties. Additionally, it has the potential to impede the functions of digestive enzymes, especially trypsin and chymotrypsin, required for the breakdown of protein (Kies et al., 2006; Denstadli et al., 2006), thereby limiting nutrient utilization. According to a previous study by Liu et al. (2014), high (4%) levels of dietary phytic acid can reduce the activity of digestive enzymes, such as protease, lipase, and amylase, thereby leading to reduced growth, feed intake, and decreased phosphorus digestibility. Furthermore, the study probed into genes responsible for appetite regulation and observed an up-regulation of cocaine- and amphetamine-regulated transcripts (CART) and cholecystokinin (CCK) – factors that regulate appetite through signal satiety and depression of feeding behaviors – in the brain. The author posited that the elevated mRNA expression of CART and CCK potentially underlies the decreased feed intake noted in grass carp (*Ctenopharyngodon idella*) fed a PA-supplemented diet (Liu et al., 2014). Similarly, the inclusion of 2.6% PA (although the actual dietary concentration is 3.0%) in high-rapeseed protein concentrate diets led to a significant reduction in feed intake and growth but no effect on the feed conversion efficiency in rainbow trout. The adverse effect observed on growth was attributed to the capability of phytic acid to chelate with mineral ions and/or phytate–protein complexes (Kaiser et al., 2021). Although the levels of other ANFs such as trypsin inhibitors, saponins, and cyanogenic glycosides are relatively lower in hempseeds compared to SBM and other plant proteins, these factors still have the potential to cause adverse effects, including gut epithelium irritation, reduced protein digestibility and weight loss, hemolysis of red blood cells, intestinal hypertrophy, and inflammation, as detailed in the review by Francis et al. (2001). Consequently, it would be beneficial to develop methods aimed at reducing the levels of PA and other ANFs present in hemp grains, which can be achieved through agronomic, enzymatic, or physicochemical approaches, particularly when considering its use in fish feed applications. Researchers have found that some processing

Table 4 Antinutritive components of hempseed reported in literature vs. other notable plant protein used in aquafeed formulation

Ingredients	Phytic acid g/100 g	Condensed tannins (mg/g)	Trypsin inhibitor unit/mg	Saponins (mg/100g)	Cyanogenic glycosides (g/100g)
Hemp cake/meal[a]	3.50–7.55	1.36–2.14	10.8–28.4	5.85–8.12	0.006–0.024
SBM[b]	1.07–2.54	0.29–1.93	41.5–96.9	66.1–180	ND
Rapeseed cake/meal[c]	3.9–7.5	1.30–4.81	1.19–4.81	ND	ND
Linseed meal[d]	2.3–4.07	0.43–0.82	12.5–24.6	ND	0.057–0.16

[a]Values for reference ranges compiled from: Galasso et al. (2016), Russo and Reggiani (2013), Mattila et al. (2018). [b]Values for reference ranges compiled from: Kwanyuen and Burton (2005), Barakat et al. (2015); Lusas and Rhee (1995). [c]Values for reference ranges compiled from: Francis et al. (2001), Mattila et al. (2018), Jiang et al. (2024). [d]Values for reference ranges compiled from: Oomah et al. (1996); Russo and Reggiani (2013); Xu et al. (2022). ND, data not found.

methods enhance nutrient composition and reduce ANFs in hemp-based products. For instance, hemp-based products that were subjected to lactic acid fermentation show significant reductions in PA levels from 60.3 g/kg to 21 g/kg, total saponin from 0.64 mg/g to 0.30 mg/g, and condensed tannins from 2.1 g/kg to undetectable levels (Nionelli et al., 2018). Additionally, hempseed meal treated with acidic soaking solutions – specifically, 100% grape vinegar and a 50% grape vinegar-50% water mix – effectively reduced the PA content of hempseed meal without compromising protein content (Kanbur, 2024).

5 Hempseed meal as an alternative ingredient for aquaculture species

Research on the use of hemp in aquaculture species is limited, and most findings are still in the laboratory stage. However, a study by Webster et al. (2000) revealed that the incorporation of 20% hempseed meal (HM) in the diet of sunshine bass (*Morone chrysops* x *M. saxatilis*) had no adverse effect on the fish performance. The weight gain (WG), specific growth rate, and feed conversion ratio (FCR) for the group fed the HM-supplemented diet (27% SBM + 27% meat-and-bone meal + 20% hempseed meal) were comparable to other dietary groups, including the control group fed 30% FM. Furthermore, the addition of 20% HM did not cause any changes in hepatosomatic index or fillet composition. Based on the diet composition table, it was noted that the AA composition of the HM-blended diet met the requirements for sunshine bass and was comparable to the fish meal diet, potentially accounting for the similar growth performance observed among the groups. The study concluded that the inclusion of 20% HM in the sunshine bass diet is feasible, indicating its suitability for aquafeed production. In a study conducted by Sample (2022), the dietary utilization of HM as a replacement for FM in striped bass (*Morone saxatilis*) diet was evaluated. Striped bass were fed graded levels of HM (5%, 10%, 20%, 30%, 40%, and 50%) for 49 days. The findings indicate that up to 50% (322.56 g/kg diet) HM can be used in place of FM without any negative effects on the percentage WG, thermal growth coefficient, and survival rate. The fish body indices, such as muscle ratio and intraperitoneal fat ratio, as well as the final body mass, did not differ significantly among the dietary groups. However, the hepatosomatic index, a bioindicator of energy reserve in fish, was higher in the group fed 10% HM compared to the control group, 20%, and 30% HM-fed groups. The body compositions of fish-fed FM and HM were not statistically different. Additionally, the apparent digestibility coefficient of protein was 85.5% for FM and 81.4% for HM diets, indicating that striped bass efficiently digest and utilize the hempseed meal protein. In a study by Lunger et al. (2007), it was observed that inclusion of 40% HM as a substitute for FM in the diet of juvenile cobia (*Rachycentron canadum*) had no significant effect on growth performance, feed efficiency, muscle ratio, viscerosomatic index, packed cell volume, and plasma protein levels. The fillet composition of the group fed HM was similar to those fed SBM, soy protein isolate, and a fish meal-based diet. However, it was noted that HM did improve the textural characteristics of the fish fillet compared to those fed FM. In a related study on common carp (*Cyrinus carpio*), it was reported that feeding 10% hempseed cake led to improved growth performance and packed cell volume. However, it also resulted in a significant reduction in the mean corpuscular concentration hemoglobin, a measure of anemia in fish, and glucose concentrations relative to the control group. There were no significant effects on other hematological variables and fatty acid composition, except for GLA (Maly et al., 2018). Furthermore, the therapeutic potential of hemp meal (HM) and hempseed oil (HO) was evaluated in Indian major carps, specifically Rohu (*Labeo rohita*) and Mrigal (*Cirrhinus mrigala*), against copper (Cu)-induced toxicity. The fish were exposed to Cu toxicity for 30 days, after which they were fed graded levels of a hemp-based diet (HM: 5%, 10%, 15%; HO: 1%, 2%, 3%) along with a control diet for 50 days. The empirical data showed that Cu exposure led to an increase in white blood cell count, hematocrit levels, mean corpuscular volume, and lymphocyte counts, while simultaneously causing a significant decrease in red blood cells and hemoglobin compared to the control group. The immune response was adversely affected, as evidenced by reduced total protein levels and elevated lysozyme activity and immunoglobulin M. Additionally, the results also revealed that the activities of enzymes, such as alanine aminotransferase, aspartate aminotransferase, and alkaline phosphatase, were increased after Cu exposure relative to the control group in both fish species. The elevated levels of these liver function enzymes are indicative of Cu toxicity and potential damage to liver cells. Moreover, the activities of antioxidant enzymes, including superoxide dismutase, catalase, glutathione peroxidase, and glutathione reductase, in the liver, brain, gills, and muscles of the Cu-exposed group were significantly higher than those observed in the control group. This elevated activity suggests that the fish were undergoing oxidative stress due to Cu exposure. However, the feeding of hemp-based diets (HM and HO) following Cu exposure effectively mitigated the negative effects, restoring all the physio-immunological indices to levels comparable to the control group (Afridi et al., 2024). This study shows the potential of hemp-based products in alleviating metal toxicity in fish.

Furthermore, the efficacy of HO as an alternate ingredient for aquaculture was conducted by Saoud et al. (2018). The study investigated the dietary effect of HO (no Δ^9-THC) relative to SO, and a blend (1:1) of SO and cannabis oil (CO; Δ^9-THC containing oil). The diets were similar in protein and energy content and the feeding lasted for 8 weeks. In their findings, the highest final body weight, total length, and best FCR value were recorded in SO group than those offered HO and SO-CO blended oil, with the SO-CO yielding a numerically lower final body weight. However, irrespective of oil fed there was no recorded effect on the hemato-immunological indices like red blood cell, white blood cell, lymphocytes, neutrophils, monocytes, hematocrit, total protein, and lysozyme activity of the fed Nile tilapia (*Oreochromis niloticus*). The whole-body compositions among the dietary groups were not different statistically, however, it was noted that mean respiration rate increased abruptly in group-fed SO-CO blended oil compared to SO. In fish, oxygen consumption is an indirect measure of metabolic rate, hence, the increase in metabolic rate observed suggests an increase in energy use and reduction in growth for the SO-CO group. The study however stressed that the indifferences observed in growth between fish-fed HO with no Δ^9-THC and those fed CO containing Δ^9-THC could be related to other cannabinoids derivative, especially CBD, which was reported to cause reduction in body weight in rat (Ignatowska-Jankowska et al., 2011). Nevertheless, the author's opinion is still subject to revalidation as more studies on HO are done. In a related study, Juries (2021) investigated the potential of hemp-derived CBD oil containing varying concentrations of CBD (0, 20, 40, and 60 mg/kg) in the diet of Nile tilapia for 10 weeks. The results showed that, compared with the control group, the final body weight (FBW), total length (TL), and feed conversion ratio (FCR) of the fish were not significantly affected by the CBD diets. Although there were no noticeable changes in blood cell counts, Fulton's condition index displayed a decreasing trend as the concentration of CBD increased. The findings of the study indicated that feeding Nile tilapia CBD-containing HO resulted in the production of leaner fish without negatively impacting their overall well-being. These findings suggest that hempseed cake/meal and HO could serve as a viable alternative ingredient for aquaculture feed production.

6 Sustainability of industrial hempseed meals for aquafeed industry and economic potential

Industrial hemp is an incredibly versatile and sustainable plant with a short growth period and requires less water and fertilizer inputs. The grains are rich in protein, fiber, and unsaturated fatty acids and low in carbohydrate, making them an ideal ingredient for aquafeed applications. In recent years, there has been a substantial increase in the cultivation of industrial hemp across various countries. The total global acreage currently stands at approximately 520 000 acres, with China, Canada, USA, France, and Chile being the leading producers (Cannabis Business Plan, 2024). Although, this data mainly convers the top five producers, other countries like Turkey, South Africa, Syria, India, Thailand, Russia, New Zealand, Japan, and Iran have also reportedly involved in the cultivation of industrial hemp (Johnson, 2013). As of 2023, the global market size of industrial hemp was valued at $5.38 billion and is projected to reach $24.30 billion by 2033, reflecting a compound annual growth rate (CAGR) of 16.27% over the forecast period (Globe Newswire, 2024). Further, the Hemp Benchmarks report for 2023 Wholesale Hemp Product Price Trends indicated that the market price of hemp grain is between US $0.55–$0.65/lb ($1.21–$1.43/kg) which is slightly higher than soybean retail price of US$0.26–$0.44/lb (US$0.57–US$0.97/kg). Even though hemp grains are currently more expensive than soybeans, it is believed that increased awareness among local farmers, favorable government policies, and the provision of incentives to farmers in the form of subsidies, rapid expansion of planting acreage, and increased investment in infrastructure will help scale up production. As demand for hemp grains and their coproducts increases, this would potentially cause prices to drop and be competitive with other ingredients currently used in aquafeed production. For example, the rapidly growing cultivation of industrial hemp in the Asia-Pacific region, driven by increased global demand and enhanced access to raw materials for planting, has heightened industry investment in research focused on the development of innovative hemp-based products for commercial purposes (Kaur and Kander, 2023). Additionally, hemp as a renewable raw material can be integrated into a crop rotation system, and this contributes to its biodiversity conservation potential. Hemp is known for its carbon sequestration efficiency and is often regarded as an eco-friendly crop with the capacity to reduce climate change because of its potential to sequester about 8 to 22 tons of carbon dioxide per hectare (Sorrentino, 2021; Ismagilov and Rusakov, 2024). A study on crops' biodiversity friendliness by Montford and Small (1999) reported that HO seed/grain is more biodiversity-friendly than other oilseed crops, such as soybeans, rapeseed, and sunflowers. These countless attributes of hemp have increased the interest of many countries around the world and prompted them to revise their laws to allow for the cultivation of industrial hemp (Głodowska and Łyszcz, 2017; Semwogerere et al., 2020). Furthermore, the health benefits of hemp, in terms of the high levels of omega-3 fatty acids and a balanced ratio of n-6/n-3, are likely to boost HO production globally. The large percentage of co-products (seed cake/meal) produced will benefit the animal feed industry. Notably, depending on the processing method, 1 kg of hempseed can yield 200–400 g/kg of protein and 300–350 g/kg of oil (Callaway et al., 2004; House et al., 2010; Alonso-Esteban et al., 2022). These and other factors would help ensure hemp remains a sustainable alternative feed ingredient for animal feeding.

7 Exploratory case study: improving hempseed meal quality for aquafeed production

Hempseed meals are being explored as a sustainable and nutrient-dense alternative to commonly used conventional ingredients in aquafeed production.

Research has shown that high dietary insoluble fiber (>10%) can reduce dry matter digestibility and nutrient utilization in carnivorous fish species (NRC, 2011; Altan and Korkut, 2011). The high indigestible fiber content in hempseeds may limit their utilization by fish. Most of the fiber of hempseed is located within the seed coat. Removal of the seed coat through a dehulling process could improve the suitability of hemp for carnivorous fish species. To address this, we utilized a laboratory-scale dehuller (Hemp Seeds Sheller, Model QLG-1, Zhengzhou Longer Machinery Co. Ltd, China) to remove the seed coats from the whole hempseed grain. The dehulled product consist of the hemp embryos and is colloquially known as 'hemp hearts.' The clean and dried hemp grain was loaded into the hopper of sheller. The machine was adjusted so the exit from the hopper was approximately 1/4 of the open size to prevent any blockages and provide a slow rate of flow for the grain to ensure thorough seed coat removal. The grain samples fall slowly against the three rubber rollers which provide shearing force to separate the hull from the embryo, colloquially known as a 'hemp heart.' With the aid of an integrated blower, the hull was blown into a waste bag, while the dehulled sample was collected. To ensure efficient dehulling with minimal breakage of the embryos, the gap between rollers was meticulously adjusted and optimized to around 0.9–1.0 mm. Observations indicated that larger (1.2 mm) and smaller (0.8 mm) roller spacing resulted in unsatisfactory dehulling, leading to a poor removal of the seed coat or a clean product with excessive waste of hemp hearts, respectively (Fig. 1). Inspection of the dehulled material revealed fiber level may still be higher than desired. As a result, the dehulled material was screened to remove excess remaining hulls. This process resulted in three distinct products for testing, the whole hempseed (WHS), dehulled hempseed (DHS), and dehulled-sieved hempseed (DSHS) materials.

Table 5 shows that the dehulling process significantly reduced the fiber content in DHS by 34.7% and by 67.7% in DSHS compared to the WHS.

Figure 1 Whole hempseed and hemp heart after dehulling. Source: courtesy of Qianna Moncur.

Table 5 Chemical composition of whole hempseed and dehulled hempseed

Hemp sample (g/100g, DM)	DM	Protein	Lipid	Fiber	Ash	NFE[a]	GE (MJ/kg)[b]
Whole hemp	92.63	25.24	29.31	31.30	5.28	8.87	19.26
Dehulled	93.82	29.61	30.05	20.44	5.49	14.41	21.57
Dehulled & sieved	94.69	34.22	42.97	10.12	5.80	6.89	26.49

[a]Nitrogen-free extracts (NFE) = 100 − (crude lipid + crude ash + crude protein + crude fiber). [b]Gross energy (GE) was calculated using the value of 23.9 kJ/g, 39.8 kJ/g, and 17.6 kJ/g for protein, lipid, and carbohydrates, respectively (Schulz et al., 2005). DM, dry matter; GE, gross energy.

Relative to the WHS, the protein content in DHS and DSHS increased by 14.8% and 26.2%, respectively. Furthermore, sieving DHS reduced fiber levels and improved the crude protein content by 13.5% but maintained a similar ash contents. The lipid content was comparable in both WHS (29%) and DHS (30%) but higher in DSHS (~43%), and this contributed to the high gross energy value recorded in DSHS. The DSHS has low fiber and carbohydrate content but a high level of lipids, which may limit the amount of hemp that can be included in the feed. Therefore, removing excess oil either mechanically or with solvents would allow for more dehulled hemp ingredients to be included in aquatic animal feed formulations. The dehulling process combined with a simple sieving step resulted in a hempseed-based ingredient with significantly lower fiber and higher protein content compared to the seed cake often obtained from the traditional screw pressing of whole hempseeds. This could potentially improve the production efficiency of fish and shellfish when included in aquafeed.

7.1 Exploratory case study: Δ^9-THC accumulation in fish tissues

Regulators are concerned about the trace levels of cannabinoids on the grain. While there are concerns about the impact of cannabinoids on animal health, the primary issue is the potential of cannabinoids bioaccumulating in the animals that enter the food supply chain. Understanding how cannabinoids may enter the food supply is imperative for the successful use of hemp as a feed ingredient. Hempseeds have a Δ^9-THC content of up to 12 mg/kg, and it was reported that the seeds' surface contained the majority of the Δ^9-THC, which was found to reduce significantly after solvent treatment. Upon further processing, less than 0.5 mg Δ^9-THC/kg was detected in the hearts after the seed coat was removed (Ross et al., 2000). Although little is known about how cannabinoids affect fish, recent findings with zebrafish (*Danio rerio*), a model fish, provided insight into the uptake and metabolism of cannabinoids by aquatic organisms. The study found that in zebrafish larvae exposed to different concentrations of Δ^9-THC and CBD, Δ^9-THC levels peaked within the first hour, while CBD levels peaked between the 60-minute and 3-hour exposure periods. For both compounds, the initial rate of absorption increases with concentration.

Table 6 Projected milligrams of cannabinoids in market size (400 g/14 oz) fish fed a 35% hemp diet (20% hempseed meal and 15% hempseed oil), assuming different rates of incorporation into the fillet

	100% cannabinoid incorporation				1% cannabinoid incorporation[a]			
Feed Conversion Ratio	100 mg/kg[a]	10 mg/kg[b]	4 mg/kg[c]	0.5 mg/kg[d]	100 mg/kg[a]	10 mg/kg[b]	4 mg/kg[c]	0.5 mg/kg[d]
2.00:1	28.0	2.80	1.12	0.14	0.28	0.028	0.0112	0.0014
1.75:1	24.5	2.45	0.98	0.1232	0.245	0.0245	0.0098	0.00126
1.50:1	21.0	2.10	0.84	0.105	0.21	0.021	0.0084	0.00112
1.25:1	17.5	1.75	0.70	0.0882	0.175	0.0175	0.007	0.00084
1.00:1	14.0	1.40	0.56	0.07	0.14	0.014	0.0056	0.0007

[a]Losses may be related to excretion, degradation, and/or localization in non-consumed tissues.
[b]Level of Δ^9-THC found in some commercial hempseed products (Yang et al., 2017. Cannabis Cannabinoid Res. 2:1 274–281).
[c]Level of Δ^9-THC tolerated for dehulled hempseed and hempseed oil products from Canada.
[c]Benchmark level of Δ^9-THC for company which obtained FDA GRAS status for shelled hempseeds.
[d]Level of Δ^9-THC found on hemp embryos (Ross et al., 2000. J Anal Toxicol 24:8 7).

However, following the larval maximum concentration peak, there was a subsequent decline through the remainder of the five-hour experiment, which is suggestive of compound metabolism and/or excretion (Achenbach et al., 2018). Furthermore, it was observed that the increase in Δ^9-THC and CBD levels in the larvae was followed by detection of their corresponding metabolites, including hydroxylated THC, CBD-glucuronide, 11-nor-THC-9-carboxylic acid, hydroxy-CBD, and THC-glucuronide. The levels of these metabolites in the media also increased with time, and their accumulation in the media correlated with the decreasing concentrations in the larvae, suggesting that the metabolites were actively excreted from the larvae. Given the minimal studies on feeding hemp to fish or shellfish, studies on ingredient utilization and zebrafish were used to conduct theoretical calculations on cannabinoid accumulation in tissues using different FCR and cannabinoid levels. We hypothesize a worst-case scenario where hemp heart is contaminated during processing and the level of Δ^9-THC increased to 100 mg Δ^9-THC/kg, which is 25 times higher than the maximum threshold limits for commercial hulled hempseed products (4 mg/kg) in the United States. With an FCR ratio of 2:1, the level of Δ^9-THC likely to be accumulated in the fish body is 28 mg/kg. However, based on reported findings, the level of Δ^9-THC in hemp hearts is less than 0.5 mg/kg, which means the tissue accumulation would be around 0.14 mg/kg for an FCR of 2:1 and 0.07 mg/kg for an FCR of 1:1 (Table 6). Notably, this does not include metabolism or excretion of cannabinoids, such as observed with zebrafish, or losses from partitioning to non-consumed tissues. Thus, the actual levels of cannabinoids are expected to be even lower. The estimates suggest that using hemp grain could produce fish with safe levels of cannabinoids. As the majority of the Δ^9-THC in industrial hemp is located in the seed coat, removing the hulls would help improve the nutritional value of hempseed meal and further reduce the Δ^9-THC level. However, this is only a prediction and must be confirmed through an in vivo feeding trial.

8 Conclusion and future trends in research

Industrial hemp has long been cultivated for a variety of purposes and is now being investigated as an alternative feed ingredient owing to its good nutritional profile. Hempseeds have easily digestible protein rich in sulfur-containing AAs, arginine, aspartic acid, and glutamic acid, and the oil is naturally enriched in PUFAs, especially α-linolenic acid and γ-linolenic acid, both associated with various health benefits. These attributes have made hemp grains promising feed ingredients. Although hemp has good attributes, it is not immune from challenges. It contains some notable anti-nutrient components and is high in fiber. For efficient use as an aquafeed ingredient, dehulling the seed could enhance the quality of the product and reduce the carbohydrate to a level appropriate for fish and shellfish. Moreover, the anti-nutrient components, especially PA, should be reduced or eliminated to a level that fish can tolerate by using existing practices. Future research efforts should be directed towards the issue of potential Δ^9-THC accumulation in fish or shellfish fillets and other notable organs to explicitly address the safety concerns. Emphasis should also be directed to how the extrusion process affects the feed's total Δ^9-THC content in order to ascertain the extent of Δ^9-THCA decarboxylation at high extrusion temperatures. Additionally, more attention should be paid to its digestibility, effects on aquatic animals' growth and health, physio-metabolic assessment, and inherent factors' impact on gut histomorphology and gut microbiota to enhance our understanding of the suitability of hempseed cake/meal for aquaculture.

9 Where to look for further information

The following articles provide a good overview of the subject:

- Mohamed, N. and House, J.D., 2024. Safety and efficacy of hemp-derived products in animal feeds–a narrative review. *Canadian Journal of Animal Science*, (ja).
- Trovato, E., et al., 2023. Hemp seed-based food products as functional foods: a comprehensive characterization of secondary metabolites using liquid and gas chromatography methods. *Journal of Food Composition and Analysis*, 117, p.105151.
- EFSA Panel on Additives and Products or Substances used in Animal Feed (FEEDAP), 2011. Scientific Opinion on the safety of hemp (Cannabis genus) for use as animal feed. *EFSA Journal*, 9(3), p.2011.
- National Research Council, 2011. *Nutrient requirements of fish and shrimp*. National Academies Press.
- Singh, P., et al., 2023. Classification, benefits, and applications of various anti-nutritional factors present in edible crops. *Journal of Agriculture and Food Research*, 14, p.100902.
- Samtiya, M., et al., 2020. Plant food anti-nutritional factors and their reduction strategies: an overview. *Food Production, Processing and Nutrition*, 2, pp.1–14.
- Madden, S.M., et al., 2022. A systems thinking approach investigating the estimated environmental and economic benefits and limitations of industrial hemp cultivation in Ireland from 2017–2021. *Sustainability*, 14(7), p.4159.
- Hassan, F.U., et al., 2023. Potential of dietary hemp and cannabinoids to modulate immune response to enhance health and performance in animals: opportunities and challenges. *Frontiers in Immunology*, 14, p.1285052.

10 References

Abdel-Warith, A.A., et al., 2019. Effect of replacing fish meal by full fat soybean meal on growth performance, feed utilization and gastrointestinal enzymes in diets for African catfish Clarias gariepinus. *Brazilian Journal of Biology*, 80, pp.535–543.

Achenbach, J.C., et al., 2018. Analysis of the uptake, metabolism, and behavioral effects of cannabinoids on zebrafish larvae. *Zebrafish*, 15(4), pp.349–360.

Adams, I.B. and Martin, B.R., 1996. Cannabis: pharmacology and toxicology in animals and humans. *Addiction*, 91(11), pp.1585–1614.

Adeoye, A.A., et al., 2020. Preliminary assessment of black soldier fly (Hermetia illucens) larval meal in the diet of African catfish (Clarias gariepinus): Impact on growth, body index, and hematological parameters. *Journal of the World Aquaculture Society*, 51(4), pp.1024–1033.

Afridi, A.J., et al., 2024. Therapeutic role of hemp (Cannabis sativa) against copper-induced toxicity in Labeo rohita and Cirrhinus mrigala. *Biological Trace Element Research*, 202(1), pp.307–318.

Ajani, E.K., et al., 2016. *Research article total replacement of fishmeal by soybean meal with or without methionine fortification in the diets of Nile Tilapia, Oreochromis niloticus*. Oreochromis Niloticus.

Alonso-Esteban, J.I., et al., 2022. Chemical composition and biological activities of whole and dehulled hemp (Cannabis sativa L.) seeds. *Food Chemistry*, 374, p.131754.

Altan, Ö. and Korkut, A.Y., 2011. Appearent digestibility of plant protein based diets by European sea bass Dicentrarchus labrax L., 1758. *Turkish Journal of Fisheries and Aquatic Sciences*, 11(1), pp.292–297. https://doi.org/10.12714/egejfas.40.4.08

Badillo-Zapata, D., et al., 2021. Total or partial replacement of fishmeal with soybean meal in the diet of the Pacific fat sleeper Dormitator latifrons juveniles. *Latin American Journal of Aquatic Research*, 49(1), pp.40–47.

Banskota, A.H., et al., 2022. Biochemical characterization and in vitro digestibility of protein isolates from hemp (Cannabis sativa L.) by-products for salmonid feed applications. *Molecules*, 27(15), p.4794.

Barakat, H., et al., 2015. Stability of saponins from chickpea, soy and faba beans in vegetarian, broccoli-based bars subjected to different cooking techniques. *Food Research International*, 76, pp.142–149.

Barbara, F., et al., 2020. The seed of industrial hemp (Cannabis sativa L.): nutritional quality and potential functionality for human health and nutrition. *Nutrients*, 12(7), p.1935. https://doi.org/10.3390/nu12071935

Bell, J.G., et al., 1995. Effects of dietary borage oil [enriched in γ-linolenic acid, 18: 3 (n-6)] or marine fish oil [enriched in eicosapentaenoic acid, 20: 5 (n-3)] on growth, mortalities, liver histopathology and lipid composition of juvenile turbot (Scophthalmus maximus). *Fish Physiology and Biochemistry*, 14, pp.373–383.

Berto, B.M., et al., 2020. Linseed oil: characterization and study of its oxidative degradation. *Grasas y Aceites*, 71(1), pp.e337–e337.

Bonini, S.A., et al., 2018. Cannabis sativa: a comprehensive ethnopharmacological review of a medicinal plant with a long history. *Journal of Ethnopharmacology*, 227, pp.300–315.

Burton, R.A., et al., 2022. Industrial hemp seed: from the field to value-added food ingredients. *Journal of Cannabis Research*, 4(1), p.45.

Callaway, J.C., 2004. Hempseed as a nutritional resource: an overview. *Euphytica*, 140, pp.65–72.

Chang, C.S., et al., 2010. Gamma-linolenic acid inhibits inflammatory responses by regulating nf-κB and AP-1 activation in lipopolysaccharide-induced RAW 264.7 macrophages. *Inflammation*, 33, pp.46–57.

Cheng, Z., et al., 2012. Dietary supplementation of arginine and/or glutamine influences growth performance, immune responses and intestinal morphology of hybrid striped bass (Morone chrysops× Morone saxatilis). *Aquaculture*, 362, pp.39–43.

Codabaccus, B.M., et al., 2011. An extended feeding history with a stearidonic acid enriched diet from parr to smolt increases n-3 long-chain polyunsaturated fatty acids biosynthesis in white muscle and liver of Atlantic salmon (Salmo salar L.). *Aquaculture*, 322, pp.65–73.

Dawson, M.R., et al., 2018. Evaluation of poultry by-product meal as an alternative to fish meal in the diet of juvenile Black Sea Bass reared in a recirculating aquaculture system. *North American Journal of Aquaculture*, 80(1), pp.74–87.

Denstadli, V., et al., 2006. Feed intake, growth, feed conversion, digestibility, enzyme activities and intestinal structure in Atlantic salmon (Salmo salar L.) fed graded levels of phytic acid. *Aquaculture*, 256(1–4), pp.365–376.

Dossou, S., et al., 2018. Growth performance, blood health, antioxidant status and immune response in red sea bream (Pagrus major) fed Aspergillus oryzae fermented rapeseed meal (RM-Koji). *Fish & Shellfish Immunology*, 75, pp.253–262.

Eriksson, M. and Wall, H., 2012. Hemp seed cake in organic broiler diets. *Animal Feed Science and Technology*, 171(2–4), pp.205–213.

FAO, 2024. The state of world fisheries and aquaculture-blue transformation in action. *The State of World Fisheries and Aquaculture*, 2024, pp.R1–232.

Fan, Y.Y. and Chapkin, R.S., 1998. Importance of dietary γ-linolenic acid in human health and nutrition. *The Journal of Nutrition*, 128(9), pp.1411–1414.

Fawole, F.J., et al., 2017. Haemato-biochemical, non-specific immunity, antioxidant capacity and histopathological changes in Labeo rohita fingerlings fed rubber protein isolate. *Fish Physiology and Biochemistry*, 43(3), pp.677–690.

Fawole, F.J., et al., 2020. Substituting fishmeal with Hermetia illucens in the diets of African catfish (Clarias gariepinus): effects on growth, nutrient utilization, haemato-physiological response, and oxidative stress biomarker. *Aquaculture*, 518, p.734849.

Fawole, F.J., et al., 2021. Insect (black soldier fly larvae) oil as a potential substitute for fish or soy oil in the fish meal-based diet of juvenile rainbow trout (Oncorhynchus mykiss). *Animal Nutrition*, 7(4), pp.1360–1370.

Fawole, F.J., et al., 2023. Housefly maggot meal complement soybean meal in a fish-free diet for hybrid catfish (Clarias gariepinus♀ x Heterobranchus longifilis♂): effect on growth, body composition, blood biochemistry and antioxidant enzyme activity. *Animal Feed Science and Technology*, 295, p.115543.

Francis, G., et al., 2001. Antinutritional factors present in plant-derived alternate fish feed ingredients and their effects in fish. *Aquaculture*, 199(3–4), pp.197–227.

Fuentes-Quesada, J.P., et al., 2018. Enteritis induction by soybean meal in Totoaba macdonaldi diets: effects on growth performance, digestive capacity, immune response and distal intestine integrity. *Aquaculture*, 495, pp.78–89.

Gakhar, N., et al., 2012. Effect of feeding hemp seed and hemp seed oil on laying hen performance and egg yolk fatty acid content: evidence of their safety and efficacy for laying hen diets. *Poultry Science*, 91(3), pp.701–711.

Galasso, I., et al., 2016. Variability in seed traits in a collection of Cannabis sativa L. genotypes. *Frontiers in Plant Science*, 7, p.688.

Gaudioso, G., et al., 2021. Processed animal proteins from insect and poultry by-products in a fish meal-free diet for rainbow trout: impact on intestinal microbiota and inflammatory markers. *International Journal of Molecular Sciences*, 22(11), p.5454.

Gill, A.R., et al., 2022. Physiological and morphological responses of industrial hemp (Cannabis sativa L.) to water deficit. *Industrial Crops and Products*, 187, p.115331.

Gimeno-Martínez, D., et al., 2023, October. Characterisation of the fat profile of different varieties of hemp seeds (Cannabis sativa L.) for food use. In *Biology and Life Sciences Forum* (Vol. 26, No. 1, p. 89). MDPI.

Głodowska, M. and Łyszcz, M., 2017. 11. *Cannabis sativa L. and its antimicrobial properties–a review*. Research and Development of Young Scientists in Poland - Agronomy and Plant Protection.

Gu, M., et al., 2016. Soybean meal induces enteritis in turbot Scophthalmus maximus at high supplementation levels. *Aquaculture*, 464, pp.286–295.

Gülck, T. and Møller, B.L., 2020. Phytocannabinoids: origins and biosynthesis. *Trends in Plant Science*, 25(10), pp.985–1004.

Gurung, R., et al., 2022. Carcass traits of growing meat goats fed different levels of hempseed meal. *Animals*, 12(15), p.1986

Halle, I. and Schöne, F., 2013. Influence of rapeseed cake, linseed cake and hemp seed cake on laying performance of hens and fatty acid composition of egg yolk. *Journal für Verbraucherschutz und Lebensmittelsicherheit*, 8, pp.185–193.

Hardy, R.W., et al., 2018. Partial replacement of soybean meal with Methylobacterium extorquens single-cell protein in feeds for rainbow trout (Oncorhynchus mykiss Walbaum). *Aquaculture Research*, 49(6), pp.2218–2224.

Hessle, A., et al., 2008. Cold-pressed hempseed cake as a protein feed for growing cattle. *Acta Agriculturae Scand Section A*, 58(3), pp.136–145.

Hofmanová, T., et al., 2014. Nutritional properties of non-traditional seeds. *Journal of Life Medicine*, 2(1), 10–14.

House, J.D., et al., 2010. Evaluating the quality of protein from hemp seed (Cannabis sativa L.) products through the use of the protein digestibility-corrected amino acid score method. *Journal of Agricultural and Food Chemistry*, 58(22), pp.11801–11807.

Ibrahim, R.E., et al., 2022. Effect of fish meal substitution with dried bovine hemoglobin on the growth, blood hematology, antioxidant activity and related genes expression, and tissue histoarchitecture of Nile tilapia (Oreochromis niloticus). *Aquaculture Reports*, 26, p.101276.

Ignatowska-Jankowska, B., et al., 2011. Cannabidiol decreases body weight gain in rats: involvement of CB2 receptors. *Neuroscience Letters*, 490(1), pp.82–84.

Ismagilov, R.R. and Rusakov, I.A., 2024. Hemp (cannabis sativa l.) Is an effective carbon-depositing crop. In *BIO Web of Conferences* (Vol. 108, p. 10001). EDP Sciences. pp. 1–6. https://doi.org/10.1051/bioconf/202410810001

Jannathulla, R., et al., 2019. Fishmeal availability in the scenarios of climate change: Inevitability of fishmeal replacement in aquafeeds and approaches for the utilization of plant protein sources. *Aquaculture Research*, 50(12), pp.3493–3506.

Jiang, W., et al., 2024. High replacement of soybean meal by different types of rapeseed meal is detrimental to rainbow trout (Oncorhynchus mykiss) growth, antioxidant capacity, non-specific immunity and Aeromonas hydrophila tolerance. *Frontiers in Nutrition*, 11, p.1363411.

Johnson, R., 2013. *Hemp as an agricultural commodity* (pp. 1–29). Library of Congress Washington DC Congressional Research Service.

Juries, T.C., 2021. *The effects of Cannabidiol oil derived from Hemp on the growth performance and wellbeing of Oreochromis niloticus* (Master of Science dissertation, Stellenbosch: Stellenbosch University).

Kaiser, F., et al., 2021. Effects of supplemented anti-nutritive substances from rapeseed on growth and health performance of rainbow trout (Oncorhynchus mykiss). *Aquaculture*, 536, p.736422.

Kanbur, G., 2024. Effect of Heat Treatment, Water and Vinegar Soaking on Protein and Phytic Acid Levels in Hemp Seed Meal. *Turkish Journal of Agriculture-Food Science and Technology*, 12(8), pp.1442–1447.

Kari, Z.A., et al., 2023. Recent advances, challenges, opportunities, product development and sustainability of main agricultural wastes for the aquaculture feed industry–a review. *Annals of Animal Science*, 23(1), pp.25–38.

Kaur, G. and Kander, R., 2023. The sustainability of industrial hemp: a literature review of its economic, environmental, and social sustainability. *Sustainability*, 15(8), p.6457.

Kies, A.K., et al., 2006. Interaction between protein, phytate, and microbial phytase. In vitro studies. *Journal of Agricultural and Food Chemistry*, 54(5), pp.1753–1758.

Kmiecik, D., et al., 2019. Degradation of tocopherol molecules and its impact on the polymerization of triacylglycerols during heat treatment of oil. *Molecules*, 24(24), p.4555.

Kroeckel, S., et al., 2012. When a turbot catches a fly: Evaluation of a pre-pupae meal of the Black Soldier Fly (Hermetia illucens) as fish meal substitute–growth performance and chitin degradation in juvenile turbot (Psetta maxima). *Aquaculture*, 364, pp.345–352.

Krogdahl, Å., et al., 2003. Effects of graded levels of standard soybean meal on intestinal structure, mucosal enzyme activities, and pancreatic response in Atlantic salmon (Salmo salar L.). *Aquaculture Nutrition*, 9(6), pp.361–371.

Kumar, A., et al., 2023. Phytic acid: a reservoir of phosphorus in seeds plays a dynamic role in plant and animal metabolism. *Phytochemistry Reviews*, 22(5), pp.1281–1304.

Kumar, V., et al., 2021. Insect (black soldier fly, Hermetia illucens) meal supplementation prevents the soybean meal-induced intestinal enteritis in rainbow trout and health benefits of using insect oil. *Fish & Shellfish Immunology*, 109, pp.116–124.

Kwanyuen, P. and Burton, J.W., 2005. A simple and rapid procedure for phytate determination in soybeans and soy products. *Journal of the American Oil Chemists' Society*, 82(2), pp.81–85.

Lefort, N., et al., 2016. Consumption of Buglossoides arvensis seed oil is safe and increases tissue long-chain n-3 fatty acid content more than flax seed oil–results of a phase I randomised clinical trial. *Journal of Nutritional Science*, 5, p.e2.

Li, X., et al., 2021. *Nutrition and functions of amino acids in fish. Amino acids in nutrition and health: amino acids in the nutrition of companion, zoo and farm animals*. Springer. pp.133–168.

Lin, Y.H. and Cheng, M.Y., 2017. Effects of dietary organic acid supplementation on the growth, nutrient digestibility and intestinal histology of the giant grouper

Epinephelus lanceolatus fed a diet with soybean meal. *Aquaculture*, 469, pp.106–111.

Liu, L., et al., 2014. Feed intake, feed utilization and feeding-related gene expression response to dietary phytic acid for juvenile grass carp (Ctenopharyngodon idellus). *Aquaculture*, 424, pp.201–206.

Lunger, A.N., et al., 2007. The effects of organic protein supplementation upon growth, feed conversion and texture quality parameters of juvenile cobia (Rachycentron canadum). *Aquaculture*, 264(1–4), pp.342–352.

Lusas, E.W. and Rhee, K.C., 1995. *Soy protein processing and utilization*. Practical Handbook of Soybean Processing and Utilization.

Maly, O., et al., 2018. Use of by-products from hemp processing in the nutrition of common carp (Cyprinus carpio L.). In *Proceedings of the MendelNet Conference* (pp. 165–170).

Matthäus, B. and Musazcan Özcan, M., 2015. Oil content, fatty acid composition and distributions of vitamin-e-active compounds of some fruit seed oils. *Antioxidants*, 4, 124–133.

Mattila, P.H., et al., 2018. Contents of phytochemicals and antinutritional factors in commercial protein-rich plant products. *Food Quality and Safety*, 2(4), pp.213–219.

Mierliță, D., 2018. Effects of diets containing hemp seeds or hemp cake on fatty acid composition and oxidative stability of sheep milk. *South African Journal of Animal Science*, 48(3), pp.504–515.

Montero, L., et al., 2023. Hemp seeds: nutritional value, associated bioactivities and the potential food applications in the Colombian context. *Frontiers in Nutrition*, 9, p.1039180.

Montford, S. and Small, E., 1999. A comparison of the biodiversity friendliness of crops with special reference to hemp (Cannabis sativa L.). *Journal of Industrial Hemp*, 6, 53–63.

Mounes, H.A., et al., 2024. Effect of substituting soybean meal in oreochromis niloticus Diets with Pumpkin (Cucurbita maxima) seed cake on water quality, growth, antioxidant capacity, immunity, and carcass composition. *Animals*, 14(2), p.195.

Mustafa, A.F., et al., 1999. The nutritive value of hemp meal for ruminants. *Canadian Journal of Animal Science*, 79(1), pp.91–95.

Nagappan, S., et al., 2021. Potential of microalgae as a sustainable feed ingredient for aquaculture. *Journal of Biotechnology*, 341, pp.1–20.

National Research Council, 2011. *Nutrient requirements of fish and shrimp*. National Academies Press.

Nionelli, L., et al., 2018. Pro-technological and functional characterization of lactic acid bacteria to be used as starters for hemp (Cannabis sativa L.) sourdough fermentation and wheat bread fortification. *International Journal of Food Microbiology*, 279, pp.14–25.

Norajit, K., et al., 2011. Effects of the addition of hemp powder on the physicochemical properties and energy bar qualities of extruded rice. *Food Chemistry*, 129(4), pp.1919–1925.

Olsen, R.E., et al., 2007. Total replacement of fish meal with plant proteins in diets for Atlantic cod (Gadus morhua L.) II–Health aspects. *Aquaculture*, 272(1–4), pp.612–624.

Oomah, B.D., et al., 1996. Phytic acid content of flaxseed as influenced by cultivar, growing season, and location. *Journal of Agricultural and Food Chemistry*, 44(9), pp.2663–2666.

Panaite, T., et al., 2017. Flaxseeds: nutritional potential and bioactive compounds. *Bulletin UASVM Food Science and Technology*, 74(2), pp.65–73.

Pang, A., et al., 2023. Differential analysis of fish meal substitution with two soybean meals on juvenile pearl gentian grouper. *Frontiers in Marine Science*, 10, p.1170033.

Parker, N.B., et al., 2022. Assessment of spent hemp biomass as a potential ingredient in ruminant diet: nutritional quality and effect on performance, meat and carcass quality, and hematological parameters in finishing lambs. *Journal of Animal Science*, 100(10), pp. 263. Website: https://academic.oup.com/jas/article-abstract/100/10/skac263/6661414.

Parker, T.D., et al., 2003. Fatty acid composition and oxidative stability of cold-pressed edible seed oils. *Journal of Food Science*, 68(4), pp.1240–1243.

Phipps, B. and Schluttenhofer, C., 2022. Perspectives of industrial hemp cultivation. In *Industrial Hemp* (pp. 1–36). Academic Press.

Pohlenz, C., et al., 2014. Effects of dietary arginine supplementation on growth, protein optimization and innate immune response of channel catfish Ictalurus punctatus (Rafinesque 1818). *Aquaculture Research*, 45(3), pp.491–500.

Ross, S.A., et al., 2000. GC-MS analysis of the total δ9-thc content of both drug-and fiber-type cannabis seeds. *Journal of Analytical Toxicology*, 24(8), pp.715–717.

Russo, R. and Reggiani, R., 2013. Variability in antinutritional compounds in hempseed meal of Italian and French varieties. *Plant*, 1(2), pp.25–29.

Sagan, A., et al., 2019. Assessment of the properties of rapeseed oil enriched with oils characterized by high content of α-linolenic acid. *Sustainability*, 11(20), p.5638.

Sample, A., 2022. Evaluation of Hemp seed meal as a fish meal replacement through growth and digestibility trials in striped bass (Morone saxatilis). https://digitalcommons.georgiasouthern.edu/etd/2405?utm_source=digitalcommons.georgiasouthern.edu%2Fetd%2F2405&utm_medium=PDF&utm_campaign=PDFCoverPages.

Saoud, I.P., et al., 2018. Effect of cannabis oil on growth performance, haematology and metabolism of Nile Tilapia Oreochromis niloticus. *Aquaculture Research*, 49(2), pp.809–815.

Sarker, P.K., et al., 2013. Sustainability issues related to feeding salmonids: a C anadian perspective. *Reviews in Aquaculture*, 5(4), pp.199–219.

Schulz, C., et al., 2005. Effects of varying dietary fatty acid profile on growth performance, fatty acid, body and tissue composition of juvenile pike perch (Sander lucioperca). *Aquaculture Nutrition*, 11(6), pp.403–413.

Semwogerere, F., et al., 2020. Bioavailability and bioefficacy of hemp by-products in ruminant meat production and preservation: a review. *Frontiers in Veterinary Science*, 7, p.572906.

Sorrentino, G., 2021. Introduction to emerging industrial applications of cannabis (Cannabis sativa L.). Rendiconti Lincei. *Scienze fisiche e naturali*, 32(2), pp.233–243.

Šťastník, O., et al., 2019. The effect of hempseed expellers on selected quality indicators of broiler chicken's meat. *Acta Veterinaria Brno*, 88(1), pp.121–128.

Sun, J., et al., 2024. Lignanamides: a comprehensive review of chemical constituents, biological activities, extraction methods and synthetic pathway. *Food Chemistry*, 460, p.140459.

Tacon, A.G., 2005. State of information on salmon aquaculture feed and the environment. Report to the WWF Salmon Aquaculture Dialogue, 80.

Tocher, D.R., 2003. Metabolism and functions of lipids and fatty acids in teleost fish. *Reviews in Fisheries Science*, 11(2), pp.107–184.

Vasilaki, A., et al., 2023. Fishmeal, plant protein, and fish oil substitution with single-cell ingredients in organic feeds for European sea bass (Dicentrarchus labrax). *Frontiers in Physiology*, 14, p.1199497.

Venold, F.F., et al., 2012. Severity of soybean meal induced distal intestinal inflammation, enterocyte proliferation rate, and fatty acid binding protein (Fabp2) level differ between strains of rainbow trout (Oncorhynchus mykiss). *Aquaculture*, 364, pp.281–292.

Vodolazska, D. and Lauridsen, C., 2020. Effects of dietary hemp seed oil to sows on fatty acid profiles, nutritional and immune status of piglets. *Journal of Animal Science and Biotechnology*, 11, pp.1–18.

Vonapartis, E., et al., 2015. Seed composition of ten industrial hemp cultivars approved for production in Canada. *Journal of Food Composition and Analysis*, 39, pp.8–12.

Wagner, B., et al., 2022. Transfer of cannabinoids into the milk of dairy cows fed with industrial hemp could lead to Δ9-THC exposure that exceeds acute reference dose. *Nature Food*, 3(11), pp.921–932.

Wang, J., et al., 2020. Effects of replacing menhaden fishmeal with cottonseed flour on growth performance, feed utilization and body composition of juvenile red drum Sciaenops ocellatus. *Aquaculture*, 523, p.735217.

Wang, S., et al., 2017. Effect of unconventional oilseeds (safflower, poppy, hemp, camelina) on in vitro ruminal methane production and fermentation. *Journal of the Science of Food and Agriculture*, 97(11), pp.3864–3870.

Wang, S., et al., 2019. Cannabisin F from hemp (Cannabis sativa) seed suppresses lipopolysaccharide-induced inflammatory responses in BV2 microglia as SIRT1 modulator. *International Journal of Molecular Sciences*, 20(3), p.507.

Wang, X.S., et al., 2008. Characterization, amino acid composition and in vitro digestibility of hemp (Cannabis sativa L.) proteins. *Food Chemistry*, 107(1), pp.11–18.

Webster, C.D., et al., 2000. Use of hempseed meal, poultry by-product meal, and canola meal in practical diets without fish meal for sunshine bass (Morone chrysops× M. saxatilis). *Aquaculture*, 188(3–4), pp.299–309.

Winders, T.M., et al., 2022. Influence of hempseed cake inclusion on growth performance, carcass characteristics, feeding behavior, and blood parameters in finishing heifers. *Journal of Animal Science*, 100(6), p.skac159.

Xin, Y., et al., 2022. Components of the endocannabinoid system and effects of cannabinoids against bone diseases: a mini-review. *Frontiers in Pharmacology*, 12, p.793750.

Xu, L., et al., 2022. Flaxseed meal and its application in animal husbandry: a review. *Agriculture*, 12(12), p.2027.

Xu, Y., et al., 2021. Hempseed as a nutritious and healthy human food or animal feed source: a review. *International Journal of Food Science & Technology*, 56(2), pp.530–543.

Xuan, X., et al., 2019. Effects of different levels of macroalga Gracilaria lemaneiformis on growth performance and feed utilization on the red sea bream, Pagrosomus major. *Journal of Applied Phycology*, 31, pp.3213–3222.

Yan, X., et al., 2015. Characterization of lignanamides from hemp (Cannabis sativa L.) seed and their antioxidant and acetylcholinesterase inhibitory activities. *Journal of Agricultural and Food Chemistry*, 63(49), pp.10611–10619.

Yang, Y., et al., 2017. Cannabis sativa (Hemp) Seeds, A 9–tetrahydrocannabinol, and potential overdose. *Cannabis and Cannabinoid Research*, 2(1), 274–281. [online]

Zając, M., et al., 2019. The quality of pork loaves with the addition of hemp seeds, de-hulled hemp seeds, hemp protein and hemp flour. *Lwt*, 105, pp.190–199.

Zarantoniello, M., et al., 2019. A six-months study on Black Soldier Fly (Hermetia illucens) based diets in zebrafish. *Scientific Reports*, 9(1), p.8598.

https://cannabusinessplans.com/global-hemp-market/#:~:text=In%202023%2C%20hemp%20acreage%20in,acreage%20at%20about%20520%2C000%20acres.

https://www.globenewswire.com/news-release/2024/06/17/2899987/0/en/Global-Industrial-Hemp-Market-Size-To-Worth-USD-24-30-Billion-By-2033-CAGR-Of-16-27.html.

https://www.selinawamucii.com/insights/prices/united-states-of-america/soya-beans/ assessed on 10/30/2024.

https://www.hempbenchmarks.com/hemp-market-insider/2023-wholesale-hemp-product-price-trends/#:~:text=Contract%20prices%20for%20conventionally%2Dgrown,%241.10%20to%20%241.30%20per%20pound.

Chapter 8

The environmental sustainability of hemp: common metrics and a consideration of cultivation geography

Christopher Dillis and Van Butsic, University of California-Berkeley, USA

1 Introduction

Hemp has been cultivated for centuries (Clarke, 1999), yet its more recent history has witnessed widespread regulatory hurdles resulting from its conflation with tetrahydrocannabinol (THC) cannabis. In the US, in particular, hemp was effectively banned as an agricultural crop for the majority of the twentieth century (Fike, 2016). Given that hemp and THC cannabis are the same plant species (*Cannabis sativa* L.), lawmakers were decidedly hesitant to allow the advancement of an agricultural sector involving hemp (Small and Marcus, 2002). For instance, even as late as 2018, growers in the US were required to obtain a permit from the Drug Enforcement Administration (DEA) in order to cultivate the plant (Dingha et al., 2019).

Significant legislative changes in recent years have revitalized the hemp industry. The 2014 Farm Bill marked the beginning of this resurgence by allowing test programs for hemp cultivation in various US states (Malone and Gomez, 2019). By 2019, 41 states had passed legislation permitting hemp cultivation, and the 2018 Farm Bill further declassified hemp from the list of controlled substances, legalizing its production as an agricultural commodity. This legislative shift has been driven in part by an increased demand for

http://dx.doi.org/10.19103/AS.2025.0161.14

non-food crops and other derivatives within the agricultural sector, leading to a progressive recovery of hemp's importance (García-Tejero et al., 2019).

One significant lingering effect of hemp's regulatory setbacks, however, is a lack of research and development, especially relative to other agricultural crops (Blandinières and Amaducci, 2022). Furthermore, the nature of hemp itself is unique in terms of the suite of end products the plant can provide. Indeed, hemp is cultivated to supply a variety of industries including agriculture, textile, automotive, construction, biofuel, cosmetics, and pharmaceuticals (Salentijn et al., 2015). The variety of end products is accompanied by a variety of cultivars and growth habits, meaning to research hemp cultivation is to research several distinct crops. Unsurprisingly, grower surveys indicate a need for research (Ellison, 2021), especially given that hemp cultivation is lacking the general research foundation afforded to almost all traditional agricultural crops that have benefited from university extension programs and federally funded research. Beyond cultivation, hemp's resurgence also suggests research on its environmental sustainability is also an immediate need, as there remain critical research gaps in this area (Aryal et al., 2023). Research on the cultivation of hemp may be limited but given the recency of hemp's resurgence, research on its environmental sustainability is also an immediate need, as there remain critical research gaps in this area (Aryal et al., 2023).

Hemp has potential for rapid expansion in cultivated area and cultivation intensity, so its environmental impact is an especially important piece of its research profile. Even conservative estimates project the global industrial hemp market to more than double between 2019 and 2025 (Burton et al., 2022). Hemp is grown for commercial or research purposes in at least 47 countries worldwide and production is legal in 46 US states (Adesina et al., 2020). The US now stands as the third-largest producer of hemp and has seen a rapid increase in both hemp acreage and the number of production licenses issued in recent years. However, the US hemp industry has already experienced its first boom-bust cycle, fueled by optimism for phytochemical derivatives (i.e. CBD) in 2019 and followed by a market collapse in 2020 (Dingha et al., 2019). The rapid expansion leading to the oversupply issues of 2019 and 2020 led to a drop off in hemp acreage (Owram, 2020a), yet the industry did eventually stabilize and resume growth (Owram, 2020b).

The capacity for hemp as an agricultural industry to expand quickly into new markets and geographies underscores that its growth must be accompanied by an understanding of its potential environmental impacts and benefits. To date, there has been a reasonable body of literature surrounding common environmental sustainability metrics, such as water use, nitrogen input, pesticide requirements, and environmental benefits; however, given the novelty of hemp and various other factors such as its potential to be grown on marginal lands, an additional sustainability metric should be considered: its cultivation geography.

This chapter will provide a brief review of the literature surrounding the aforementioned environmental sustainability metrics. This will be accompanied by a case study exploring its potential for geographic sustainability. That is, we ask whether hemp is more likely than other crops to be grown in areas with heightened capacity for environmental impacts. For instance, hemp has been given attention as an ideal crop for marginal lands (Scalabrin et al., 2024), including in mountainous regions (Giupponi et al., 2020; Desanlis et al., 2013). Additionally, recent research has demonstrated that cultural and political factors have shaped the geography of legal cultivation of hemp's conspecific, THC cannabis, in that it has been excluded from traditional agricultural zones and instead ushered into environmentally sensitive areas (Dillis et al., 2024). This case study will therefore explore whether hemp has been susceptible to the same geographic forces and whether this represents a broader concern for the environmental sustainability of the hemp industry.

1.1 Definitions

- Cultivation Geography: The spatial extent and density of farming, with particular attention to the relative proximity of natural landscape features.
- THC Cannabis: Cultivars of (*Cannabis sativa* L.) whose concentration of Delta-9-tetrahydrocannabinol (THC) exceeds the legal threshold of 0.3%.
- Environmental Benefits: Positive effects for natural resources resulting either directly or indirectly from cultivation practices or crop characteristics.
- Generalized Additive Models: A class of flexible statistical models that combine linear relationships with non-linear functions to capture complex patterns in data.

2 Environmental sustainability metrics

2.1 Water use

The most common water demand comparison for hemp (or any textile) has been with cotton, as the world's most prominent fiber crop (Baffes, 2005). The result of the comparison is a rather resounding consensus that hemp requires less water than cotton (Averink, 2015; Adesina et al., 2020; Blandinières and Amaducci, 2022; Gill et al., 2023; Wise et al., 2023), by as much as 70% (Schultz et al., 2020). In fact, hemp is considered to be more sustainable than cotton not only in terms of water use but also its general ecological footprint throughout production (Schumacher et al., 2020). The water efficiency of hemp (i.e. water demand for optimal yield) surpasses many of the most common agricultural crops, including corn (Amaducci et al., 2000) and most fiber crops, in particular (Gill et al., 2023). As such, hemp is especially suited to arid climates (Gill et al., 2023) given its physiology and morphology.

Hemp has received attention as an especially water stress-tolerant crop for both fiber (Bourdot et al., 2017; Gill et al., 2023) and phytochemical (CBD and CBG; García-Tejero et al., 2020) varieties. Hemp is also known to demonstrate increases in water-use efficiency under drought conditions (Cosentino et al., 2013), likely due at least in part to its ability to develop a deep root system (Amaducci et al., 2008). As a result, hemp does well as long as plants are propagated from seed and sufficiently irrigated while they are still establishing, as they are able to access lower soil strata with reliable water thereafter (Herppich et al., 2020). Because the plants are able to quickly establish deep root systems, hemp is capable of producing economically viable yields with relatively minimal irrigation (Di Bari et al., 2004).

Hemp is nonetheless not entirely immune to water stress for all varieties or under all conditions. For instance, seed production is negatively affected by both water stress (Baldini et al., 2020) and high temperatures (Ferfuia et al., 2021). Water deficits can also lead to changes in morphological characteristics, chlorophyll content, dry matter, and oil yield, as well as increases in carotenoids and antioxidant enzymes (Bahador and Tadayon, 2020). Blandinières and Amaducci (2022) also report reductions of 20–25% in general aboveground biomass between well-irrigated and water-stressed crops. However, some varieties of hemp are more responsive to irrigation and demonstrate changes in water-use efficiency under various growing conditions (Campbell et al., 2019) and choosing location-appropriate varieties can result in low irrigation demand (Satriani et al., 2021). Therefore, further applied research on hemp cultivars, end products, and their respective water demands is especially important for matching crops to their ideal water-use environments and capitalizing on the capacity for hemp to be particularly water efficient.

2.2 Nitrogen input

Hemp's nutrient requirements are relatively modest compared to many other crops, as hemp requires relatively little nutrient input throughout its life cycle (Donaldson, 2022), with relatively low nitrogen requirements specifically (Finnan and Burke, 2013; Prade et al., 2011). For instance, hemp has been demonstrated to require less nitrogen than many crops, including linseed, maize, sorghum, oilseed rape, and most of all, cotton (58% less; Tang et al., 2017). However, research on the effects of fertility management on hemp, particularly regarding phytochemical concentrations, remains limited (Wylie et al., 2021).

Nitrogen requirements for hemp can vary based on the intended end product (Aubin et al., 2015). As with most crops, soil health plays a crucial role in determining nitrogen input needs. In nutrient-rich soils, hemp exhibits negligible yield increases with additional nitrogen, while showing considerable responses in nutrient-poor soils (Amaducci et al., 2015). This adaptability makes hemp a viable option for cultivation on marginal soils with poor nutrient quality, yet current guidelines for nitrogen input are not as well-researched or standardized as those for other crops, highlighting a need for further study.

2.3 Pesticides

Hemp is often touted for its minimal pesticide requirements (Van der Werf et al., 1995); however, this narrative has historically been limited to specific hemp varieties (Struik et al., 2000). Namely, hemp planted for fiber production is particularly resistant to weeds, as its dense planting configuration causes the crop to outcompete and suppress them (Kaiser et al., 2015). For this reason, there has been a tradition of growing hemp without the need for herbicides in regions, such as Europe, where densely planted fiber hemp has been most common (Flessner et al., 2020). However, weed control is commonly employed in hemp cultivation worldwide, and research on the effects of various herbicides on hemp is still catching up (Flessner et al., 2020). There is evidence that hemp may be particularly sensitive to herbicides commonly used for other crops such as corn and soybean (Ortmeier-Clarke et al., 2022), yet progress on suitable herbicides continues to be made (Gitsopoulos et al., 2024).

Currently, France, the world's largest producer of hemp, has authorized 13 pesticide products, including those containing Tefluthrin, although these are reported not to be widely used (Marquardt and Savage, 2023). In the US, the Environmental Protection Agency (EPA) has approved nearly 100 pesticide products, all but one of which are biological. Despite this guidance, pesticide regulation has been somewhat fraught. In particular, the 2014 Farm Bill's lack of guidance on pesticide use led to a chaotic regulatory environment, with states making individual decisions regarding pesticide regulations (Cranshaw et al., 2019). Furthermore, the dynamic nature of hemp introduces further complications as pesticide needs vary based on the intended end product (Ely et al., 2022; Cranshaw et al., 2019).

2.4 Environmental benefits

Hemp's environmental benefits are well-documented, including phytoremediation of contaminated soils, converting high amounts of atmospheric CO_2 to biomass through bio-sequestration, and its suitability for bioenergy production (Adesina et al., 2020). Additionally, hemp is regarded as an ideal crop for organic agriculture due to its high root biomass, especially in deep soil layers, underscoring its potential in sustainable cropping systems (Amaducci et al., 2008).

Hemp's capacity for carbon sequestration, in particular, has been known for decades (Pervaiz and Sain, 2003). The carbon uptake of hemp significantly outweighs its carbon footprint, making it a valuable crop for CO_2 mitigation (Scrucca et al., 2020). The use of hemp as a powerful tool against climate change continues to be an active area of research (Madden et al., 2022). Hemp's negative carbon footprint combined with its high cellulose content has also established it as an effective candidate for bioethanol production (Zhao et al., 2020). The crop is also a much more efficient bioenergy feedstock than most annual energy crops (Finnan and Styles, 2013) and can be integrated into food crop rotations to promote sustainable bioenergy supply chains with minimal transport requirements.

Remediation of contaminated soils is another well-known environmental service of hemp (Adesina et al., 2020). When cultivated for single-use fiber, in particular, hemp can be grown on soils high in heavy metals, which do not appear to have a significant impact on yield (Blandinières and Amaducci, 2022). The use of hemp for phytoremediation has been shown to be effective for lead (Xu et al., 2021), as well as cadmium, zinc, and arsenic (Pietrini et al., 2019; Canu et al., 2022).

The environmental benefits of hemp are indeed broad, well documented, and continuing to be explored. For instance, researchers have also highlighted the potential for hemp to ease land use pressures due to its capacity to produce more fiber per hectare than most common fiber crops (Alden et al., 1998), and note that this opportunity remains mostly unexamined (Aryal et al., 2023). In total, the benefits of hemp, coupled with its relatively lower impact metrics, make this a particularly environmentally sustainable crop. Given its novelty and capacity for expansion, however, one must also consider its geography – a topic which has received almost no attention thus far.

3 Cultivation geography: case study

3.1 Background

The geography of a crop can influence its potential impact on the environment. Perhaps the most salient example of this is THC cannabis in California, USA. Licensed THC cannabis in the state has been demonstrated to be grown in remote areas, even more so than other common 'rural' crops (Dillis et al., 2022; 2024). These areas have a heightened sensitivity to impacts, as they are home to a large amount of biodiversity and protected freshwater aquatic species (Carah et al., 2015).

There are several metrics that can be used to assess the potential impact of agriculture on a landscape, including the average slope of the terrain, the proximity to or presence of natural water courses, and the necessity of sourcing irrigation water from natural surface flows. Agricultural activity on steep slopes requires grading and clearing not only for the cultivation site but often for infrastructure needed for access (Carah et al., 2015). As such, the potential for erosion, including the potential for runoff to degrade water quality of nearby watercourses, rises as average slope increases. Proximity to natural watercourses is therefore an additional metric of impact potential, not only with regard to sediment input but any other agricultural runoff (Bauer et al., 2015). Finally, sourcing of irrigation water, particularly in regions with extended dry periods, is a key factor in determining potential impact to streamflow quantity (Dillis et al., 2023). When farms do not have access to reservoir-fed irrigation networks or large groundwater aquifers, they must source irrigation water from natural surface flows. While onsite water storage can be used to mitigate irrigation needs in dry periods (Dillis et al., 2020), the process of obtaining and using storage is often fraught with logistical, regulatory, and financial hurdles that introduce complications and incentivize drawing water directly from surface streams (Dillis et al., 2023).

The current geography of licensed THC cannabis in California is strongly influenced by social and political factors. When the state of California created a legal cannabis industry in 2018, individual counties and municipalities were afforded the ability to decide whether or not to allow cultivation (Padon et al., 2022). As a result, much of the traditional agricultural regions of the state decided to maintain prohibitions on licensed cannabis cultivation and much of the production has remained relegated to marginal lands where the potential for environmental impacts has been greater. The social and political factors at play range from stigmas surrounding cannabis to the novelty of the crop as a new competitor for land, water, and labor (Dillis et al., 2024). While hemp has the benefit of federal legalization, it is not necessarily immune to many of the socio-political forces shaping the spatial distribution of THC cannabis. Therefore, this case study will compare the geography of hemp to both that of THC cannabis and all other 'traditional' agricultural crops in California. It will use three metrics of environmental impact (average slope, presence of natural watercourse, and access to reservoir/aquifer irrigation sources) to make this comparison and ultimately assess whether hemp may be prone to forces that affect its geographic sustainability.

3.2 Methods and results

Spatial hemp and THC cannabis (referred to hereafter as *cannabis*) datasets were obtained via public records requests to the California Department of Agriculture (on 3/12/2024) and the California Department of Cannabis Control (on 12/23/2023), respectively. Spatial data for all other crop types (referred to hereafter as *general crops*) were downloaded from the California Department of Water Resources 2018 i15 Crop Mapping Layer (CDWR, 2018). The *general crops* dataset was randomly sampled so that the number of observations (n = 2500) approximated that of *cannabis* (n = 2501) and *hemp* (n = 2581) farms (Fig. 1). Because spatial farm data were available only as centroids, a buffer of 500 m was applied to each in order to provide a conservative spatial sample of each farm.

The following spatial variables were calculated for each farm type, including *average slope*, presence of natural *watercourse*, and presence of *irrigation* network/aquifer. *Average slope* was calculated using Digital Elevation Model (DEM) data from the National Elevation Dataset (USGS, 2020). Presence of natural *watercourse* on each farm was determined using the National Hydrography Dataset (NHD; USGS, 2019). Any farms that intersected a perennial, ephemeral, or intermittent stream were considered as having a *watercourse*. Farms were determined to have access to *irrigation* if they either had an underlying groundwater aquifer – identified using data from the California Department of Water Resources (State of California, 2019) or were within 1 km of an irrigation canal/ditch, as identified by the NHD (USGS, 2019).

Generalized Additive Models (GAMs) for each environmental impact metric were fit using the mgcv package (Wood et al., 2016) in R Statistical Software (R Core Team, 2018). The use of GAMs for comparisons between farm types was necessary to account for spatial autocorrelation. The first GAM used a log link function to estimate a beta distribution for *average slope* (A_i) based on farm *type* (T_i), while the remaining GAMs used logit link functions to estimate binomial distributions for *watercourse* (W_i) and *irrigation* (R_i):

$log(A_i) = + {}_tT_i + {}_9 +$ (1)

$logit(W_i) = + {}_tT_i + {}_9 +$ (2)

$logit(R_i) = + {}_tT_i + {}_9 +$ (3)

Log estimates of A_i, and log-odds estimates of W_i and R_i, for each farm type are produced by adding the coefficient for farm type ($_t$) and the sum of

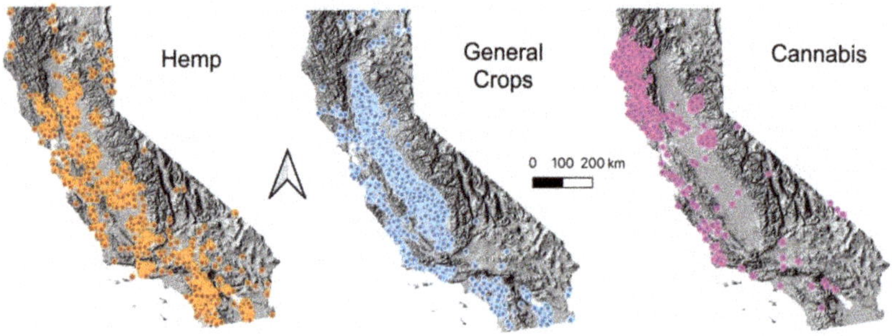

Figure 1 Study area map. The spatial distributions of hemp farms (n = 2581), randomly sampled farms of general crops (n = 2500), and cannabis farms (n = 2501) are depicted in orange, blue, and purple, respectively.

coefficients for nine basic functions ($_9$) to the intercept (), with hemp serving as the reference level. For all GAMs, coefficient estimates were considered reliable in cases where 95% confidence intervals constructed from the standard errors did not overlap zero.

Raw data and model estimates for all three environmental impact metrics suggested that the impact potential of hemp was elevated relative to *general crops*, but still significantly lower than the impact potential of *cannabis*. The average slope of *hemp* farms (Mean = 3.73%, Std Dev = 5.55) exceeded that of *general crops* (Mean = 1.98%, Std Dev = 3.99), but both were considerably lower than that of *cannabis* (Mean = 12.66%, Std Dev = 8.33). The model estimates for *hemp* (Intercept MLE = −3.19, Std Err = 0.02, Estimate = 3.93%) were also reliably greater than those of *general crops* (MLE = −0.13, Std Err = 0.03, Estimate = 3.48%), but reliably less than *cannabis* (MLE = 0.61, Std Err = 0.04, Estimate = 7.03%; Fig. 2), indicating that the results were robust to spatial autocorrelation (i.e. spatial clustering of farm types).

The percentage of *hemp* farms with an onsite *watercourse* (55.21%) was greater than that of *general crops* (40.48%), but far lower than *cannabis* farms (90.52%). These results were robust to spatial autocorrelation, in that model estimates for *hemp* (Intercept MLE = 0.69, Std Err = 0.06, Estimate = 66.60%) were reliably greater than those of *general crops* (MLE = −0.35, Std Err = 0.06, Estimate = 58.51%), but reliably less than *cannabis* (MLE = 0.89, Std Err = 0.11, Estimate = 82.88%; Fig. 3).

The percentage of *hemp* farms with access to *irrigation* (76.09%) was lower than for *general crops* (92.16%), but substantially higher than *cannabis* farms (32.15%). These results were robust to spatial autocorrelation, in that model estimates for *hemp* (Intercept MLE = 1.26, Std Err = 0.08, Estimate = 77.98%) were reliably lower than those of *general crops* (MLE = 1.45, Std Err = 0.11, Estimate = 93.81%), but reliably greater than *cannabis* (MLE = -0.84, Std Err = 0.11, Estimate = 60.37%; Fig. 4).

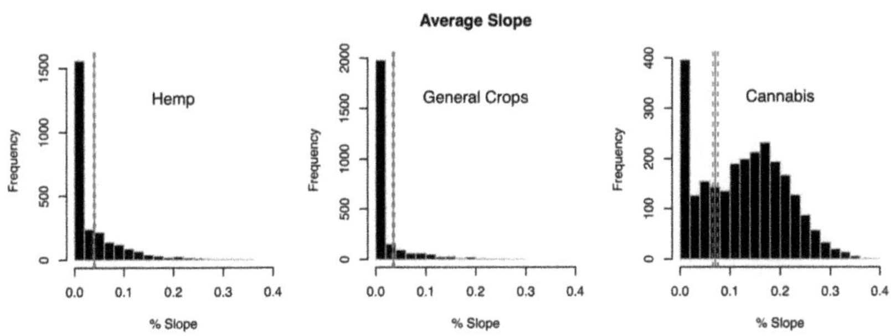

Figure 2 Average slope by farm type. Raw data are depicted in black, while Generalized Additive Model estimates are overlaid in red. Solid red lines depict maximum likelihood estimates, while dashed red lines depict the 95% confidence intervals.

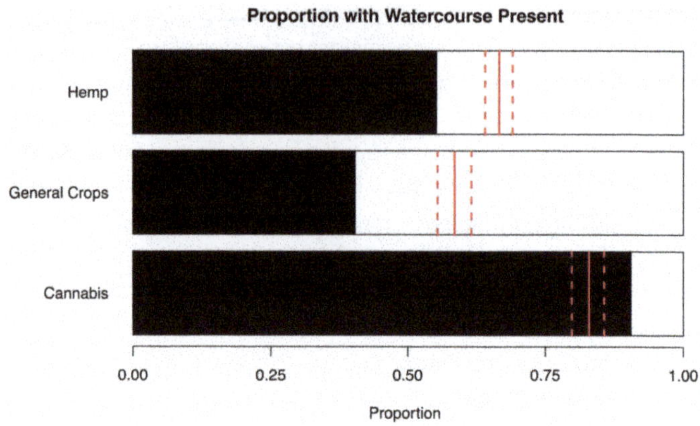

Figure 3 Watercourse presence by farm type. Raw data of proportions of watercourses per farm type are depicted in black, while Generalized Additive Model estimates are overlaid in red. Solid red lines depict maximum likelihood estimates, while dashed red lines depict the 95% confidence intervals.

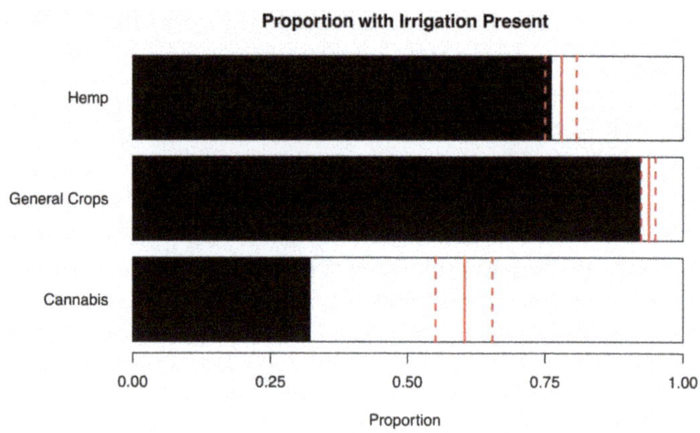

Figure 4 Irrigation presence by farm type. Raw data are depicted in black, while Generalized Additive Model estimates are overlaid in red. Solid red lines depict maximum likelihood estimates, while dashed red lines depict the 95% confidence intervals.

4 Case study discussion

This case study provides a comparative analysis of three geographic and environmental sustainability metrics for hemp, cannabis, and general crops in California. The findings offer important insights into the potential environmental impacts of hemp cultivation, particularly when juxtaposed with cannabis and other traditional agricultural crops. While model estimates indicated hemp farms are situated on terrain that is less flat than farms of general crops, the practical difference appears to be negligible and the rugged terrain more typical of cannabis far exceeded the average slope of both. Similarly, the likelihood of natural watercourses on hemp farms was somewhat higher than that on farms with general crops, but substantially lower than that on cannabis farms. Finally, in terms of access to either reservoir water or groundwater aquifers, hemp appears to be much more similar to general crops than is cannabis, which is instead much more often reliant on natural surface flow or the groundwater that feeds it. While there are statistically significant differences between hemp and general crops, hemp appears to be more aligned with the latter than with cannabis in a pragmatic examination of all three environmental impact metrics considered herein. It should be noted, however, that hemp in California is still distinct from general crops in terms of these metrics, representing an opportunity for the industry to transition toward a more sustainable geography.

The sociopolitical factors that have directed cannabis toward areas with higher environmental impact potential do not appear to have exerted the same pressures on hemp, at least in California. Still, given the growth potential of the hemp industry, its status as a newcomer in many agricultural regions, and potential stigmas against hemp as a cannabis conspecific, consideration should be given to the cultivation geography of this crop outside of California as well. Areas with a long tradition of hemp cultivation are likely to demonstrate a more extensive geographic and regulatory integration of hemp with traditional crops. However, in regions where hemp is a novel crop, potentially regarded with skepticism from traditional agriculturalists, the environmental sustainability of hemp would benefit from an assertion of its cultivation in traditional agricultural zones rather than marginal lands with heightened potential for impact.

5 Conclusion

The environmental sustainability of hemp cultivation is well supported. Although it is still catching up with other crops in terms of research and development, hemp's agronomic characterization continues to demonstrate its substantial upside for the environment. Hemp uses water efficiently, has low fertilizer and pesticide requirements, and offers a relatively large suite of environmental benefits. Its potential for carbon sequestration and suitability for bioenergy production highlight its ecological benefits and makes hemp a valuable crop for climate change mitigation. Hemp's high root biomass supports sustainable cropping systems and organic agriculture. Its capacity for non-detrimental absorption of heavy metals further makes it a powerful tool for phytoremediation of contaminated and marginal soils.

While hemp may be suitable for marginal soils, directing cultivation toward marginal lands may come with consequences for environmental sustainability. Luckily, while hemp may still be in the process of integrating itself with traditional agriculture, its geography is far from problematic in terms of several environmental impact metrics considered herein. However, policymakers, farmers, and environmental stakeholders should remain attentive to where hemp cultivation is occurring and resilient to any exclusion from traditional agricultural zones. In order for hemp to fulfill its potential for environmental sustainability, it should be guaranteed access to cultivation in low-impact environments.

6 Where to look of further information

- USDA Hemp Program (https://www.ams.usda.gov/rules-regulations/hemp).
- National Hemp Association (https://nationalhempassociation.org/).

7 References

Adesina, I., et al., 2020. A review on the current state of knowledge of growing conditions, agronomic soil health practices and utilities of hemp in the United States. *Agriculture, 10*(4), p.129.

Alden, D.M., et al., 1998. Industrial hemp's double dividend: a study for the USA. *Ecological Economics, 25*(3), pp.291–301.

Amaducci, S., et al., 2000. Crop yield and quality parameters of four annual fibre crops (hemp, kenaf, maize and sorghum) in the North of Italy. *Industrial Crops and Products, 11*(2–3), pp.179–186.

Amaducci, S., et al., 2008. Characterisation of hemp (Cannabis sativa L.) roots under different growing conditions. *Plant and Soil, 313*, pp.227–235.

Amaducci, S., et al., 2015. Key cultivation techniques for hemp in Europe and China. *Industrial Crops and Products, 68*, pp.2–16.

Aryal, K., et al., 2023. Knowledge mapping for a secure and sustainable hemp industry: a systematic literature review. *Case Studies in Chemical and Environmental Engineering, 8*, p.100550.

Aubin, M.P., et al., 2015. Industrial hemp response to nitrogen, phosphorus, and potassium fertilization. *Crop, Forage & Turfgrass Management, 1*(1), pp.1–10.

Averink, J., 2015. *Global water footprint of industrial hemp textile* (Master's thesis). Enschede: University of Twente.

Baffes, J., 2005. Cotton: Market setting, trade policies, and issues. Aksoy, M.A. and Beghin, J.C. eds., 2004. *Global agricultural trade and developing countries*. World Bank Publications. Washington, DC.

Bahador, M. and Tadayon, M.R., 2020. Investigating of zeolite role in modifying the effect of drought stress in hemp: antioxidant enzymes and oil content. *Industrial Crops and Products, 144*, p.112042.

Baldini, M., et al., 2020. Suitability assessment of different hemp (Cannabis sativa L.) varieties to the cultivation environment. *Industrial Crops and Products, 143*, p.111860.

Bauer, S., et al., 2015. Impacts of surface water diversions for marijuana cultivation on aquatic habitat in four northwestern California watersheds. *PLoS One*, 10(3), p.e0120016.

Blandinières, H. and Amaducci, S., 2022. Adapting the cultivation of industrial hemp (Cannabis sativa L.) to marginal lands: a review. *GCB Bioenergy*, *14*(9), pp.1004–1022.

Bourdot, A., et al., 2017. Characterization of a hemp-based agro-material: influence of starch ratio and hemp shive size on physical, mechanical, and hygrothermal properties. *Energy and Buildings*, *153*, pp.501–512.

Burton, R.A., et al., 2022. Industrial hemp seed: from the field to value-added food ingredients. *Journal of Cannabis Research*, *4*(1), p.45.

Campbell, B.J., et al., 2019. Genotypex environment interactions of industrial hemp cultivars highlight diverse responses to environmental factors. *Agrosystems, Geosciences & Environment*, *2*(1), pp.1–11.

Canu, M., et al., 2022. Hemp cultivation in soils polluted by Cd, Pb and Zn in the Mediterranean area: sites characterization and phytoremediation in real scale settlement. *Applied Sciences*, *12*(7), p.3548.

Carah, J.K., et al., 2015. High time for conservation: adding the environment to the debate on marijuana liberalization. *BioScience*, *65*(8), pp.822–829.

CDWR, 2018. *California division of water resources, division of regional assistance regional offices: Northern, North Central, South Central and Southern Regional offices, and water use efficiency branch*. i15 Crop Mapping Layer.

Clarke, R.C., 1999. *Botany of the genus Cannabis* (pp. 1–19). Binghamton, NY: Haworth Press.

Cosentino, S.L., et al., 2013. Evaluation of European developed fibre hemp genotypes (Cannabis sativa L.) in semi-arid Mediterranean environment. *Industrial Crops and Products*, *50*, pp.312–324.

Cranshaw, W., et al., 2019. Developing insect pest management systems for hemp in the United States: a work in progress. *Journal of Integrated Pest Management*, *10*(1), p.26.

Desanlis, F., et al., 2013. Hemp agronomics and cultivation. In S. Allegret, et al. (Eds.), *Hemp: industrial production and uses* (pp. 98–124). Wallingford, UK: Cabi.

Di Bari, V., et al., 2004. Potential productivity of fibre hemp in southern Europe. *Euphytica*, *140*, pp.25–32.

Dillis, C., et al., 2020. Water storage and irrigation practices for cannabis drive seasonal patterns of water extraction and use in Northern California. *Journal of Environmental Management*, *272*, p.110955.

Dillis, C., et al., 2022. The threat of wildfire is unique to cannabis among agricultural sectors in California. *Ecosphere*, *13*(9), p.e4205.

Dillis, C., et al., 2023. Water demands of permitted and unpermitted cannabis cultivation in Northern California. *Environmental Research Communications*, *5*(2), p.025005.

Dillis, C., et al., 2024. A theory of geo-social marginalization: a case study of the licensed cannabis industry in California. *Journal of Environmental Management*, *355*, p.120396.

Dingha, B., et al., 2019. Industrial hemp knowledge and interest among North Carolina organic farmers in the United States. *Sustainability*, *11*(9), p.2691.

Donaldson, L., 2022. From illegal to ideal: The necessity of hemp production. *Natural Sciences Education*, *51*(2), p.e20090.

Ellison, S., 2021. Hemp (Cannabis sativa L.) research priorities: opinions from United States hemp stakeholders. *GCB Bioenergy*, *13*, pp.562–569.

Ely, K., et al., 2022. Cannabis/hemp: Sustainable uses, opportunities, and current limitations. In *Cannabis/hemp for sustainable agriculture and materials* (pp. 59–87). Springer. Singapore.

Ferfuia, C., et al., 2021. Performance and stability of different monoecious hemp cultivars in a multi-environments trial in north-eastern Italy. *Agronomy*, *11*(7), p.1424.

Fike, J., 2016. Industrial hemp: renewed opportunities for an ancient crop. *Critical Reviews in Plant Sciences*, *35*(5–6), pp.406–424.

Finnan, J. and Burke, B., 2013. Nitrogen fertilization to optimize the greenhouse gas balance of hemp crops grown for biomass. *GCB Bioenergy*, *5*(6), pp.701–712.

Finnan, J. and Styles, D., 2013. Hemp: a more sustainable annual energy crop for climate and energy policy. *Energy Policy*, *58*, pp.152–162.

Flessner, M.L., et al., 2020. Evaluating herbicide tolerance of industrial hemp (Cannabis sativa L.). *Crop Science*, *60*(1), pp.419–427.

García-Tejero, I.F., et al., 2019. Seeking suitable agronomical practices for industrial hemp (Cannabis sativa L.) cultivation for biomedical applications. *Industrial Crops and Products*, *139*, p.111524.

García-Tejero, I.F., et al., 2020. Yield of new hemp varieties for medical purposes under semi-arid Mediterranean environment conditions. *Comunicata Scientiae*, *11*, pp.e3264–e3264.

Gill, A.R., et al., 2023. The potential of industrial hemp (Cannabis sativa L.) as an emerging drought resistant fibre crop. *Plant and Soil*, *493*(1), pp.7–16.

Gitsopoulos, T., et al., 2024. Response of industrial Hemp (Cannabis sativa L.) to herbicides and weed control. *International Journal of Plant Biology*, *15*(2), pp.281–292.

Giupponi, L., et al., 2020. Overview on Italian hemp production chain, related productive and commercial activities and legislative framework. *Italian Journal of Agronomy*, *15*(3), pp.194–205.

Herppich, W.B., et al., 2020. Effects of drought and heat on photosynthetic performance, water use and yield of two selected fiber hemp cultivars at a poor-soil site in Brandenburg (Germany). *Agronomy*, *10*(9), p.1361.

Kaiser, C., et al., 2015. Industrial hemp production. *Center for Crop Diversification - University of Kentucky*, *27*, pp.101–106.

Madden, S.M., et al., 2022. A systems thinking approach investigating the estimated environmental and economic benefits and limitations of industrial hemp cultivation in Ireland from 2017–2021. *Sustainability*, *14*(7), p.4159.

Malone, T. and Gomez, K., 2019. Hemp in the United States: a case study of regulatory path dependence. *Applied Economic Perspectives and Policy*, *41*(2), pp.199–214.

Marquardt, S. and Savage, L., 2023. *Growing Hemp for the future: a global fiber guide*. Textile Exchange.

Ortmeier-Clarke, H.J., et al., 2022. Dose–response screening of industrial hemp to herbicides commonly used in corn and soybean. *Weed Technology*, *36*(2), pp.245–252.

Owram, K., 2020a. Hemp prices plunge as CBD demand falls short: Cannabis weekly. *Bloomberg*, January, 27.

Owram, K., 2020b. CBD supply seen shrinking as farmers flee hemp: Cannabis weekly. *Bloomberg*, August, 10.

Padon, A.A., et al., 2022. Local laws regulating cannabis in California two years post legalization: assessing incorporation of lessons from tobacco control. *Cannabis*, *5*(3), p.47.

Pervaiz, M. and Sain, M.M., 2003. Carbon storage potential in natural fiber composites. *Resources, conservation and Recycling*, *39*(4), pp.325–340.

Pietrini, F., et al., 2019. Morpho-physiological and metal accumulation responses of hemp plants (Cannabis Sativa L.) grown on soil from an agro-industrial contaminated area. *Water*, *11*(4), p.808.

Prade, T., et al., 2011. Biomass and energy yield of industrial hemp grown for biogas and solid fuel. *Biomass and Bioenergy*, *35*(7), pp.3040–3049.

R Core Team, 2018. *R: a language and environment for statistical computing*. Vienna, Austria: R Foundation for Statistical Computing.

Salentijn, E.M., et al., 2015. New developments in fiber hemp (Cannabis sativa L.) breeding. *Industrial Crops and Products*, *68*, pp.32–41.

Satriani, A., et al., 2021. The cultivation of industrial hemp as alternative crop in a less-favoured agricultural area in southern Italy: the Pignola case study. *Pollutants*, *1*(3), pp.169–180.

Scalabrin, E., et al., 2024. Hemp cultivation opportunities for marginal lands development. *PLoS One*, *19*(3), p.e0299981.

Schultz, C.J., et al., 2020. Consumer and health-related traits of seed from selected commercial and breeding lines of industrial hemp, Cannabis sativa L. *Journal of Agriculture and Food Research*, *2*, p.100025.

Schumacher, A.G.D., et al., 2020. Industrial hemp fiber: a sustainable and economical alternative to cotton. *Journal of Cleaner Production*, *268*, p.122180.

Scrucca, F., et al., 2020. Energy and carbon footprint assessment of production of hemp hurds for application in buildings. *Environmental Impact Assessment Review*, *84*, p.106417.

Small, E. and Marcus, D., 2002. Hemp: a new crop with new uses for North America. *Trends in New Crops and New Uses*, *24*(5), pp.284–326.

State of California, 2019. California State Geoportal. 2019 SGMA Basin Prioritization Map.

Struik, P.C., et al., 2000. Agronomy of fibre hemp (Cannabis sativa L.) in Europe. *Industrial Crops and Products*, *11*(2–3), pp.107–118.

Tang, K., et al., 2017. A comprehensive study of planting density and nitrogen fertilization effect on dual-purpose hemp (Cannabis sativa L.) cultivation. *Industrial Crops and Products*, *107*, pp.427–438.

USGS, 2019. U.S. Geological Survey, National Hydrography Dataset (NHD).

USGS, 2020. U.S. Geological Survey, National Elevation Dataset (NED).

Van der Werf, H.M., et al., 1995. Agronomic research on hemp (Cannabis sativa L.) in the Netherlands, 1987–1993. *Journal of the International Hemp Association*, *2*(1), pp.14–17.

Wise, K., et al., 2023. Comparative study of water requirements and water footprints of fibre crops hemp (Cannabis sativa) and cotton (Gossypium hirsutum L.). *Journal of Agrometeorology*, *25*(3), pp.392–396.

Wood, S.N., et al., 2016. Smoothing parameter and model selection for general smooth models. *Journal of the American Statistical Association*, *111*(516), pp.1548–1563.

Wylie, S.E., et al., 2021. Fertility management for industrial hemp production: current knowledge and future research needs. *GCB Bioenergy*, *13*(4), pp.517–524.

Xu, Y., et al., 2021. Accumulation and subcellular distribution of lead (Pb) in industrial hemp grown in Pb contaminated soil. *Industrial Crops and Products*, *161*, p.113220.

Zhao, J., et al., 2020. Bioconversion of industrial hemp biomass for bioethanol production: a review. *Fuel*, *281*, p.118725.

Part 3

Hemp diseases, pests and weeds

Chapter 9

Integrated pest management in hemp cultivation

Marguerite Bolt, Purdue University, USA

1 Introduction

Hemp production is an ever-evolving landscape. As acreage increases, records of arthropods found in the crop will also increase. This will provide more opportunities to evaluate damage caused by pests, identify beneficial organisms, and investigate management strategies. Several insects and mites have been identified as pests of hemp either by directly damaging the plant (i.e. chewing buds, leaves, or stems) or indirectly by transmitting pathogens and altering cannabinoid levels (McPartland and Hillig, 2003). Economic injury levels and economic thresholds are yet to be developed for hemp. At this point, it has been a challenge both to determine the extent of crop loss caused by specific pests and to assign a monetary value to the crop. This does not mean that economic loss caused by insects and mites has not been observed in hemp. There have been scenarios where pests have caused significant damage to the crop, causing a reduction in yield (Bakro et al., 2018; Cranshaw et al., 2019; Hu et al., 2021). There are key approaches needed to understand pest damage in hemp and to improve pest management in hemp. Identifying

http://dx.doi.org/10.19103/AS.2025.0161.20

pests and associated damage is a necessary starting point that will guide pest management decisions.

- Pest: organism that causes damage or interferes with plant production. Pests can also cause damage to physical structures and impact human or animal health;
- Integrated pest management (IPM): a multifaceted approach to manage pests while minimizing negative impacts on the environment and people;
- Cultural control: management practices that disrupt the life cycle of the pest (establishment, survival, reproduction, dispersal) by making the environment less favorable;
- Mechanical/physical control: management practices that physically block the pest, knock pests off of plants, or kill pests without the use of pesticides;
- Biological control: organisms that reduce a pest population through presence, predation, or parasitism; and
- Chemical control: use of pesticides to manage a pest population.

2 Identification of pests

Proper identification of a pest by inspecting and monitoring plants is the first step in integrated pest management. The presence, diversity, and pest damage will differ based on the geographic location and the growing environment. In some locations, data on the diversity of arthropods in hemp is available, but in other areas, hemp production has been minimal, and such surveys have not been conducted. One of the most comprehensive resources, *Hemp Diseases and Pests* by McPartland et al. (2005), describes 150 different insects and mites associated with hemp grown in North America, Europe, and Asia. There are more recent observational pest resources available for regions within the United States (Cranshaw et al., 2019; Munir et al., 2023; Ahmed et al., 2024). For example, agricultural outreach and education programs have published observational data on their websites in the form of guides and fact sheets. These resources provide the foundation for identifying common pests and are necessary for developing management plans. Observational data continues to be important, and growers should keep records of what they notice.

Many different arthropod species are observed on hemp plants every year. They may be feeding but causing minimal damage. For example, the saltmarsh caterpillar, *Estigmene acrea* (Drury), is consistently observed on hemp in some locations, but damage is minimal (Cranshaw, 2017). Other leaf feeders like flea beetles may cause ugly-looking damage, but the significance to the plant is typically minor (Cranshaw et al., 2019). It is still important to record observations of pests. A minor pest one year could be important in subsequent years.

There are generalist pests that feed on hemp and many other plant species, and there are specialist pests that primarily feed on *C. sativa*. Photographs of pests provide a time-stamped record and can be useful for identification, especially if a physical sample cannot be collected. Physical samples are important if the pest needs to be sent to a diagnostic clinic for accurate identification. Sometimes the pest is not present or is difficult to observe, so the damage they cause is used for identification. Pests also leave behind signs of their presence, like frass (insect excrement) or shed skins. Region-specific resources are useful for pest identification, and information from agricultural outreach and education specialists, researchers, crop consultants, and farmers can provide valuable insight into what pests are typically present. This is especially important for those planting hemp for the first time. Becoming familiar with the most common pests in the hemp-growing environment will aid in identification and selection of appropriate management tools.

3 Integrated pest management

Integrated pest management (IPM) uses multiple techniques and an understanding of pest life cycles to reduce or prevent damage to plants and minimize negative impacts on the ecosystem. This is not to eradicate pests, which would be futile, but to minimize economic loss caused by pests. IPM relies on setting thresholds, monitoring pests, prevention strategies, and then responding to pests with a variety of management techniques. There is limited research on IPM in hemp, but many of the insects and mites observed in this system are also found on other crops. This provides an opportunity to apply the same or similar monitoring, prevention, and pest management strategies to hemp.

Hemp is grown for its metabolites, seeds, and stalks in a variety of environments, including warehouses, glasshouses, and high tunnels as well as outdoors. The production environment plays a critical role in which pests are going to be the most abundant. Indoor producers will have more control over the environment but that can provide ideal conditions for pest populations to increase rapidly. Outdoor producers will have less control over the growing environment, but there are many natural enemies of pests found outdoors, and conditions may not be optimal for pest populations to increase rapidly. Understanding the environment and geography of production will allow for selection of appropriate pest management tools.

The current literature on economically important pests in hemp is limited, and economic thresholds have not been developed yet. Key pests have been identified in some regions; the next step is developing economic thresholds and economic injury levels, both of which are necessary in determining what management practices should be used and when.

3.1 Economic injury and pest monitoring

Economic injury level (EIL) was first proposed by Stern et al. (1959) and is still used in crop production today. EIL is the lowest population density of a specific pest that causes economic injury. This requires understanding how much damage a pest causes to the plant and how much damage causes yield loss. Additionally, EIL must also factor in the cost of managing the pest and the market value of the crop. The life cycle of specific pests feeding on hemp and the yield loss they cause is going to require years of observation. Hemp markets are still developing in many parts of the world, and the value of the crop is variable depending on the type of hemp being grown and the end use. The cost of managing the pest below a set threshold also needs to be developed for key pests of hemp.

Stern et al. (1959) also defined a more practical approach to pest management timing, which is the economic threshold. This is the point in which the density of a specific pest triggers a management decision to prevent injury and reduce the pest population. This can be thought of as an action threshold, a time when treatment should be used to prevent the pest population from increasing to the economic injury level. The economic threshold is set below the economic injury level. There are several components of developing economic thresholds that have been the focus of research, which include understanding the pest and the damage they cause, the type of hemp being grown, and life stage of the plant when damage occurs. Data collection on monitoring pests in different types of hemp and assessing damage is an ongoing process in which researchers, educators, and growers participate.

Monitoring for pests is a critical aspect of IPM. Growers should regularly scout a subset of plants for pests throughout the growing cycle or until pests are no longer present. The method and the frequency of observation are important for identifying pest presence and density. Scouting technique should aim to cover the growing area by observing plants in a set pattern of inspection. This pattern is usually a zig-zag, or 'X' across the entire growing area. There may be parts of the field or facility that are of greater concern; problem areas should also be scouted. Pest distribution in an area can be uniform, within the field along the edges, or randomly clumped. If scouting aims to look for specific pests, understanding the distribution of the pest can aid in how to walk the field and the method of scouting. Scouting should begin early in the growth cycle and should continue throughout the life of the plant or until pests are no longer causing damage. The frequency will vary by operation, but once a week is a common frequency. Weather events, like strong southern winds, may cause more frequent scouting of pests that migrate from warmer regions. Pest monitoring using traps may trigger more frequent scouting as well. Having a set scouting pattern and frequency will aid in consistent observations and hopefully prevention of economic injury to the crop.

The number of plants observed will depend on how large the growing area is, if there are variations in the growing environment (microclimates), if there are multiple cultivars being grown, and the amount of time available to observe plants. Ground truthing is still the standard for plant observation, but the use of drones may help in the identification of problem areas in a field, allowing for more targeted scouting and management. The number of plants observed should be a representative sample for the growing area. In low-density plantings, 25 plants may be enough to get a representative sample (McKeag, 2019). In field crops that are planted more densely, which is the case for grain and fiber hemp, 50–100 plants are sufficient (Anon, 2022).

Tools including sweep nets and beat sheets are useful for visual observations of plants. Certain tools will be better suited for specific types of hemp or certain times of year. Scouting techniques will vary based on if the hemp crop is for fiber, grain, or cannabinoids. For example, it will be easier to use a sweep net in a fiber field when the plants are young, but as the plants increase in height, this approach is not as feasible. In a cannabinoid crop where plants are large and bushy, a beat sheet will be a useful technique to assess pests. In a grain crop, the use of a sweep net should be avoided once grain is maturing to prevent knocking seeds off the plants. For some pests, visual observations may require the use of a hand lens or jewelers loupe of 15–20x magnification. There are additional items that should be taken when scouting including: collection vials or bags, note taking device, camera, a device with GPS, flag tape, and a field guide for pest identification. Smartphones can act as a camera and note taking tool, are equipped with GPS, and can have downloadable guides.

The use of sticky cards and pheromone traps should be considered in hemp production systems. These are important monitoring tools that can trigger management responses when trap catches reach a specific level. Yellow sticky cards are often used indoors to monitor for fungus gnats, thrips, winged aphids, and whiteflies. These should be changed weekly so an accurate number of pests is recorded (Pundt, 2019). Pheromone lures are often used in outdoor traps. There are pheromone lures available for several of the notable pests in hemp, including corn earworm, European corn borer, and several other common lepidopteran pests. There is currently no pheromone lure developed for Eurasian hemp borer. Pheromone lures use chemicals that attract a specific species into the trap. This allows for a population estimate, with higher catches triggering management in the crop. A major drawback of this monitoring method is that counts are based only on insects that come to the trap, which can underestimate the pest density.

3.2 Cultural control

Cultural control practices aim to reduce pests by disrupting their life cycle through modification of the growing environment. Cultural control begins before planting, with the selection of appropriate genetics. Cultivars need to be suited for the growing environment. Cultivars that are not adapted to the growing environment are not going to perform as well compared to cultivars bred for that region. Selection of genetics that have resistance to specific pests is another strategy. One example of trait selection for pest management is alfalfa that is resistant to potato leafhopper. In resistant alfalfa cultivars, glandular trichomes on the plant create a physical barrier to the nymphs and adults and prevent egg laying on the stems (McCaslin and Whalen, 2017). There is an opportunity to investigate traits for pest resistance in hemp. This will be further discussed in Section 5.

Site selection is an important decision for both overall crop health and also to reduce infestation of certain pests. Pests in weedy field edges and neighboring crops can move into the hemp field when the other host plants are removed or harvested. One example of this is common stalk borer, *Papaipema nebris* (Guenée), which can move into crops when weedy host plants die, typically due to the use of herbicides in the spring (Rice and Davis, 2010). Common stalk borer has been observed in hemp in some Midwestern US states. While neighboring crops and weeds should be considered when selecting a hemp field, growers may have limited options available.

Crop rotation can disrupt the life cycles of pests by using non-host crops, eliminating the food source. This method of pest management works best on pests that are specific to one crop, have a longer life cycle, and do not have a widespread distribution. Another technique, intercropping, is the use of other plant species within the field for the purpose of increasing environmental diversity and manipulating insect behavior. Hemp has been intercropped with cowpeas and squash to improve pollinator abundance for squash yields, but the hemp yields were also improved due to intercropping (Dingha et al., 2024). While this study does not focus on pests, it does show that hemp yields can be improved by intercropping. When hemp was intercropped with marigold, basil, cilantro, dill, and sage, many pests were collected, but there was very little damage to the hemp plants (Connelly, 2023). However, the most abundant pests observed in this study were plant bugs, leafhoppers, and leaf-miner flies, which are not typically associated with extensive damage to hemp in the Midwest region of the United States where the study was conducted. Intercropping can also increase predators and parasitoids of pests by providing shelter and supplemental and alternative food sources (Gurr et al., 2017; Gontijo, 2019).

Phenological asynchrony in pest management is when a pest is not present at the same time as when the crop is vulnerable to attack (Messina and

Jones, 1990). This is achieved by altering planting and harvesting dates of the crop. Phenological asynchrony is used in sweet corn to reduce corn earworm damage, wheat to reduce Hessian fly injury to young plants, and in alfalfa to reduce alfalfa weevil populations. Phenological asynchrony is less understood with hemp but could be a useful tool for certain pests that attack at specific times of year. Selecting faster maturing genetics or planting earlier could reduce damage caused by pests like corn earworm, which are often observed on hemp after sweet corn is done silking in the Midwest United States.

Sanitation should not be overlooked as an important strategy for pest management in hemp. Plants entering a facility or field need to be inspected for pests. This allows for the treatment or isolation of the plants before being introduced to clean spaces. In indoor production, there should be a separate quarantine area for incoming plants. Heavily infested plants may need to be destroyed if treatment options are not available. For pests that remain in hemp residue after harvest, like Eurasian hemp borer, destroying stems and branches can reduce the overwintering population. Chopping or chipping would kill the larvae and can be an important sanitation method for those that struggle with Eurasian hemp borer.

Maintaining healthy plants is an important pest management strategy. Over-fertilization of nitrogen can increase susceptibility to pests (Altieri and Nicholls, 2003; Bala et al., 2018). Overwatering, a common issue for indoor producers, can increase populations of fungus gnats, which need a warm and moist environment.

3.3 Mechanical control

Mechanical control options include cultivation of soil, hand-picking off pests, spraying pests off plants with water, and physical barriers. Each method comes with its own benefits and downsides, and feasibility will vary based on the type of hemp produced, the scale of operation, and the growing environment.

Cultivating fields where pests overwinter may be necessary in certain locations. Pests or pest eggs that overwinter in the crop residue or in the upper soil layers can be destroyed through cultivation. In fields where hemp was planted the previous year, cultivation can kill early season volunteers. Removal of hemp volunteers is important for cannabis aphid management. These aphids need a spring food source before hemp is planted, and these volunteers serve that purpose. However, soil cultivation can negatively impact beneficial organisms and alter soil structure.

Removing pests by hand is a time-consuming method and is more common in small-scale production. This is a useful method of pest control for caterpillars in cannabinoid hemp fields. Plants are inspected; caterpillars are removed and destroyed. This is a challenge for large operations that have many plants to individually scout. It is also a challenge to detect early instars of caterpillars because of their small size.

Similar to hand-picking, spraying off plants with water can be a useful technique to remove pests. Adequate water pressure is needed to knock pests off without damaging the plant itself. This technique is better suited for pests like aphids, mites, and even small caterpillars which are more easily knocked off the plants and struggle to get back on. The feasibility of this at a large scale makes this a technique suited for smaller operations.

Physical control methods include barriers that prevent pests from reaching the plants. This can include netting, sealing up cracks in indoor facilities, and putting sticky traps up to catch pests. In hemp, the type of barrier installed will depend on the growing environment. High tunnel producers can install fine mesh barriers along the open sides of the structure to prevent pest movement. Floating row covers may be useful if there are early season pests in cannabinoid production. Smaller operations may be able to use exclusion netting for pests like corn earworm, Eurasian hemp borer and other insects that attack the bud tissue. This method would be labor intense since each plant would require a barrier. Additional types of barriers that have been suggested for hemp include copper screens or foil for snails and diatomaceous earth for leafhoppers (Zuefle and Grab, 2023).

3.4 Biological control

Biological control techniques involve the use of beneficial organisms to reduce pests. These organisms exist in nature and can be leveraged in outdoor and indoor hemp production. There are many predatory and parasitic arthropods and nematodes that will attack pests of hemp. There are also pathogens that can attack pests, including fungi, bacteria, and viruses. There are three general strategies for using biological control: augmentative, classical, and conservation.

Augmentative biological control is used to inundate the growing environment with a predator or parasitoid. This requires releasing large numbers of beneficial organisms into the environment to provide pest suppression. It is important to know which pests are present so the appropriate agent can be released. This is achieved through scouting and correct identification of the pests causing damage. If there are high pest populations before natural enemies are released, an application of an insecticide or miticide may be necessary to knock down the population. Once natural enemies are released, product selection becomes very important because broad-spectrum pesticides can harm beneficial insects and mites. There are many commercially available predators and parasitoids from insectaries. Many insectaries are familiar with hemp production and are well versed on which predators and parasitoids are going to work best.

Classical biological control requires the introduction of a host-specific natural enemy from the same place as the exotic pest. Exotic pests typically have few predators or parasitoids in their new environment, making management with biological control agents a challenge. A suitable control agent is identified and released, where it establishes in the environment. This method requires investigating the species that attack the pest in their native region and evaluating if those natural enemies can be released in the new environment for control with minimal impact on native species. One of the most well-known examples is the release of the vedalia beetle for the management of cottony cushion scale in California. There are no examples of classical biological control being used in hemp.

Conservation biological control practices help to maintain predators and parasitoids in the environment. It is a more ecological approach to pest management and may be done in conjunction with augmentative biological control. The goal is to alter or conserve the growing environment in a way that increases the presence of biological control agents. This is done by providing continuous food, habitat that provides shelter during the growing season and over the winter, and the right microclimate (Rabb et al., 1976). One way to achieve this is to plant a diverse mix of vegetation near or within the field or to plant cover crops after the primary crop is harvested. The effort and cost of conservation can reduce insecticide inputs if there is a healthy population of natural enemies.

Hemp plants provide refuge and in grain and fiber production, their pollen is a food source for insects. Habitat and additional food supply are important for natural enemies; it is no surprise that there is a diversity of predators and parasitoids found in hemp fields. A review by Cranshaw et al. (2019) noted lady beetles, lacewings, syrphids, chloropids, spiders, and other generalist predators such as minute pirate bugs, damsel bugs, and big-eyed bugs were common in hemp fields. Conservation biological control practices should be used to preserve the diversity of natural enemies in outdoor grown hemp.

In indoor systems, beneficial insects and mites are introduced to manage common greenhouse pests. These are often released regularly to keep pest populations low. Commercially available beneficial control agents used in *C. sativa* production include predatory mites, midges, green lacewing, rove beetles, minute pirate bug, nematodes, and parasitic wasps. Selection of the appropriate biological control agent will depend on what pests are present.

Parasitoids of hemp pests include wasps and flies. These organisms use insects as a source of food for their offspring. They lay eggs on or inside the host, where the larvae feed. Some species of parasitoids are commercially available and can be released into the growing environment. One example of a hemp pest that has several parasitoids is corn earworm. Corn earworm has both wasp and fly parasitoids in the United States (Reay-Jones, 2019). In

Kentucky-grown hemp, 45% of observed corn earworms were parasitized by two different species of tachinid flies (*Winthemia rufopicta* and *Lespesia aletiae*) but caused only 45.9% mortality (Falcon-Brindis et al., 2022). This reduction in control could be due to the caterpillars molting, shedding their skin and the fly eggs with it.

Pests in hemp can be infected by entomopathogenic fungi, viruses, and bacteria (Cranshaw et al., 2019). These microbes occur in nature, and several of these agents can be purchased as commercial products. In indoor cannabis production, *Beauveria bassiana*, *Isaria fumosorosea*, *Lecanicillium lecanii*, and *Metarhizium anisopliae* are all entomopathogenic fungi used to control different pests (Lemay and Scott-Dupree, 2022). There are also viruses belonging to the family Baculoviridae that are highly host specific, but have not gained widespread adoption (Rohrmann, 2019). These products are sprayed on plants, where they infect the target pest, causing cessation of feeding, resulting in an insect that appears discolored and flaccid. Nuclear Polyhedrosis Virus (NPV) for corn earworm control in hemp is commercially available. The bacterium *Bacillus thuringiensis* (Bt) is used in hemp production to control caterpillar pests and flies. Bt subspecies *kurstaki* (Btk) and Bt subspecies *aizawa* (Bta) work on certain species of butterfly and moth larvae where Bt subspecies *israelensis* (Bti) works on certain species of fly larvae. It is important to read the product label to determine which active ingredient is going to be useful for the management of specific pests.

4 Chemical control

Chemical control is typically viewed as the last resort approach in integrated pest management. Pesticides can be powerful tools that are necessary for pest management in many crops but can also present risks. These risks include harming beneficial organisms, environmental contamination, chemical exposure to the applicator, damage to the plant, and increased cost of production. However, there are products available that have reduced environmental risks and are specific to certain pests.

In the United States, pesticides registered under section 25(b) of the Federal Insecticide, Fungicide Rodenticide Act (FIFRA) are an option for hemp producers. These products contain ingredients deemed as minimum risk. This exempts the product from an EPA label, and efficacy data is not required. A list of the active and inert ingredients allowed in these minimum risk products can be found at the following webpage: https://www.epa.gov/minimum-risk-pesticides/active-ingredients-allowed-minimum-risk-pesticide-products.

Hemp acreage is minor compared to most crops. There has not been a push from large chemical companies to register conventional pesticides for use in hemp. However, in the United States, a program exists that helps support registering conventional pesticides for use on specialty crops. The Interregional Research Project No. 4 (IR-4) has been conducting research with land grant universities to obtain pesticide product registration for minor crops. There are ongoing insecticide trials through the IR-4 project, but it is unclear when these products will be registered for use on hemp in the United States.

The list of available pesticides for use in hemp is limited compared to other crops. Most pesticides registered in the United States are biological-based products. These products are typically more effective when pest pressure is low. Some products are more effective on earlier instars of insects. One major advantage of these products is their specificity on certain pests. Many of these pesticides are also registered for use in certified organic systems and have a short re-entry interval and a short pre-harvest interval. Many of these pesticides have less residual activity, requiring more frequent applications. Combining these biological-based insecticides with multiple modes of action can be an effective management tool. Britt (2021) found that combining two pesticides, one containing a virus and the other containing an entomopathogenic fungus, provided similar control of corn earworm larvae compared to chlorantraniliprole. The use of multiple biological-based insecticides could be a useful tool for hemp producers because they lack conventional insecticide options.

5 Genetics and breeding efforts

Breeding resistant cultivars is an important tool for managing pests. Several different mechanisms can be targeted in traditional breeding practices. Plants can be bred to have tolerance to pests, where damage can be sustained, but there is no reduction in yield. Plants can be bred to have resistance to pests, where a trait or traits influence pest interactions with the crop. Breeding for resistant crops includes antixenosis traits and antibiosis traits. Antixenosis is when pests do not prefer a plant where antibiosis is when a pest's biology is altered by a plant.

Antixenosis has been observed in *Cannabis sativa*. Both morphological and chemical traits can reduce preference for this crop. For example, the morphological structure of hemp flowers alters both oviposition and settling behavior of corn earworm, a major pest of hemp (Shrestha et al., 2025b). Trichomes on plants also provide a defense mechanism against pests. The structural trichomes of *C. sativa* are suggested to provide protection against pests, wind, and light (Dayanandan and Kaufman, 1976). However, structural trichomes may influence the mobility of green lacewing (*Chrysoperla carnea*), a beneficial insect (Carretero et al., 2022). Plant chemistry can influence feeding and oviposition. Rothschild and Fairbairn (1980) found that *Pieris brassicae* L. can distinguish cannabidiol (CBD) extracts from THC extracts. Host plants sprayed with CBD extract had more eggs than those sprayed with THC. It is unclear if this difference would be observed on high THC cannabis plants versus high CBD hemp plants. Extracts from cannabis plants have been used as insecticides and repellants for certain pests, including houseflies and green peach aphids (McPartland, 1997; Benelli et al., 2018). There is a need to understand how the many different chemicals produced by *C. sativa*, including terpenes and cannabinoids, influence feeding and oviposition.

Antibiosis has been observed in *C. sativa*. Antibiosis can be due to morphological and chemical traits that impact feeding, reduce fecundity, or cause mortality. Glandular trichomes can physically stop the movement of certain insects, including rice root aphids that get stuck on trichomes as they crawl from the roots to the upper surfaces of leaves of hemp plants (Cranshaw and Wainwright-Evans, 2020). Unfortunately, Carretero et al. (2022) found that the mobility of two aphid predators was reduced on the inflorescences, potentially due to the glandular trichomes.

Despite insects lacking cannabinoid receptors (McPartland et al., 2001), there are several examples of antibiosis demonstrated through *Cannabis sativa* feeding trials. Tobacco hornworm (*Manduca sexta*) will preferentially consume *C. sativa* tissue that contains lower amounts of cannabidiol (CBD) compared to tissue with high CBD, and when these caterpillars are fed a diet spiked with higher amounts of CBD, there is reduced growth and increased mortality (Park, et al., 2019). Reduced growth and increased mortality have also been observed in cabbage looper (*Trichoplusia ni*) larvae fed diets spiked with cannabinoids (Stack et al., 2023). CBD reduces fall armyworm feeding (*Spodoptera frugiperda*) and decreases production of the detoxification enzyme CYP450 (Abendroth et al., 2023). While cannabinoids have a negative effect on some insects, there is also the possibility of insects harnessing these compounds. The moth *Arctia caja* can successfully consume cannabinoids and store them within the cuticle, which may provide protection against mammalian predators (Rothschild et al., 1977).

The use of modern breeding technology can accelerate trait selection in plants, which has created opportunities for crops resistant to pests. Using gene-editing tools is an option in *C. sativa* and has been investigated to increase trichome production (Roscow, 2019) and to alter cannabinoid content. RNA interference (RNAi) shows promise for pest management in *C. sativa* (Willow et al., 2023). There are also opportunities to use technology like CRISPR (clustered regularly interspaced short palindromic repeats) to improve cannabis plants (Shiels et al., 2022). One potential application of CRISPR could be focused on insect vectored diseases. The use of gene editing tools for resistance to beet curly top has been successfully done in sugar beets using CRISPR-Cas9 (Yıldırım et al., 2023). This would be an important advancement for hemp growers in locations that have beet leafhoppers and struggle with beet curly top disease.

6 Pests of concern

While many different arthropods have been observed in hemp production, there are certain species of greatest concern. Some pests cause direct feeding damage, leading to yield loss, while others spread pathogens when they feed or cause wounds that allow for secondary infection. Economically important pests will vary by region. For example, beet leafhopper is not found in the Eastern United States but is an important virus vector in the Western United States. Understanding which pests are common in a specific growing region will aid in proper identification and management.

6.1 Chewing pests

Corn earworm (*Helicoverpa zea*) is found across the United States and has many different host plants, including cotton, corn, and tomatoes. Corn earworm overwinters in warmer locations and seasonally migrates north. This pest causes damage to hemp flower buds and the grain and is more problematic in cannabinoid hemp production. Feeding damage from corn earworm causes an open wound on the plants, which can be a site for infection. Corn earworm adults can be monitored using pheromone traps. This is useful when timing insecticide applications, especially for biopesticides, many of which are more effective on young larvae. There are natural enemies of this pest, including wasps that parasitize the eggs of corn earworm and tachinid flies that parasitize the larvae (Falcon-Brindis et al., 2022). There are also predators of corn earworm eggs, including the minute pirate bug, which is commonly found in hemp fields. Management practices include manually picking off caterpillars and the use of insecticides. Products that contain Nuclear Polyhedrosis Virus (NPV), *Beauveria bassiana*, *Bacillus thuringiensis* subspecies *kurstaki*, and *Bacillus thuringiensis* subspecies *aizawai* can be used for corn earworm management (Britt, 2021; Shrestha et al., 2025a).

Eurasian hemp borer (*Grapholita delineana*) is found in Europe, Asia, and North America. In North America, Eurasian hemp borer has been a major pest in some Midwestern states. Illinois and Wisconsin have seen high populations in cannabinoid hemp production. Eurasian hemp borer can be found in 'ditchweed,' the feral populations of hemp found along fields and roadways throughout the region. It is unclear if this pest has alternate hosts, but Japanese knotweed has been suggested as a possible overwintering host for the larvae (Anon, 2023). There are currently no pheromone lures available for monitoring adult moths. The larvae remain in the stems and branches of the hemp plant, making insecticides inefficient for control. Removing infested plant material, especially after harvest, is currently the best management approach. After the flower buds are removed from stems and branches, the woody material should be destroyed if Eurasian hemp borer has been observed.

Grasshoppers (*Melanoplus* spp. and *Dissosteira carolina*) can be a significant pest when conditions are dry. The most problematic species observed on hemp in the United States belong to the genus *Melanoplus*. Large populations can destroy many small plants and damage larger plants by chewing on the stems. In Colorado, the two-striped grasshopper (*Melanoplus bivittatus*) causes significant damage to hemp plants (Schreiner, 2019). In Kentucky, Carolina grasshoppers (*Dissosteira carolina*) were problematic in high tunnels, where small plants were destroyed (Villanueva, 2022). In other locations, grasshoppers are a minor pest. In areas that have experienced loss from grasshoppers, tillage in and around the field, as well as weed management, can destroy the eggs in the ground or expose them to predators (Anon, 2020b).

Fungus gnats (*Bradysia* spp.) are commonly found in indoor production and are often a consequence of overwatering. The fly larvae feed on decaying material and roots, causing damage to the plants. Yellow sticky cards are used to monitor for and trap the adult flies. A potato slice can be placed on the growing media for 48 hours and inspected for larvae that have adhered to the bottom of the slice (Sadof, 2020). Management includes reducing the frequency of watering, removing decayed plant tissue, and selecting growing media that is well drained. There are several biological control options, including the nematode (*Steinernema feltiae*), predatory mite (*Stratiolaelaps scimitus*), and *Bacillus thuringiensis* var. *israelensis* (Bti).

6.2 Sucking pests

Whiteflies (*Trialeurodes vaporariorum* and *Bemisia tabaci*) are common greenhouse pests that have been observed on hemp. Whiteflies have many different hosts, including crops and weedy plants. While the greenhouse whitefly (*Trialeurodes vaporariorum*) is more common on indoor grown cannabis, the sweetpotato whitefly (*Bemisia tabaci*) has been observed on outdoor grown cannabis in southern Europe, Brazil, and the southern United States (McPartland et al., 2000; Cranshaw et al., 2019). Adult and immature whiteflies suck fluid from plants and cause stippling damage, yellowing of leaves, stunting, and reduced plant vigor. This damage is similar to other piercing-sucking pests. Whiteflies can also spread pathogens, including lettuce chlorosis virus (LCV), which has been detected in *C. sativa* in Israel (Hadad et al., 2019). Whitefly adults are often noticed when plants are disturbed, causing the pests to fly off the plants. Yellow sticky cards can be used for monitoring. There are several biological control agents used for whiteflies, including the parasitic wasps *Encarsia Formosa* and *Eretmocerus eremicus*. The predatory mite, *Amblyseius swirskii*, will also feed on whiteflies. Application of entomopathogenic fungi, *Beauveria bassiana* or *Isaria fumosorosea*, is also an option for whitefly management.

Thrips (*Frankliniella occidentalis* and *Thrips tabaci*) are another common pest in greenhouse production. Western flower thrips (*Frankliniella occidentalis*) and onion thrips (*Thrips tabaci*) have been observed on *C. sativa*. Thrips puncture plant tissue and suck out fluid, which causes stippling, leaf yellowing, and leaf drop. Symptoms are similar to aphids, mites, or whiteflies. However, stippling damage appears larger because thrips scrape the epidermal tissue and suck fluid, and they do not have piercing mouthparts. Black frass, rather than honeydew, is important for identification. Damage caused by thrips can reduce yield through feeding and through the spread of pathogens. McCune et al. (2021) found that an initial infestation of five onion thrips per cannabis plant can cause a reduction of yield in indoor systems. Thrips spread viruses, including tomato spotted wilt virus (TSWV), which have been shown to infect *C. sativa* in experimental settings (Kegler and Spaar, 1997). However, thrips vectored viruses have not been observed outside of experimental settings. Management of thrips includes predatory mites, minute pirate bugs, predatory nematodes, and the entomopathogenic fungus *Beauveria bassiana*. Some predators may be more effective for certain thrips species. *Amblyseius swirskii*, *Neoseiulus cucumeris*, and minute pirate bugs (*Orius insidiosus*) consumed more onion thrips than western flower thrips in choice assays (Summerfield et al., 2024). Chemical control options are limited, but soaps and oils are used for thrips, including those containing azadirachtin.

Rice root aphid (*Rhopalosiphum rufiabdominate*) is a pest of grain, vegetables, and ornamental crops throughout North America and has become a problematic pest on indoor cannabis production. They can survive in many types of growing media and spend much of their life cycle near the roots until plants flower and become mature (Cranshaw and Wainwright-Evans, 2020). Infestations can lead to generalized decline and slowing of plant growth; lower leaves may also turn yellow. Monitoring for this pest includes the use of sticky tape on the lower stem, which traps aphids as they crawl up. If this pest is found, growing media should not be reused, and containers should be sanitized. The life cycle of the aphid makes selection of biological control agents challenging. The soil-dwelling mite, *Stratiolaelaps scimitus* may be recommended for rice root aphids, but there is no published evidence of successful suppression (Lemay and Scott-Dupree, 2022). Management options include the entomopathogenic fungi *Beauveria bassiana*, *Isaria fumosorosea*, and *Lecanicillium lecanii* (Cranshaw and Wainwright-Evans, 2020). The limited management options highlight the importance of sanitary practices to prevent infestations of this pest.

Cannabis aphids (*Phorodon cannabis*) have been found globally in indoor and outdoor grown hemp, and they appear to feed only on cannabis plants (Cranshaw et al., 2018). Like other aphid species, populations can rapidly increase if unchecked, especially in indoor production. In outdoor crops, cannabis aphid populations often reach higher numbers later in the season, where they will lay eggs to overwinter. No obvious plant wilting or yellowing occurred when large populations of cannabis aphid were observed in Colorado hemp fields (Cranshaw et al., 2018). Lab studies have found that cannabis aphids are able to transmit Potato Virus Y (PVY), but this has not been observed in the field (Pitt et al., 2022). Because cannabis is the only known host for this aphid, removing infested plants indoors and removing early season hemp volunteer outdoors could be an effective method to reduce populations. There are predators and parasitoids of cannabis aphid. Commonly observed predators include the aphid midge larvae (*Aphidoletes aphidimyza*), green lacewing larvae (*Chrysoperla* spp.), brown lacewing larvae (*Micromus variegatus*), minute pirate bug nymphs and adults (*Orius insidiosus*), multiple species of syrphid fly larvae, and multiple species of lady beetle larvae and adults (Lemay and Scott-Dupree, 2022). There are three species of parasitic wasps in the genus *Aphidius* and *Aphelinus abdominalis* that will parasitize cannabis aphid, but *Aphidius ervi* and *Aphidius matricariae* are likely to provide better control (Lemay and Scott-Dupree, 2022). The entomopathogenic fungus, *Beauveria bassiana* is affective at controlling cannabis aphids (Lopez Restrepo and Kovalchuk, 2025). Spraying off plants or the use of insecticidal soaps or oils are other management practices that may be useful in indoor systems.

Beet leafhopper (*Circulifer tenellus*) is a pest found in arid and semi-arid regions around the world and feeds on a wide variety of weedy plants and crops (Cook, 1967; Bennett, 1971). Beet leafhopper pierce the plant cell and suck out fluids, salivating in the process of feeding on the phloem (Stafford et al., 2009). This salivation allows for the transmission of beet curly top virus, BCTV (Curtovirus). BCTV can infect many different crops, including *C. sativa*. BCTV has been identified in hemp in Colorado, Arizona, California, Washington, and New Mexico (Creamer et al., 2024). This virus can cause a range of symptoms in cannabis, which is dependent on the infection stage and plant growth stage (Hu et al., 2021). Hu et al. (2021) observed early symptoms of new growth that were light green to yellow and blotchy mosaic mottling. They also observed shortened internode length and the curling and twisting of new leaves. Late-stage symptoms include severe leaf curling, stunting, and leaf necrosis. However, Hu et al. (2021) observed that symptoms can be isolated to individual branches of plants, an important indication of the virus. Yield loss and plant mortality are observed in infected plants. It appears that CBD and fiber hemp are preferred by the beet leafhopper compared to grain hemp (Creamer et al., 2024). Yellow sticky cards can be placed in the field to monitor for the leafhoppers, but few management options are available. Management of the beet leafhopper is challenging because it is highly mobile and can spread the virus within several hours (Anon, 2020a). In CBD hemp, reflective surface mulches could be an effective tool to deter leafhoppers, but this has not been tested. Breeding plants that are not attractive to the leafhopper or are resistant to BCTV infection should be investigated.

Potato leafhopper (*Empoasca fabae*) is a pest primarily associated with potatoes, alfalfa, and beans but can feed on a wide variety of plants. This is a migratory species that moves from southern to northern locations in the eastern United States. In hemp, noticeable damage has occurred in some locations. The damage, called hopperburn, causes the leaf tips to appear burned or scorched and yellow. Symptoms may be confused with nitrogen deficiency; however, the scorching from potato leafhopper will occur all over the plant, where nitrogen deficiency is most apparent in the lower canopy and can be remedied by fertilizing the plants. Additionally, leaves may curl, and plants may be stunted. The leafhoppers move quickly and can be difficult to spot. When observing symptomatic plants, remain still for several minutes and the insects often can be observed. A sweep net can also be used if plants are small. Management of potato leafhopper in hemp is not well defined. Selecting fields that do not border alfalfa or other susceptible crops may reduce movement of this pest into the hemp field.

Two-spotted spider mites (*Tetranychus urticae*) are associated with indoor cultivation but can also be found outdoors during dry periods. These mites are ubiquitous in greenhouse production; most producers struggle with this pest at some point. When populations are high, apparent webbing occurs on the plants, which may first be mistaken as a spider web. They cause stippling damage, which appears as white dots on the leaves. Monitoring should include inspecting leaves or shaking branches over a white sheet of paper to observe mites. Removing infested plants and maintaining clean growing spaces is important to reduce movement to healthy plants. Forceful spraying of water can knock mites off the plants and could be an effective strategy in some growing environments. There are many predators of spider mites, including green and brown lacewings, minute pirate bugs, predatory thrips (*Scolothrips sexmaculatus*), the spider mite destroyer lady beetle (*Stethorus picipes*), and the midge (*Feltiella acarisuga*). There are some chemical control options available, including soaps and oils.

Broad mites (*Polyphagotarsonemus latus*) are common in greenhouse environments and feed on many different plant species. Broad mite infestations can cause stunting and severely malformed terminal growth. Detection before damage is challenging because the mites are small and require a hand lens or dissecting microscope for identification. Follow sanitation practices and thoroughly inspect plants before moving out of quarantine. There are several species of predatory mites that feed on broad mites, including *Amblyseius swirskii* and *Neoseiulus cucumeris*.

Hemp russet mite (*Aculops cannabicola*) is an eriophyid mite found indoors and outdoors. It is difficult to see, requiring at least 15x magnification for identification. This pest often goes undetected until damage to the plant occurs, which can include bronzing, small leaves, and brittle tissue. Management options should include proper sanitation and starting with clean material and removal of infested plants. There are few commercially available biocontrol agents that are known to attack this mite, but *Amblysieus andersoni* is a predatory mite that will feed on russet mites (Anon, 2018). Chemical control options include horticultural oils.

7 Case study: hemp russet mites

Hemp russet mites (HRM), *Aculops cannabicola* (Farkas), are a widely distributed pest observed in indoor and outdoor grown cannabis. This pest makes an interesting case study because it is ubiquitous, is host-specific, can cause considerable damage to plants, and is difficult to manage.

HRM has global distribution and was first found in Hungary in 1958 and partially described in 1960 by Farkas. HRM was found in other Eastern European countries and Central Asia, as well as in the United States on feral hemp populations in the Midwest region (McPartland and Hillig, 2003). A more complete description was provided by Petanović et al. (2007) using mites found in Serbia. It is a year-round pest in controlled environments (Villanueva et al., 2022). In outdoor production, populations increase throughout the season (Cranshaw et al., 2019). *C. sativa* is the only known host for HRM (Petanović et al., 2007).

HRM belongs to the family Eriophyidae, which consists of small, cigar-shaped mites. Mites are 110–210 μm in length, can range in color from pale yellow to white, have four legs near the mouthparts, and the body is tapered at the rear (Villanueva et al., 2022). This pest can complete its life cycle in 1 week when there is high relative humidity and temperature (Anon, 2018). Unlike spider mites, HRM does not produce silk that facilitates movement to neighboring plants. HRM will hang from the edges of leaves, which allows the mites to become airborne as a breeze from the wind or a fan takes them to a new plant (Anon, 2018). This behavior allows for the mites to readily spread throughout the growing environment (Fig. 1).

HRM feeds by piercing and extracting the contents of plant cells with its mouthparts. Damage caused by HRM can include leaf margin curling or distortion, brittle leaves, reduced leaf size, and russeting or bronzing on the leaves. It is important to note that some cultivars produce leaves that have an upward curled leaf margin in the absence of HRM (Cranshaw et al., 2019). The yield loss caused by this pest has been variable depending on what type of hemp is being grown. HRM can reduce flower size in cannabinoid hemp, causing yield loss (Cranshaw et al., 2019). Mites have even been observed feeding on the glandular trichomes, causing a reduction in resin yields and

Figure 1 Russet mites (*Aculops cannabicola*) observed on the abaxial surface of a hemp (*Cannabis sativa* L.) leaf. Individual mites are circled to highlight their small size and translucent, elongate morphology.

extractable cannabinoids (Petanović et al., 2007; McPartland and Hillig, 2003). Small plants are especially susceptible to HRM damage and can rapidly decline as the mites feed (Szczepaniec et al., 2024).

Leaves that have curled margins, are brittle, distorted, or show russeting should be observed for mites. Mites are often found next to the midrib of the leaves or around glandular trichomes (McPartland and Hillig, 2003). As the population increases, they can be observed on the tops of leaves, petioles, and even on the flower buds (Villanueva et al., 2022). Populations often increase to large numbers because the mites can go unnoticed. Their small size requires a hand lens of at least 15x magnification or a dissecting microscope. Typically, diagnosis occurs after damage to plants is observed due to the small size of the mites. Diagnosing HRM when populations are low is critical to prevent significant damage to the plants. Carter et al. (2024) developed a rapid method of identifying HRM presence that is a useful tool when populations are very low and would likely be missed through scouting alone. The use of real-time polymerase chain reaction (PCR) assays resulted in accurate identification of HRM without amplifying non-target pests like two-spotted spider mites and western flower thrips (Carter et al., 2024). With this method, plants could be quarantined and tested for HRM and treated as needed, preventing introduction of the pest to the facility.

Management of HRM has been a challenge since the pesticide options are limited, and biological control agents have not been well studied. The best approach for the management of this pest is to prevent infestation by inspecting plants coming to a facility or field and quarantining newly arrived plants. Biological control options are limited. One species of predatory mite, *Amblyseius andersoni*, can be used for HRM (Anon., 2018). *Hirsutella thompsonii*, an entomopathogenic fungus, was detected in HRM in Kentucky when a large number of darkened mites were observed on leaves (Ricciardi et al., 2024). When Ricciardi et al. (2024) conducted laboratory assays, *H. thompsonii* killed 28% of mites 14 days after inoculation. *H. thompsonii* is pathogenic to other mite species and shows promise as a biocontrol tool in *C. sativa*.

Several miticides have been tested on HRM, and some effectively reduce mite populations. Hayes et al. (2023) found that the application of sulfur reduced HRM counts and resulted in higher biomass and an increase in multiple cannabinoids. They also found that hot water immersion and surfactant dips reduced HRM during clonal propagation with no risk to the cuttings. Szczepaniec et al. (2024) found abamectin had the greatest control of mites in a greenhouse experiment, but populations were reduced across all treatments compared to the control. The other treatments included etoxazole, fenpyroximate, 1% mineral oil, 2% mineral oil, rosemary oil, and sulfur, all of which were applied once. In field experiments, rosemary oil had the greatest reduction in mite densities (Szczepaniec et al., 2024). While conventional miticides are not currently labeled for use in hemp as of 2025, rosemary oil is a miticide option approved in many locations.

8 Conclusion and future trends in research

Hemp is still considered a relatively new crop, despite its historic production in many parts of the world. There is a need to better understand economic injury caused by pests and evaluate management strategies for specific pests. This is difficult to do when the economic value of hemp is evolving as new products are developed. Pest presence and pressure are likely to change as acreage increases around the world. This will require further research on pest management practices for emerging pests in hemp.

Research on IPM strategies for economically important pests is necessary, especially with so few chemical control options available. There must be a multifaceted approach to minimize the damage caused by insects and mites. Continuous monitoring of pests in each growing cycle will aid in our understanding of the insects and mites that damage hemp and will be important for selecting management tools and when to use them.

9 Where to look for further information

9.1 Further reading

- Fact sheets on different hemp pests 'Colorado State Hemp Insects Website.'

- Comprehensive book with hemp pests from multiple countries *Hemp Diseases and Pests* by McPartland, Clark, and Watson and Hemp Diseases and Pests 2nd Edition by McPartland.
- The book chapter 'Management of Insect Pests on Cannabis in Controlled Environment Production' by Lemay and Scott-Dupree (2022) within the book *Handbook of Cannabis Production in Controlled Environments*.
- Illustrated guide for common pests *Field Guide to Hemp Arthropod Pests* by Osborne et al. (2024).
- Seminal article on integrated pest management in hemp, 'Developing Insect Pest Management Systems for Hemp in the United States: A Work in Progress' by Cranshaw et al. (2019).

9.2 Key journals, conferences, and working groups

- The Cannabis Research Conference is an annual event and includes a variety of subject areas. The event is hosted by the Institute for Cannabis Research (ICR), the Global Hemp Innovation Center (GHIC), and the National Interdisciplinary Cannabis Symposium (NICS).
- Many hemp pest articles are found in the *Journal of Integrated Pest Management*.
- A national project in the United States includes priorities on insect pests, Building Hemp Research Initiatives & Capacity in the U.S., USDA Hemp Multistate Project S1084 and features a website hosted through Cornell University.

10 References

Abendroth, J.A., et al. (2023), 'Cannabidiol reduces fall armyworm (*Spodoptera frugiperda*) growth by reducing consumption and altering detoxification and nutritional enzyme activity in a dose-dependent manner', *Arthropod-Plant Interactions*, 17(2), 195–204.

Ahmed, M.Z., et al. (2024), 'Arthropod and mollusk pests of hemp, *Cannabis sativa* (Rosales: Cannabaceae), and their indoor management plan in Florida', *Journal of Integrated Pest Management*, 15(1), 1.

Altieri, M.A. and Nicholls, C.I. (2003), 'Soil fertility management and insect pests: harmonizing soil and plant health in agroecosystems', *Soil and Tillage Research*, 72(2), 64–67.

Anon. (2018), 'Pest management of hemp in enclosed production hemp russet mite (*Aculops cannibicola*)', Colorado State University, https://webdoc.agsci.colostate.edu/hempinsects/PDFs/Hemp%20Russet%20Mite_New_6-28-18.pdf (accessed 26 December 2024).

Anon. (2020a), 'Beet leafhopper and beet curly top virus - agricultural biology', Colorado State University, https://agsci.colostate.edu/agbio/ipm-pests/beet-leafhopper-beet-curly-top-virus/ (Accessed 30 May 2025).

Anon. (2020b), 'Grasshoppers', Colorado State University, https://webdoc.agsci.colostate.edu/hempinsects/PDF/Grasshoppers.

Anon. (2022), 'Agronomy Guide for Field Crops - Chapter 10: Field Scouting', Ontario Ministry of Agriculture, Food and Rural Affairs (OMAFRA), https://www.ontario.ca/files/2022-10/omafra-agronomy-guide-for-field-crops-chapter-10-en-2022-10-13.pdf (accessed 18 February 2025).

Anon. (2023), 'Eurasian hemp borer, *Grapholita delineana*', Colorado State University, https://agsci.colostate.edu/agbio/ipm-pests/eurasian-hemp-borer/ (accessed 7 January 2025).

Bakro, F., et al. (2018), 'An overview of pathogen and insect threats to fibre and oilseed hemp (*Cannabis sativa* L.) and methods for their biocontrol', *Journal of Natural Fibers*, 15(1), 26–51.

Bala, K., et al. (2018), 'Effect of plant nutrition in insect pest management: a review', *Journal of Pharmacognosy and Phytochemistry*, 7(4), 2737–2742.

Benelli, G., et al. (2018) 'The essential oil from industrial hemp (*Cannabis sativa* L.) by-products as an effective tool for insect pest management in organic crops', *Industrial crops and products*, 122, 308–315.

Bennett, C.W. (1971), 'The curly top disease of sugarbeet and other plants', Monograph No 7. St. Paul (MN) American Phytopathological Society.

Brendan Hayes, C., et al. (2023), 'Biology and management of hemp russet mite (Acari: Eriophyidae)' *Journal of Economic Entomology*, 116(5), 1706–1714.

Britt, K.E. (2021), 'Insect pest management in hemp in Virginia', PhD Thesis, Virginia Polytechnic Institute and State University, https://hemp.ucdavis.edu/sites/g/files/dgvnsk12601/files/media/documents/Britt_KE_D_2021.pdf (accessed 26 December 2024).

Carretero, P.L., et al. (2022), 'Glandular trichomes affect mobility and predatory behavior of two aphid predators on medicinal cannabis', *Biological Control*, 170, 104932.

Carter, O., et al. (2024), 'Development of a real-time PCR assay for detection of hemp russet mite (*Aculops cannabicola*)', *Environmental Entomology*, 53(1), 34–39.

Connelly, G. (2023), 'Evaluating ecosystem and agronomic services provided by companion cropping in hemp (*Cannabis sativa* L.)', Master's thesis, University of Wisconsin-Madison, https://agroecology.wisc.edu/wp-content/uploads/sites/75/2023/08/Grace-Connelly-Masters-Thesis_-1.pdf (accessed 18 February 2025).

Cook, W.C. (1967), 'Life history, host plants, and migrations of the beet leafhopper in the western United States (No. 1365)' Agricultural Research Service, US Department of Agriculture.

Cranshaw, W.S. (2017), 'Saltmarsh Caterpillar', https://webdoc.agsci.colostate.edu/hempinsects/PDFs/Saltmarsh%20Caterpillar%20with%20Photos%20NEW%207-3-17.pdf (accessed 17 February 2025).

Cranshaw, W.S. and Wainwright-Evans, S. (2020), '*Cannabis sativa* as a host of rice root aphid (Hemiptera: Aphididae) in North America', *Journal of Integrated Pest Management*, 11(1), 15.

Cranshaw, W.S., et al. (2018), '*Phorodon cannabis* Passerini (Hemiptera: Aphididae), a newly recognized pest in North America found on industrial hemp', *Insecta mundi*.

Cranshaw, W.S., et al. (2019), 'Developing insect pest management systems for hemp in the United States: a work in progress', *Journal of Integrated Pest Management*, 10(1), 1–10.

Creamer, R., et al. (2024) 'Interactions of beet leafhopper (Hemiptera: Cicadellidae), vector of beet curly top virus, and hemp in New Mexico', *Environmental Entomology*, 53(1), 11–17.

Dayanandan, P. and Kaufman, P.B. (1976), 'Trichomes of *Cannabis sativa* L. (Cannabaceae)', *American Journal of Botany*, 63(5), 578–591.

Dingha, B.N., et al. (2024), 'Intercropping industrial hemp and cowpea enhances the yield of squash—a pollinator-dependent crop', *Agriculture*, 14(4), 636.

Falcon-Brindis, A., et al. (2022), 'Parasitism of corn earworm, *Helicoverpa zea* (Boddie) (Lepidoptera: Noctuidae), by tachinid flies in cultivated hemp', *Insects*, 13(6), 519.

Gontijo, L.M. (2019), 'Engineering natural enemy shelters to enhance conservation biological control in field crops', *Biological Control*, 130, 155–163.

Gurr, G.M., et al. (2017), 'Habitat management to suppress pest populations: progress and prospects', *Annual Review of Entomology*, 62(1), 91–109.

Hadad, L., et al. (2019), 'Lettuce chlorosis virus disease: a new threat to cannabis production', *Viruses*, 11(9), 802.

Hu, J., et al. (2021), 'First report of Beet curly top virus infecting industrial hemp (*Cannabis sativa*) in Arizona' *Plant Disease*, 105(4), 1233–1233.

Kegler H. and Spaar D. (1997), 'On the virus susceptibility of varieties of *Cannabis sativa* L. *Archives of Phytopathology and Plant Protection*, 30(5),457–464. https://doi.org/10.1080/03235409709383198.

Lemay, J. and Scott-Dupree, C. (2022), 'Management of insect pests on Cannabis in controlled environment production' In Zheng, Y. (Eds.), *Handbook of Cannabis production in controlled environments*, CRC Press, Boca Raton, FL, pp. 271. https://doi.org/10.1201/9781003150442.

Lopez Restrepo, D. and Kovalchuk, I. (2025), 'Entomopathogenic fungi effectively control *Phorodon cannabis* aphid population in *Cannabis sativa* plants', *Plants*, 14(6), 931.

McCaslin, M. and Whalen, D. (2017), 'An Update on potato leafhopper resistance in Alfalfa', https://fyi.extension.wisc.edu/forage/an-update-on-potato-leafhopper-resistance-in-alfalfa/ (accessed 18 February 2025).

McCune, F., et al. (2021), 'Onion thrips, *Thrips tabaci* (Thysanoptera: Thripidae), reduces yields in indoor-grown cannabis', *Phytoprotection*, 101(1), 14–20.

McKeag, L. (2019), 'Pest scouting and using the UMass scouting sheets', University of Massachusetts Extension, https://ag.umass.edu/vegetable/fact-sheets/pest-scouting-using-umass-scouting-sheets (accessed 18 February 2025).

McPartland, J.M. (1997), 'Cannabis as repellent and pesticide', *Journal of the International Hemp Association* 4(2), 89–94.

McPartland, J.M. and Hillig, K.W. (2003), 'The hemp russet mite', *Journal of Industrial Hemp*, 8(2), 107–112.

McPartland, J.M., et al. (2005), *Hemp diseases and pests: management and biological control: an advanced treatise*, CABI, Wallingford, UK.

McPartland, J.M., et al. (2001), 'Cannabinoid receptors are absent in insects', *Journal of Comparative Neurology*, 436(4), 423–429.

Messina, F.J. and Jones, V.P. (1990), 'Relationship between fruit phenology and infestation by the apple maggot (Diptera: Tephritidae) in Utah', *Annals of the Entomological Society of America*, 83(4), 742–752.

Munir, M., et al. (2023), 'Occurrence and distribution of common diseases and pests of US cannabis: a survey', *Plant Health Progress*, *24*(4), 498–507.

Osborne, L., et al. (2024), *Field guide to hemp arthropod pests* (1st ed., 42 pp.), University of Florida IFAS Bookstore, Gainesville, FL.

Park, S.H., et al. (2019), 'Contrasting roles of cannabidiol as an insecticide and rescuing agent for ethanol–induced death in the tobacco hornworm *Manduca sexta*', *Scientific Reports*, 9(1), 10481.

Petanović, R., et al. (2007), 'The hemp russet mite *Aculops cannabicola* (Farkas, 1960)(Acari: Eriophyoidea) found on *Cannabis sativa* L. in Serbia: supplement to the description', *Archives of Biological Sciences*, 59(1), 81–85.

Pitt, W.J., et al. (2022), 'Virus infection and host plant suitability affect feeding behaviors of cannabis aphid (Hemiptera: Aphididae), a newly described vector of Potato Virus Y', *Environmental Entomology*, 51(2), 322–331.

Pundt, L. (2019), 'Using sticky cards to monitor for greenhouse insects', UConn Extension, Department of Plant Science and Landscape Architecture, https://ipm.cahnr.uconn.edu/wp-content/uploads/sites/3216/2022/12/2019stickycardsfactsheet-1.pdf (accessed 18 February 2025).

Rabb, R.L., et al. (1976), 'Conservation and augmentation of natural enemies'. In Huffaker, C.B. (Eds.), *Theory and practice of biological control*, Academic Press, New York, NY, pp. 233–254.

Ricciardi, M.M., et al. (2024), 'First report of *Hirsutella thompsonii* as an entomopathogenic fungus of hemp russet mite (*Aculops cannabicola*) in Kentucky', *Plant Health Progress*, 25(1), 92–94.

Rice, M.E. and Davis, P.M. (2010), 'Stalk Borer (Lepidoptera: Noctuidae) ecology and integrated pest management in Corn', *Journal of Integrated Pest Management*, 1(1), C1–C6.

Reay-Jones, F.P. (2019), 'Pest status and management of corn earworm (Lepidoptera: Noctuidae) in field corn in the United States', *Journal of Integrated Pest Management*, 10(1), 19.

Rohrmann, G.F. (2019), 'Baculoviruses as insecticides: four examples', in Baculovirus Molecular Biology 4th edn., National Center for Biotechnology Information Bethesda, MD, USA, Available at: https://www.ncbi.nlm.nih.gov/books/NBK543459/ (accessed 18 February 2025).

Roscow, R.F. (2019), *Enhanced Cannabis plants and methods of making and using the same*, U.S. Patent Application 16/093,066.

Rothschild, M., et al. (1977), 'Storage of cannabinoids by *Arctia caja* and *Zonocerus elegans* fed on chemically distinct strains of *Cannabis sativa*', *Nature*, 266(5603), 650–651.

Rothschild, M. and Fairbairn, J.W. (1980), 'Ovipositing butterfly (*Pieris brassicae* L.) distinguishes between aqueous extracts of two strains of *Cannabis sativa* L. and THC and CBD', *Nature*, 286(5768), 56–59.

Sadof, C. (2020), 'Fungus gnats and shore flies', Purdue University, https://extension.entm.purdue.edu/publications/E-111/E-111.html (accessed 25 May 2025).

Schreiner, M. (2019), *'A survey of the arthropod fauna associated with hemp (Cannabis sativa L.) grown in eastern Colorado'*, Master's thesis, Colorado State University.

Shiels, D., et al. (2022), 'Hemp genome editing–Challenges and opportunities', *Frontiers in Genome Editing*, 4, 823486.

Shrestha, G., et al. (2025a), 'Evaluation of biological and synthetic insecticides for control of Lepidopteran caterpillar complex in CBD hemp, 2024' *Arthropod Management Tests*, 50(1), 032.

Shrestha, G., et al. (2025b), *How far have we come in understanding host plant resistance for corn earworm management in hemp?'* Entomological Society of America, Salt Lake City, UT.

Stack, G.M., et al. (2023), 'Cannabinoids function in defense against chewing herbivores in *Cannabis sativa* L.', *Horticulture Research*, 10(11), p. 207.

Stafford, C.A., et al. (2009), 'Stylet penetration behavior resulting in inoculation of beet severe curly top virus by beet leafhopper, *Circulifer tenellus*', *Entomologia Experimentalis et Applicata*, 130(2), 130–137.

Stern, V.M., et al. (1959), 'The integrated control concept', *Hilgardia*, 29, 81–101.

Summerfield, A., et al. (2024), 'Laboratory investigations on the potential efficacy of biological control agents on two thrips species, Onion thrips (*Thrips tabaci* Lindeman) and western flower thrips (*Frankliniella occidentalis* (Pergande))', *Insects*, 15(6), 400.

Szczepaniec, A., et al. (2024), 'Suppression of hemp russet mite, Aculops cannabicola (Acari: Eriophyidae), in industrial hemp in greenhouse and field', *Environmental Entomology*, *53*(1), 18–25.

Villanueva, R.T. (2022), 'Carolina grasshopper consumed hemp seedlings in high tunnels', *Kentucky Pest News*, https://kentuckypestnews.wordpress.com/2022/07/12/carolina-grasshopper-consumed-hemp-seedlings-in-high-tunnels/ (accessed 23 May 2025).

Villanueva, R.T., et al. (2022), Hemp russet mite, *ENTFACT-162*, University of Kentucky College of Agriculture, Food and Environment, https://entomology.ca.uky.edu/files/ef162.pdf (accessed 26 December 2024).

Willow, J., et al. (2023), 'Towards dsRNA-integrated protection of medical Cannabis crops: considering human safety, recent-and developing RNAi methods, and research inroads', *Pest Management Science*, 79(4), 1267–1272.

Yıldırım, K., et al. (2023), 'Development of highly efficient resistance to beet curly top Iran virus (Becurtovirus) in sugar beet (*B. vulgaris*) via CRISPR/Cas9 system', *International Journal of Molecular Sciences*, 24(7), 6515.

Zuefle, M.E. and Grab, H. (2023), *2023 Crop profile for hemp in New York state*. National IPM Database, https://ipmdata.ipmcenters.org/source_report.cfm?view=yes&sourceid=2485 (accessed 26 December 2024).

Chapter 10

Integrated disease management in hemp cultivation

N. Gauthier and H. Smith, University of Kentucky, USA

1 Introduction

Since hemp reintroduction in the US (Agricultural Act of 2014; Agricultural Improvement Act 2018), over 50 diseases have been confirmed to cause yield loss in hemp (Munir et al. 2023, Gauthier et al. 2019, 2022, Kesheimer et al. 2021, Kesheimer et al. 2023, Gauthier and Thiessen 2022, Thiessen et al. 2020, Rivedal et al. 2022, Ellison 2020). Unfortunately, the hemp industry has missed opportunities for scientific advances resulting from over 80 years of prohibition. Pathology publications from the prohibition era are limited to literature reviews from decades past and general observations and anecdotal reports from the illegal cannabis industry (McPartland et al. 2000). During that time, however, the

http://dx.doi.org/10.19103/AS.2025.0161.21

discipline of plant pathology has advanced in terms of taxonomy, diagnostics, and disease management.

Unfortunately, legal restrictions prevented scientists from bringing these technologies to the hemp industry. Research (government and industry) remained at a standstill while advances continued in other crop areas. Hemp prohibition relaxed in Canada and some European and other countries (hemp was never prohibited in China or France) before the US, and research resumed as early as 2014 in some US states. Still, pathology research was slow to recover. Since 2014, pathologists in the US have begun to make up lost ground, but efforts are not enough to catch up with mainstream crop advancements. Meanwhile, pathologists in Israel and Canada have taken leading roles in understanding host-pathogen interactions in high-THC (Δ-9 tetrahydrocannabinol) cannabis, particularly greenhouse production (Jerushalmi et al. 2020, 2022, Pepin et al. 2018, Punja et al. 2019, Punja 2021).

Plant pathology plays a vital role in hemp production and crop advancement. Diseases can become severe and lead to significant yield loss. Risk for disease loss can vary across regions and with rotational crop choices. In dry regions, fiber and grain hemp are grown with little-to-no disease management. Conversely, when hemp is introduced to farms with a long history of cereal crop production, particularly in regions with moderate-to-high rainfall, disease can be severe. The popularity of floral hemp (grown for cannabinoids) introduced a horticultural production system that mimicked high-THC cannabis production. Practices that were conducive to disease development were overlooked by farmers unfamiliar with this new production system. By 2019, acreage of floral hemp increased rapidly across the US, and hemp was grown in creek bottoms, former pastured lands, and abandoned fields. Susceptible cultivars and production in a wide range of conditions and production systems likely led to disease outbreaks that did not represent pre-prohibition hemp production or grain and fiber hemp in Canada and France.

Though cannabis varies across use types and cultivars, high- and low-THC cannabis are generally similar in terms of disease susceptibility. Based on recent research, there are some cultivars with varying degrees of susceptibility, but these differences are yet to be deemed significant or correlated with cannabinoid concentrations (Munir et al. 2024); resistance studies are ongoing (Mihalyov and Garfinkel 2021, Schwartz et al. 2025a; Stack et al. 2024; Weldon et al. 2020). The most distinct differences are indoor versus outdoor production systems in which the different environmental conditions drive disease outbreaks.

Despite the prevalence of cannabis-specific diseases such as Septoria leaf spot, rust of hemp, and hemp downy mildew, most cannabis diseases have broad host ranges. In short, ubiquitous diseases such as powdery mildew, Pythium root rot, and Sclerotinia white mold affect a wide range of agricultural and horticultural crops, as well as wild hosts. We are in the early stages of modern cannabis research, so this field and these observations are changing rapidly.

2 Basic concepts in plant pathology

It is important to understand some general plant pathology before key diseases and their management are presented. Plant diseases are caused by microorganisms in five main categories:

- bacteria and phytoplasma;
- fungi (true fungi);
- fungus-like water molds (oomycetes);
- nematodes; and
- viruses.

Abiotic abnormalities (disorders) are sometimes referred to as diseases, particularly if symptoms mimic those caused by plant pathogens. However, in the realm of plant pathology and disease management, the term 'plant disease' will refer only to maladies caused by the biotic plant pathogens referenced above. Common terms used are:

- Plant pathogen – microorganism that infects plants and causes disease;
- Causal agent – the pathogen causing observed disease symptoms; and
- Inoculum – pathogen propagule that is capable of causing disease (fungal spores, virus particles, bacterial cells), also referenced as a quantitative measure 'inoculum load.'

Plant diseases are plants' responses to infection by plant pathogens; they are symptoms. Symptoms such as blight, rot, spot, and wilt are considered plant diseases. Examples of names referring to causal agents include Pythium root rot, Septoria leaf spot, and Fusarium wilt. It is also common for symptoms to be part of disease names and can refer to any of several pathogens. Examples include damping off, powdery mildew, and tip blight. Different pathogens can cause similar symptoms. For example, leaf spots can be caused by several fungal or bacterial agents, many of which look alike. Root rots can be caused by fungal or oomycete pathogens, and nematode infestations can cause similar above-ground symptoms. Given all this complexity, it is important to focus on identifying the causal agent when diagnosing plant diseases.

Accurate diagnoses cannot be made based on symptoms, although symptoms may be part of the investigative process. Pathologists often use symptoms to narrow down possibilities and determine next steps in confirming causal agents. Methods for confirming causal agents can range from microscopy to immunology to DNA sequencing. Modern molecular diagnostic techniques can provide species level accuracy but also add to the time and cost of diagnoses. Ultimately, disease management is the end goal. Pathogen confirmation is necessary for understanding disease cycles and knowing sources of inoculum, host range, yield loss potential, and overwintering capacity. With this information, management is more effective and more efficient.

Three things must be present in order for disease to develop:

- a pathogen;
- a susceptible host; and
- a conducive environment.

As stated above, diseases are caused by pathogens, so without a pathogen infection does not occur. Therefore, pathogen exclusion and removal of infected tissue are a means to break the disease cycle. Pruning infected tissue can reduce inoculum load. The presence of a susceptible host (genus, species, cultivar) or susceptible tissue (plant part, age, wound/stress) must be available for the pathogen to infect. An example is Fusarium head blight (FHB) in which the primary mode of infection is through flower styles.

Even when pathogen and host are available, environmental conditions must be met. Most pathogens require moderate temperatures (60°–80°F) and free water (RH >70%) in order to infect and reproduce. Keep in mind that some diseases such as southern blight favor hot conditions, while others such as Sclerotinia white mold favor cool conditions. Likewise, some powdery mildew species can infect in low humidity conditions, while downy mildew requires high humidity or prolonged leaf wetness to complete its life cycle. In some cases, environmental conditions affect disease vectors. Beet leafhopper, the beet curly top virus (BCTV) vector, is native to arid conditions. Environmental conditions can affect regional diseases, as well as outbreaks in microclimates at specific farms or greenhouses.

3 Key diseases of hemp: overview

Pathogens of hemp can have broad host ranges, or they can be host specific. Some pathogens are indiscriminate, while many are tissue or age specific. This section organizes key hemp diseases by host growth stage for quick reference. Note that some diseases may appear at multiple growth stages and/or on different tissue types than described here. Nuances are mentioned in general, but cultivar, climate, and other external factors may affect disease development.

Diseases featured in the following sections reflect the most prevalent cannabis diseases in North America since hemp reintroduction in the US

(Munir et al. 2023; Gauthier and Thiessen 2022). Both historical literature and recent research were considered when determining the key diseases listed below. During the past decade, scientists have disproved or corrected causal agents. Key diseases listed in this section have been confirmed using molecular diagnostic tools and phylogenetic analyses. Based on reports from Asia, Europe, and the Middle East, many of the diseases listed may also be prevalent outside of North America.

Detailed information on the key hemp diseases listed below can be found in the *Compendium of Cannabis Diseases* published by the American Phytopathological Society (Gauthier and Thiessen 2022). Disease profiles were written by experts in their respective fields, assuring accuracy. The compendium is available in a 'progressive' format to keep up with advances in the cannabis industry.

4 Diseases of seedlings and transplants

Establishment of hemp is often dependent on end use. Typically, grain and fiber hemp are direct-seeded in open fields, while hemp intended for floral production can be direct-seeded (feminized seed) or transplanted. Transplants are produced in greenhouses by direct seeding or by cuttings (known as clones in the cannabis industry). Thus, production system and mode of establishment have a major influence on early-season disease. Direct-seeded hemp is more susceptible to soilborne pathogens that have established in individual fields. Pre- and post-emergent damping off can be caused by oomycetes or fungal pathogens and influenced by temperature and soil moisture. Greenhouse-produced transplants are limited in exposure to soilborne pathogens, except where contamination of soilless media occurs and if the water source is contaminated.

There are several key diseases that can occur in hemp seedlings and transplants; two of the most significant diseases are caused by fungal pathogens. Botrytis gray mold (*Botrytis cinerea*) is a gray, fuzzy fungal growth that often develops on succulent and necrotic tissue. It has a broad host range and is more common in greenhouse-cultivated hemp production or under wet or humid conditions (RH >85%). Airborne spores can infect above-ground plant parts to cause damping off or leaf/tip blight. This then results in the colonization of damaged and necrotic tissue, which can serve as initiation of disease outbreaks (Fig. 1).

Powdery mildew (*Golovinomyces* ambrosiae, *G. cichoracearum*, *G. spadiceous*) is a powdery white fungal growth that occurs on upper leaf surfaces and on succulent tissue. Similar to Botrytis gray mold, powdery mildew is more common in greenhouse-cultivated hemp production. Airborne spores can move long distances via air currents. *Golovinomyces* spp. are easily misidentified or

Figure 1 Botrytis gray mold causing dieback.

confused with one another. *Golovinomyces* spp. affecting cannabis have broad host ranges, including agronomic crops, ornamentals, and weeds (Fig. 2).

Another key disease of hemp seedlings and transplants is the oomycete disease Pythium root rot and damping off (*Pythium aphanidermatum*, *P. myriotylum*, *Globisporangium irregulare*, *G. ultimum*). Root and stem infections cause stunting, dieback, or plant girdling and collapse. Roots become soft and decayed, often slimy. Infection can occur during seed germination (pre-emergent damping off) or after seedlings emerge (post-emergent damping off). Bases of unrooted cuttings rot, and cuttings fail to root. *Pythium* and *Globisporangium* spp. are associated with wet and poorly drained soils (available water capacity >70%) and can be prevalent in field soils and in contaminated greenhouses. *Pythium* spp. have broad host ranges. While *P. aphanidermatum* favors warm soils, *P. myriotylum* and *G. ultimum* are commonly active in cooler soils (Fig. 3).

Figure 2 Powdery mildew on leaf surfaces.

Figure 3 Pythium root rot causing necrosis and root dieback.

Damping off (*Fusarium* spp., *Pythium* spp., *Rhizoctonia* spp.) is another disease that can affect seedlings and transplants of hemp. Infected seedlings either fail to emerge or collapse soon after germination. Areas of infection include stem girdling, stem rots, and root rots. Damping off is more common in cool, damp soils where plants are slow to establish (Fig. 4).

Erwinia soft rot (*Erwinia* spp.) is caused by a bacterial pathogen in which cuttings develop a slimy, smelly rot and fail to root. Infection may spread to

Figure 4 Damping off of seedlings.

stems or leaves. The pathogen favors warm, wet conditions and spreads readily via soil/media and water.

5 Diseases of the vegetative stage

Hemp plants spend several weeks in the vegetative stage, especially when planted early in the season. Long exposure to the environment can result in higher risk for infection by airborne, soilborne, and waterborne pathogens. Considering that most hemp is produced in open fields, this section will focus on field production. However, diseases that are also of concern for greenhouse-grown cannabis are noted.

5.1 Leaf diseases

Leaf diseases are often the most noticeable diseases of hemp, potentially resulting in their being the most commonly reported. Leaf diseases include spots, blights, and molds/mildews, and severity can vary with cultivar susceptibility and environmental conditions. However, some pathogens can cause more severe disease symptoms than others. Typically, leaf diseases begin in the lower and inner canopy where temperatures are more moderate and humidity is higher. Denser canopies are more susceptible to disease. Spread within canopies, often to upper plant parts, is also dependent on susceptibility and environmental conditions.

Some of the most notable leaf diseases include those already highlighted in Section 4 of this chapter. Powdery mildew (*Golovinomyces ambrosiae*,

G. cichoracearum, *G. spadiceous*, *Podosphaera macularis*), a white fungal growth, also appears on upper leaf surfaces and on succulent tissue. Similar to its occurrence at the seedling and transplant stage, it is more common in greenhouse-cultivated hemp production. Powdery mildew can also occur in fields in mild and temperate climates, often in the inner plant canopy. *Podosphaera macularis* is known to cross-infect from hops from adjacent fields (Fig. 2).

Another fungal leaf disease known to affect hemp at the vegetative stage is Septoria leaf spot (*Septoria cannabis*), which appears as brown to black irregular spots with a wide yellow halo. Spots rapidly coalesce and cause leaves to turn yellow and drop. Disease progresses from lower canopy upward and develops early in the growing season and tapers off in hot conditions. Despite high severity on some floral cultivars, yield loss has not been documented. No other hosts have been identified (Fig. 5).

Bipolaris leaf spot (also known as hemp leaf spot) (*Bipolaris gigantea*) is a fungal disease that causes circular spots with white centers that fall out. Spots remain small but may coalesce to cause leaf blighting. All leaf tissue is susceptible. Severe losses have been documented in a limited group of floral hemp cultivars, but most cultivars are slightly to moderately susceptible. Host range includes several gramineous hosts and a limited number of broadleaf weed species (Fig. 6).

Figure 5 Septoria leaf spot in the early-mid stages of development.

Figure 6 Hemp leaf spot, advanced development.

Downy mildew (*Pseudoperonospora cannabina*) can affect hemp in the field or greenhouse. This is an oomycete disease that causes water-soaked spots that are yellow-green, eventually progress to bright yellow, and become angular and vein-delimited. Sporulation on undersides of leaves is gray to black and 'downy.' The pathogen requires cool, wet conditions (RH >85%) to occur. Host range is limited to *C. sativa*. Spore movement occurs through splashing water or wet tools, and long-distance spread is through movement of plant material (Fig. 7).

Other leaf diseases worth discussing include several fungal diseases, such as anthracnose (*Colletotrichum coccodes*, *C. fioriniae*, *C. dematium* complex). Leaf lesions appear sunken and circular or irregularly shaped with gray centers and brown to black borders. Symptoms resemble other common leaf spots. *Colletotrichum* spp. can infect other crops and weeds; host range varies by species.

Cercospora leaf spot (*Cercospora flagellaris*) is another fungal disease that causes irregular spots with tan to brown centers, broad purple borders, and yellow halos. Disease usually appears late in the season when temperatures are warm. *Cercospora flagellaris* has a broad host range including soybean (Fig. 8).

Rust on hemp (*Uredo kriegeriana*) is a fungal disease that causes yellow spots with bright halos on the upper sides of leaves. Entire leaves eventually turn yellow, and prominent, orange pustules on undersides of leaves are usually the best identifier for diagnosis. *Uredo kriegeriana* is known to infect only *C. sativa* (Fig. 9).

Figure 7 Downy mildew on sugar leaves.

Figure 8 Cercospora leaf spot, coalescing of lesions.

Figure 9 Rust pustules on undersides of leaves.

Bacterial blight (*Pseudomonas cannabina*, formerly *P. syringae* pv. *cannabina*) is a bacterial disease that causes water-soaked leaf spots that are angular and vein-delimited with yellow halos. Spots eventually become necrotic, and centers fall out (shot-hole). Disease is more prevalent in tropical and subtropical climates and is known to affect only *C. sativa*.

Phytoplasma (*Candidatus phytoplasma*) is a bacterial disease that causes plants to become yellow ('yellows' disease) with elongated nodes and witches' broom shoots. Plants become stunted, and flowers fail to develop properly. Phytoplasma is vectored by leafhoppers, planthoppers, and psyllids.

5.2 Root, crown and stem diseases

Root, crown, and stem diseases are more difficult to detect than leaf diseases and often are not directly identified. Root rots can be low to moderately severe and lead to generally low plant vigor. Root loss and stem cankers result in reduced nutrient and water uptake, leading to nutrient deficiency and drought symptoms (chlorosis, stunting, and wilt) in upper plant parts. Severe root loss and expanding stem lesions can cause rapid death, particularly with temperature spikes or dry conditions. Root, stem, and crown diseases are often caused by fungal and oomycete pathogens.

Fusarium wilt (*Fusarium oxysporum*) is a fungal disease that causes chlorosis and wilting, often on one side of the plant, which is followed by defoliation and quick death. Discoloration of vascular tissue (cross section) is diagnostic. *Fusarium oxysporum* is a species complex with special forms that are specific to hosts or host groups. Several *formae speciales* have been associated with Fusarium wilt on cannabis, but research is ongoing (Fig. 10).

Figure 10 Fusarium wilt causing whole plant dieback.

Southern blight (*Agroathelia rolfsii*, formerly *Sclerotium rolfsii*) is another fungal disease that causes sudden, whole plant wilting and dieback under hot conditions. It also results in white fungal mats with round sclerotia that colonize stems at the soil level. Sclerotia are round and tan to brown colored. Southern blight is prominent in regions with warm, wet summers such as the southeastern US. *Agroathelia rolfsii* has a broad host range (Fig. 11).

White mold (*Sclerotinia sclerotiorum*) symptoms begin as whole plant wilting and chlorosis, followed by sudden plant death. Stem cankers may 'shred' or crack, and plants lodge. White, fluffy mycelia colonize lesions under cool, wet conditions. Sclerotia are hard, black, and irregularly shaped. Sclerotinia white mold is more prevalent in cool climates of northern US and Canada. *Sclerotinia sclerotiorum* has a broad host range (Fig. 12).

There are several species of nematode (*Meloidogyne* spp., *Globodera* spp., *Heterodera* spp., *Pratylenchus* spp., *Ditylenchus* sp.) that can affect hemp. Hemp-specific reports of nematode infestations are limited, likely resulting from limited nematology programs in the US. Many nematode species have broad

Figure 11 Southern blight causing stem girdling; mycelial mats and sclerotia at the soil surface.

host ranges, so one can assume that nematodes can become problematic at sites where nematode populations are high.

There are several other root, crown, and stem diseases that can occur at the vegetative stage. Bot canker (Botryosphaeriaceae spp: *Neofusicoccum parvum*, *Lasiodiplodia theobromae*) is another fungal disease that can occur at the root, crown, and stem of hemp. Canker and terminal dieback diseases by Botryosphaeriaceae spp. appear as wilting and necrosis of stems resulting from stem cankers. Species favor warm, wet conditions and have broad host ranges (Fig. 13).

Phytophthora crown rot (*Phytophthora* spp.) is an oomycete disease that results in girdling and decay of crown tissue. Sudden wilt or blighting can occur when weather turns hot or dry. *Phytophthora* spp. can readily establish in wet or poorly drained soils. Pythium root and crown rot (*Pythium* spp., *Globisporangium* spp.) is another oomycete disease, in which root infections cause chlorosis, stunting, and plant dieback. Roots develop brown to black growth tips, especially new roots. Cortex easily sloughs from the stele, resulting

Figure 12 White mold on stems, with mycelial mats.

in 'rat tail' ends. Established plants may suffer from reduced vigor but are rarely killed. Pythium root rot is prevalent in wet or poorly drained soils (Fig. 3).

5.3 Viral diseases

Viral diseases of cannabis are reported primarily from floral hemp production systems (both high and low THC). Viruses that infect cannabis are systemic; they spread through the entire plant including asymptomatic tissues. Viral infections often require molecular or immunological testing for confirmation.

Beet curly top virus (BCTV; *Curtovirus*) symptoms can include stunting, leaf distortion, twisting of new growth, and mosaic/mottling. Internodes are shortened with increased lateral branching. Yields are reduced because of plant decline, but inflorescences are not affected. BCTV is transmitted by beet leafhoppers, which are present in the western and southwestern US. Host range includes sugar beets, cucurbits, legumes, and cannabis (Fig. 14).

Cannabis cryptic virus (CanCV; *Betapartitivirus*) symptoms include yellow to light green interveinal chlorosis or streaking and leaf wrinkling.

Figure 13 Bot canker causing stem girdling.

Leaf margins may curl upward, and plants may be stunted. CanCV has similar symptoms to what is described as 'hemp streak virus' syndrome. CanCV is transmitted via cuttings and seed and can be prevalent among floral hemp cultivars (Fig. 15).

Two commonly reported virus syndromes, hemp mosaic virus (HMV) and hemp streak virus (HSV), are commonplace in cannabis industry literature and websites. However, to date, no causal agent has been confirmed for either virus syndrome. Thus, virologists suggest that these viruses may be combinations of abiotic and biotic conditions. They are not recognized as viruses.

6 Diseases of the reproductive stage

Hemp plants spend various amounts of time in the reproductive stage. Fiber hemp is harvested soon after bloom (before seed set), so the diseases below are rarely problematic. Grain hemp is harvested as grain matures, making flowers and grain susceptible to many diseases. Floral hemp, which is limited to female plants that are not fertilized, remain in flower for longer periods and

Figure 14 Beet curly top virus causing leaf distortion.

Figure 15 Cannabis cryptic virus resulting in yellow streaking.

are harvested as flowers. Differences in reproductive tissues of grain and floral hemp result in different primary diseases. Environmental conditions, including microclimates due to floral architecture, play a major role in disease severity.

6.1 Flower and grain diseases

Botrytis gray mold (*Botrytis cinerea*) is a fungal disease that results in gray, fuzzy fungal growth that can colonize dense flowers and grain heads. Gray mold can also affect vegetative growth or colonize damaged or dead tissue. *Botrytis cinerea* requires cool temperatures and high moisture, so disease is more prevalent in fields in the northern US and Canada and in greenhouses with poor air flow.

Hop latent viroid (HLVd; *Pospiviroidae* viroid) causes symptoms to develop at bloom. This is known as 'dudding' in which inflorescences develop poorly and trichome development is limited. Other symptoms include stunting, leaf curling, and brittle stems. Infected plants can be asymptomatic. HLVd is transmitted mechanically via propagation stock and by tools during pruning. It is most commonly spread via infected mother plants. HLVd has a wide host range including herbaceous and woody plants (Fig. 16).

Web blight (*Rhizoctonia solani*) is a fungal disease that results in fungal webbing in inner canopies, which can then lead to necrosis. Mycelial webbing attaches leaves and stems together. It is more common during hot, wet weather, especially in dense plantings (Fig. 17). Web blight is caused by the same pathogen that causes Rhizoctonia damping off.

Figure 16 Hop latent viroid resulting in bud distortion and "dudding."

Figure 17 Web blight expanding to the upper canopy.

Fusarium head blight (*Fusarium graminearum*, *Fusarium* spp.) causes extensive blighting of buds and seed heads. Mycelial colonization begins in tissue closest to the main stem. It is more common in dense flowers and seed heads and may be confused with budworm damage. The disease is common on wheat, cereal grains, and gramineous weeds. Fusarium head blight is most severe in crops adjacent to or following wheat and corn production. It is most prevalent on grain hemp but has also been confirmed in floral hemp (Fig. 18).

7 Postharvest and storage diseases

After harvest, most hemp enters short or long-term storage. End products vary as do storage methods. Fiber hemp is retted either naturally in the field or chemically, while grain and floral hemp are dried by varying methods to reduce moisture and then stored at specified temperatures. Postharvest and storage diseases can range from those that evolve from field infections to mold contaminants acquired during storage. As previously mentioned, environmental conditions play a major role in the length of disease cycle, inoculum load, and potential for pathogen reproduction. During the hemp 'gold rush' between 2018 and 2019, small and/or inexperienced growers dried floral material in barns, abandoned greenhouses, and closed storage facilities where air circulation is limited, and humidity is high. Floral material and grain are frequently stored in super sacks (FIBCs; flexible intermediate

Figure 18 Fusarium head blight with colonization inside flower heads.

bulk containers), and if stored before properly dried, moisture is trapped, promoting growth of mold and rot pathogens.

Botrytis gray mold (*Botrytis cinerea*), which can occur in fields and greenhouses, is typically seen at low levels in storage. However, stored tissue that is infected or contaminated can develop severe moldy growth under humid conditions. The fungus can survive as a saprophyte and colonize stored tissue.

Fusarium head blight (*Fusarium graminearum*, *Fusarium* spp.) is another fungal disease that can develop in the field on flowers or grain. Infected tissue moved into storage can worsen, especially under humid conditions. *Fusarium* spp. can survive as saprophytes, so colonization of harvested tissue can occur under humid conditions. The fungi that cause head blight can also produce mycotoxins that affect humans and animals (Fig. 18).

Saprophytes (*Alternaria* sp., *Epicoccum* sp., *Gliocladium* sp., *Penicillium* sp.) are fungal contaminants that result in moldy tissue. They are common in stored floral and grain hemp when the crop is not dried and stored properly. Airborne spores are prevalent, and colonization is favored by high humidity and moisture.

8 Disease management in field and greenhouse-cultivated hemp: an overview

Plant diseases can have major impacts on hemp crop yields. Disease management begins with prevention by exclusion or plant protection practices. Once infection occurs, diseases can be managed by eliminating or suppressing causal agent spread. As mentioned in Section 2 of this chapter, plant diseases require (1) pathogen, (2) susceptible host, and (3) conducive environment to complete their life cycle. By manipulating each of these components, disease can be prevented, suppressed, or eliminated. Not all management strategies will eliminate disease. However, using a combination of tactics from the disease triangle (Fig. 19) can reduce disease severity and yield loss. The following image is a representation of the disease triangle:

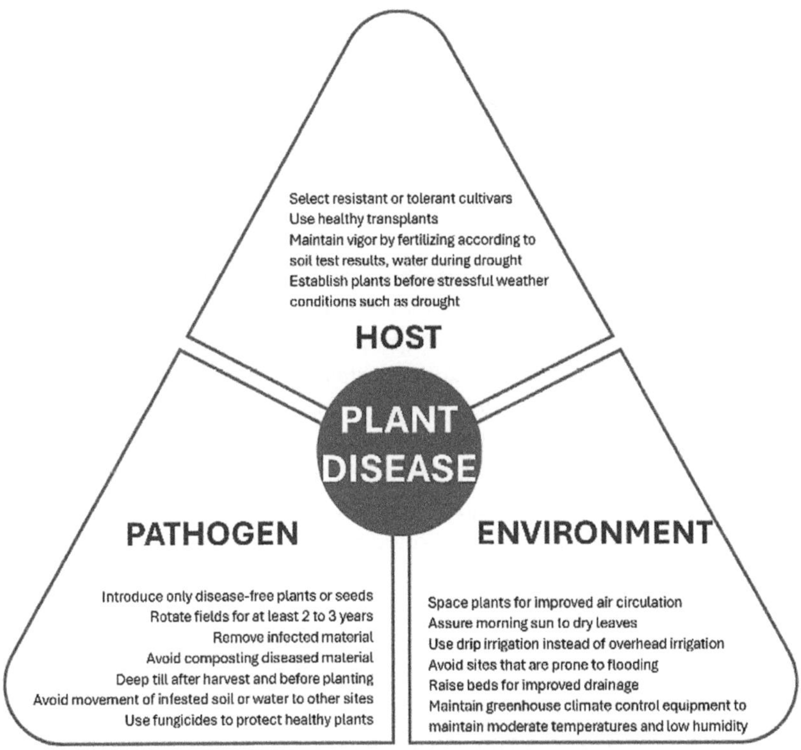

Figure 19 Disease triangle which represents the conditions required for disease development: pathogen, susceptible host, and conducive environment. Included are approaches for manipulating these components in order to manage disease in an integrated way.

9 Cultural practices for hemp disease management

Cultural practices refer to methods that change or alter production systems. These practices help prevent or delay infections or mitigate disease symptoms to an acceptable level (Buirs 2024). Often, a single cultural practice does not have the power to make major changes in a disease management program. However, when used in combination, an arsenal of cultural practices can have significant impacts on disease levels and crop health. Cultural practices can be used alone or in combination with fungicides. They include the following (Table 1):

- Rotation away from susceptible crops for 2 to 3 years, dependent on causal pathogen's life cycle and host range.
- Resistant or tolerant cultivars can reduce infection or delay disease development until later in the season. Limited information is available for resistant traits in available cultivars. Breeding is in progress.
- Air circulation helps improve drying and reduce humidity inside plant canopies. Plant and row spacing, pruning, and weed management can all help improve air movement in fields and greenhouses. Fans and open vents improve air circulation in greenhouses.
- Weed removal can reduce or eliminate alternative hosts. Weed management is most important in hemp fields, but managing certain weed hosts in adjacent fields can also reduce the inoculum of certain pathogens.
- Delayed planting can help protect seeds and seedlings from soilborne pathogens that favor cool, wet soils. Proper planting time is dependent on region and climate, as well as the problematic pathogen.
- Raised beds help improve surface drainage and raise plant roots and crowns from wet or waterlogged soils. Raised beds warm more quickly in spring which can slow soilborne pathogens that favor cooler soils.
- Maintaining plant health helps prevent infections.
- Wounds and damage by equipment, insects, or other abiotic factors create entry points for opportunistic pathogens.
- Fallowing greenhouses and indoor facilities between growing seasons breaks pathogen and vector life cycles. Sanitation and disinfection of all surfaces, including walls, benches, floors, and irrigation systems, remove residual contaminants.
- Rogueing dead, dying, and diseased plants throughout the season prevents inoculum buildup and spread to healthy plants.
- Deep tilling at the end of the season buries fine debris and residual inoculum.

Table 1 Efficacy of cultural practices for management of hemp diseases

	Leaf spots	Damping off	Downy mildew	Fusarium crown rot	Nematodes	Phytophthora root/crown rot	Powdery mildew	Pythium root rot	Rhizoctonia	Southern blight	Viruses	White mold
Clean seed/cuttings		x	x	x		x		x	x		x	
Crop rotation		x		x	x	x		x	x	x		x
Deep till		x		x		x		x	x	x		x
Drainage and site selection						x		x	x	x		
Drip irrigation	x		x									
Insect management				x				x			x	
Minimize splashing	x											
Plant spacing	x		x	x		x	x		x	x		x
Plant vigor												
Raised beds		x		x		x		x	x			
Reduced wounding/injury												
Removal of diseased tissue												
Resistant cultivars	x						x				x	
Sanitation (tools/equipment)		x		x	x	x		x	x	x	x	x
Weed management	x		x		x			x	x	x	x	

10 Use of biological, biorational, and conventional fungicides in disease management

Biological fungicide products are made from microorganisms such as bacteria or fungi. In general, biocontrol products have one of three modes of action: antagonism, competition, or induced plant resistance (SAR, systemic acquired resistance). There is limited efficacy data for biological products on cannabis, but data from trials in other crops was considered for this section (Akinrinlola and Hansen 2023, Balthazar et al. 2022, Dixon et al. 2022a, Nesbitt 2023, Scott and Punja 2020; Schwartz et al. 2025b). Some of the most reliable sources for fungicide trial data are *Plant Disease Management Reports* (plantmanagement network.org) and *Plant Health Progress* (apsjournals.apsnet.org/toc/php/cur rent).

It is important to keep in mind that biologicals have different efficacies than conventional products. They work to improve plant health and alter the microbiome of the plant. Thus, biologicals are used as part of a larger disease management program; they have no curative or quick-action activity. Building a disease management program using biologicals should be considered a holistic approach, should be combined with cultural practices, and may not be suitable with conventional products.

Pesticide sale and use in cannabis is determined by state regulations, which can be stricter than EPA-FIFRA regulations. In most states, approved cannabis pesticides fall into one of three categories:

- Products with active ingredients that are considered minimum risk pesticides;
- Botanical products; and
- Products with language that does not prohibit use on cannabis (biological and biorational product labels often have this type of language).

For example, labels may list 'indoor and outdoor plants' instead of listing specific crops. Many states consider this broad, non-prohibitive language a satisfactory approach to legal use of products.

While biological fungicides are dependent on the activity of the microbial agent, there can be significant differences across products. Manufacturers often select specific strains of these species based on efficacy, resiliency, or other traits. Thus, products with the same active ingredient (microbial agent) may have different ranges of disease management efficacy. The following list includes common biological agents manufactured for fungicidal use and their modes of action:

- *Bacillus amyloliquefaciens* – cell degradation, induced host resistance;
- *Bacillus subtilis* – cell degradation;

- *Bacillus mycoides* – induced host resistance;
- *Clonostachys rosea* (formerly *Gliocladium catenulatum*) – cell wall degradation, induced host resistance;
- *Gliocladium virens* (current name *Trichoderma virens*) – parasitism, competition;
- *Pseudomonas chlororaphis* – cell degradation;
- *Reynoutria sachalinensis* – induced host resistance;
- *Streptomyces lydicus* – cell degradation;
- *Swinglea glutinosa* – cell membrane disruption;
- *Trichoderma harzianum* – colonization/competition, cell degradation, induced host resistance, parasitism;
- *Trichoderma* spp. – colonization/competition, cell degradation, parasitism; and
- *Ulocladium oudemansii* – colonization/competition.

Besides biological control fungicides, there are many other biorational products on the market (Table 2). These products are made of materials that are not considered a threat to humans or animals and can include botanical, microbial, mineral oil, or soap. Microbial products (biocontrol agents) are covered above. The following list includes additional biorational products that are often used for management of hemp diseases:

- Acibenzolar – induced host resistance;
- Botanical oils such as garlic, capsaicin – cell degradation;
- Citric acid – cell wall disruption;
- Hydrogen peroxide – cell degradation, host resistance induction;
- Neem oil – inhibits spore germination;
- Potassium bicarbonate – inhibits mycelial development; and
- Silicon, calcium silicate – strengthens plant cell walls.

Conventional fungicides refer to any fungicidal product that is not of biological origin. While conventional pesticides are often synthetic, they can also be derived from inorganic compounds (copper, sulfur). As of this printing, there are limited conventional fungicides labeled for use on hemp or other cannabis. As described above, pesticides used for cannabis must have label language that does not prohibit use on the crop.

Copper and sulfur are the most commonly used conventional fungicides. Formulations and production across these products vary, so each label should be read carefully for appropriateness and legal parameters. Although copper and sulfur are not organic products, some formulations can be approved by the Organic Materials Review Institute (OMRI) for use when an alternative organic

Table 2 Biorational fungicides with potential to manage common diseases of hemp. Not all fungicides have been tested on cannabis; data compiled from trials with other crops

	Alternaria leaf spot	Anthracnose	Bacterial spot	Botrytis gray mold	Cercospora leaf spot	Damping off	Downy mildew	Fusarium crown rot	Phytophthora root and crown rot	Powdery mildew	Pythium root rot	Rhizoctonia	Rust	Septoria leaf spot	Southern blight	White mold
Acibenzolar			x				x			x						
Bacillus amyloliquenfaciens	x	x	x	x	x	x	x	x	x	x	x	x	x	x	x	x
Bacillus mycoides	x	x		x	x									x		
Bacillus pumilus			x				x			x						
Bacillus subtilis	x	x	x	x	x		x	x	x	x	x	x	x		x	
*Clonostachys rosea**				x		x	x	x	x	x	x	x				
Citric acid	x	x		x	x		x							x		x
Copper	x	x	x		x		x									
*Gliocladium virens**																
Hydrogen peroxide		x	x				x									
Neem oil										x						
Potassium bicarbonate					x		x			x						
Pseudomonas chlororaphis	x	x		x			x	x		x	x	x				x
Reynoutria sachalinensis	x		x	x	x		x			x				x		x
Silicon, calcium silicate				x	x									x		
Streptomyces lydicus	x		x	x	x	x	x	x	x	x	x	x	x	x		x
Sulfur										x						
Swinglea glutinosa			x							x			x			x
Trichoderma spp.				x			x	x	x		x	x			x	x
Ulocladium oudemansii				x												x

*Note: *Gliocladium virens* current name is *Trichoderma virens*; *Clonostachys rosea* is formerly known as *Gliocladium catenulatum*

product is not available. Approval can also vary across these products; check labels.

The IR-4 program is perhaps the most promising pathway for EPA (US Environmental Protection Agency) approval of conventional fungicides for hemp (ir4project.org). The program supports registration of pesticides for use on many specialty crops, facilitating data generation required by the EPA for approval. This prevents small acreage crops such as hemp from being omitted when industry focuses on approval for large acreage crops. At the time of this printing, IR-4 has completed or is in the process of facilitating Section 3 labels for a limited number of pesticides, including a broad-spectrum fungicide, one for powdery mildew, and another for oomycetes.

Postharvest residue testing is required in many US states. Depending on state laws, pesticide residues are tested, some with an extensive list of requirements https://gentechscientific.com/cannabis/state-regulations/. Among the most commonly prohibited fungicides is myclobutanil, which produces cyanide when burned.

11 Case studies

11.1 Leaf spot disease in floral hemp cultivars

Leaf spots are perhaps the most prominent diseases of hemp, as they are easily detected and most often reported. Severe foliar blighting and defoliation are known to affect photosynthesis and general plant health. A two-year, two-site study of Bipolaris, Cercospora, and Septoria leaf spots in Kentucky examined time of infection, weekly and cumulative disease severity at mid- and upper-canopy, and disease effects on floral biomass and CBD (cannabidiol) yield (Munir et al. 2024). The study investigated floral cultivars from four relatedness groups with the following treatments: inoculated plots to induce high disease severity, fungicide treatments to simulate disease-free plots, and a natural infection control.

Early season disease Septoria leaf spot developed in June and caused severe defoliation as early as July (early-mid vegetative stage). Bipolaris leaf spot developed in July and increased throughout the season, causing minimal to moderate disease severity. Cercospora leaf spot developed at mid- to late-reproductive stage, and symptoms were difficult to differentiate from senescence. Disease severity was highest among cultivars in the Trump group and lowest in the Otto group, suggesting that there is genetic diversity within the hemp germplasm. In both years and at both locations, floral biomass and CBD yield of the heavily infected plots (inoculated) were not significantly different from low disease plots (fungicide treated), even with early onset disease and defoliation of nearly 90%.

These results supported similar results by Britt (2021) in which hemp leaves were removed to mimic insect herbivory; removal of leaf tissue of up to 75% did not significantly reduce grain yield, and 50% leaf removal failed to significantly reduce cannabinoid yield. These studies suggest that hemp's photosynthetic efficiency is sufficient to overcome leaf loss while maintaining the ability to produce optimal yield. Munir et al. (2024) concluded that the use of fungicides to manage foliar diseases may not be warranted. As hemp cultivars are expanded and improved, however, traits may change. Future studies are needed to monitor thresholds and detect changes in the effects of foliar diseases on yield.

11.2 Understanding Fusarium head blight and its management

Fusarium head blight (FHB) is becoming a major concern for grain and floral hemp in the US. *Fusarium graminearum* has been documented as the most prominent causal species of FHB, but *F. avenaceum*, *F. sporotrichioides*, and *F. incarnatum-equiseti* species complex have also been confirmed to cause disease (Smith et al. 2023, 2024, Yulfo-Soto et al. 2022, Bergstrom et al. 2020). These are the same species that cause FHB of wheat and cereals (also known as scab). A national cannabis census from 2020 included reports of FHB in four of five regions across the US (Munir et al. 2023).

A 2022 Kentucky survey of FHB infections by toxigenic members of the *Fusarium sambucinum* species complex (*F. graminearum* and *F. sporotrichioides*) saw an average isolation rate from symptomatic tissue of 8.5% across 25 sites (Smith et al. 2023). Disease symptoms occurred predominantly in September and October as plants were in the mid to late reproductive stages (11–15 weeks after planting). A phylogenetic analysis of recovered *F. graminearum* isolates grouped them with wheat isolates, while a distinct clade of *F. sporotrichioides* was identified. Closer examination of *F. graminearum* from the same survey confirmed that isolates were predominantly 15-DON chemotype (95.6%). A separate survey of postharvest hemp in Kentucky found that 16.3% of sampled lots were colonized by *Fusarium* spp. belonging to the *F. sambucinum* complex (Szarka et al. 2023).

To better understand the disease cycle, Munir et al. (2026) conducted a two-year field study in Kentucky. Latent infections of *F. graminearum* were detected as early as mid-April in early planted fields when plants were in the vegetative stages, with the highest detection of latent infections beginning in mid-June. Infections were correlated with rain events and flowering but not plant age or time of planting. This study focused on latent infections, and sampling was terminated once latent infection was detected. FHB symptoms include flower and grain head blighting, as well as brittle grain. Additionally, *Fusarium* spp. within the *F. sambucinum* species complex are major producers of trichothecene mycotoxins. In the US, DON contamination is regulated in cereal grains; T2 does not have an advisory level in the US, but the EU regulates the toxin. Currently, there are no required testing or advisory levels for trichothecene mycotoxins in hemp. Immuno-testing of grain and floral material from FHB symptomatic tissue in Kentucky has resulted in the detection of DON and T2 toxins above cereal grain thresholds in the US and EU (Kagen et al. 2025; Smith 2026). Studies are ongoing.

11.3 The use of root-applied silicon for powdery mildew management in greenhouse-cultivated hemp

Powdery mildew is a major disease of cannabis, particularly in controlled systems such as greenhouses and indoor operations. Powdery mildew can also be prevalent under field conditions in temperate climates. There are limited management options for diseases of cannabis crops, including powdery mildew, especially since the EPA has only recently considered conventional products for labeling.

Silicon has been documented to decrease powdery mildew on barley, cucumber, grape, pumpkin, and rose by improving plant strength and cell rigidity. It has also been documented to change the plant's response against pathogen infection at the physiological, biochemical, and molecular levels. Research has shown that root applications of silicon are more effective at reducing powdery mildew development than when applied to leaves.

A University of Kentucky study evaluated root-applied silicon at 300, 600, and 900 kg/ha Si (268, 535, 803 lbs/acre) (Dixon et al. 2022b). At week 6, root-applied silicon resulted in 82.7% and 82.0% reduction in disease severity for 'HK' and 'Victoria' cultivars, respectively. The 300 kg/ha rate was sufficient to suppress disease in the upper canopy, and the 600 kg/ha rate was the most efficacious for the mid-canopy. The 900-kg/ha treatment did not provide disease suppression beyond that of the 600 kg/ha rate. The study concluded that 600 kg/ha total Si is a good recommendation rate for application to soilless mixes to suppress powdery mildew. In addition to powdery mildew, silicon has been reported to reduce or suppress many fungal and bacterial diseases, including *Bipolaris* spp., *Cercospora* spp., *Fusarium* spp., and *Septoria* spp., all of which have been reported as important diseases of hemp. Thus, consideration of silicon as a disease management option in field-grown cannabis should be explored. These findings suggest that silicon should be considered as part of a powdery mildew management program, particularly in greenhouse and indoor environments where powdery mildew is the most limiting factor, and where hemp is grown in soilless mixes that are low or limiting in soluble silicon.

12 Future trends

Since reintroduction of hemp to the US in 2014 and 2018, markets have experienced volatility (Mark et al. 2020, NASS 2022). After the 'gold rush' of 2018–2019, markets began to stabilize, and production of grain and fiber hemp continues to increase with market demands (NASS 2022). This increased grain and fiber acreage is likely to result in rotations with agronomic crops, while small acreage farms and greenhouse production are likely to decrease. The trend may also induce a shift toward pathogens that affect common agronomic crops. For example, important hemp diseases, southern blight and Sclerotinia white mold, also cause significant losses in soybean and many vegetable crops, while Fusarium head blight is a major concern in cereal production. Inoculum buildup of these common diseases is likely to become problematic as hemp is introduced to intensely managed farms.

Diseases can be climate-specific, and thereby region-specific, as particular pathogens favor particular temperatures, rainfall, or the flora that are associated with them. However, climate change may shift those climate patterns and the pathogens associated with them. Warming climates are already resulting in warm-climate diseases such as southern blight appearing in more temperate climates as far north as New York and Vermont (Munir et al. 2023). Warming temperatures are also likely to lead to shifts in vectors, as well as longer growing seasons that can lead to more generations per season. Vectors can also be affected by climate extremes such as drought. For example, beet leafhopper, which favors arid climates, is likely to expand its range, bringing BCTV to the central and eastern regions of the US.

Movement of pathogens and vectors can also be driven by global commerce. Expansion of hemp acreage and markets, as well as marijuana legalization, has led to shifting production systems. Pathogens can move with seeds and transplants, as well as in harvested materials such as floral material and grain. For example, HLVd has been limited to the western region of the US, particularly in transplant production (high- and low-THC cannabis). With interstate movement of infected material, the pathogen is likely to spread to regions such as the eastern US.

13 References

Akinrinlola, R. J. and Hansen, Z. R. 2023. Efficacy of organic fungicides against hemp powdery mildew caused by *Golovinomyces ambrosiae* in a greenhouse in Tennessee. *Plant Dis*. 107(6):1867–1873.

Balthazar, C., et al. 2022. Biocontrol activity of *Bacillus* spp. and *Pseudomonas* spp. against *Botrytis cinerea* and other cannabis fungal pathogens. *Phytopathology*. 112(3): 549–560.

Bergstrom, G., et al. 2020. Diseases affecting hemp in New York (2019 version). Cornell Hemp Blog. https://hemp.cals.cornell.edu/resource/diseases-affecting-hemp-new-york-2020/.

Britt, K. E. 2021. Insect Pest Management in Hemp in Virginia. Ph.D. diss. Ch 5. Virginia Polytechnic Institute, Blacksburg, VA. http://hdl.handle.net/10919/103014.

Buirs, L. and Punja, Z. K. 2024. Integrated management of pathogens and microbes in *Cannabis sativa* L. (Cannabis) under greenhouse conditions. *Plants*. 13(6):786.

Dixon, E., et al. 2022a. Suppression of hemp powdery mildew using silicon as a root-applied amendment. *Plant Health Prog.* 23:260–264.

Dixon, E., et al. 2022b. Prescreening of biological and biorational fungicides against common hemp pathogens using in vitro analyses. *Plant Health Prog.* 23:250–255.

Ellison, S. 2020. Hemp (*Cannabis sativa* L.) research priorities: opinions from United States hemp stakeholders. *GCB Bioenergy.* 13:562–569.

Gauthier, N., et al., eds. 2019. Science of hemp: production and pest management meeting proceedings, vol. 1. Science of Hemp Annual Meeting. https://digitalcommons.murraystate.edu/soh/vol1/iss1/1/.

Gauthier, N., et al., eds. 2022. Science of hemp: production and pest management meeting proceedings, vol 3. Science of Hemp Annual Meeting. https://digitalcommons.murraystate.edu/soh/vol3/iss1/1.

Gauthier, N. and Thiessen, L. D., eds. 2022. *Compendium of Cannabis Diseases*. American Phytopathological Society, Minneapolis, MN. https://apsjournals.apsnet.org/doi/book/10.1094/9780890546284.

Jerushalmi, S, et al. 2020. Fungal pathogens affecting the production and quality of medical cannabis in Israel. *Plants*. 9(7):882.

Jerushalmi, S., et al. 2022. Members of the *Fusarium oxysporum* complex causing wilt symptoms in medical cannabis in Israel, Italy, and North America comprise a polyphyletic assemblage. *Plant Dis.* 106:2656–2662.

Kagan, I. A., et al. 2025. Separation and quantification of mycotoxins in floral tissue of hemp (*Cannabis sativa*) infected with *Fusarium graminearum*. *JSFA Rep.* 5(4), 140–148.

Kesheimer, K., et al., eds. 2021. Pages 1–61 in: *Science of Hemp: Production and Pest Management Meeting Proceedings*, vol. 2. Science of Hemp Annual Meeting. https://digitalcommons.murraystate.edu/soh/vol2/iss1/1/.

Kesheimer, K., et al., eds. 2023. Pages 1–47 in: *Science of Hemp: Production and Pest Management* Meeting Proceedings, vol 4. Science of Hemp Annual Meeting. https://digitalcommons.murraystate.edu/soh/vol4/iss1/1/.

Mark, T., et al. 2020. *Economic Viability of Industrial Hemp in the United States: A Review of State Pilot Programs*. United States Department of Agriculture, EconomicResearch Service. https://www.ers.usda.gov/publications/pub-details?pubid=95929.

McPartland, J. M., et al. 2000. *Hemp Diseases and Pests: Management and Biological Control: An Advanced Treatise*. CABI, Wallingford, UK.

Mihalyov, P. D. and Garfinkel, A. R. 2021. Discovery and genetic mapping of *PM1*, a powdery mildew resistance gene in Cannabis sativa L. *Front Agron*. 3:720215.

Munir, M., et al. 2023. Occurrence and distribution of common diseases and pests of U.S. Cannabis: a survey. *Plant Health Prog.* 24:498–507.

Munir, M., et al. 2024. Leaf spot disease development and its effect on yield of essential oil-producing hemp cultivars in Kentucky. *Plant Dis.* 108(6):1621–1631.

Munir, M., et al. 2026. Detection and quantification of laten infection by *Fusarium graminearum*, causal agent of Fusarium head blight on hemp (*Cannabis sativa*) fields in Kentucky. *Plant Dis.* https://doi.org/10.1094/PDIS-04-25-0774-RE.

National Agricultural Statistics Service (NASS). 2022. National Hemp Report. Agricultural Statistics Board, United States Department of Agriculture (USDA).

Nesbitt, J. 2023. Efficacies of biological control agents for controlling *Fusarium* spp. in soilless cannabis cultivation. Thesis. University of Guelph.

Pepin, N., et al. 2018. Occurrence of powdery mildew caused by *Golovinomyces cichoracearum* sensu lato on *Cannabis sativa* in Canada. *Plant Dis.* 102(12):2644.

Punja, Z. K. 2021. Emerging diseases of *Cannabis sativa* and sustainable management. *Pest Manag Sci*. 77:3857–3870.

Punja, Z. K., et al. 2019. Pathogens and molds affecting production and quality of *Cannabis sativa* L. *Front Plant Sci*. 10:1120.

Rivedal, H. M., et al. 2022. An overview of pathogens associated with biotic stresses in hemp crops in Oregon, 2019 to 2020. *Plant Dis.* 106:1334–1340.

Schwartz, J. A., et al. 2025a. Efficacy of organic fungicides for management of hemp Septoria leaf spot. *Plant Health Prog*.26:602–608.

Schwartz, J. A., et al. 2025b. Screening of *Cannabis sativa* germplasm for susceptibility to Septoria leaf spot. *Plant Dis.* 66:213–218. https://doi.org/10.1094/PDIS-07-25-1367-RE.

Scott, C. and Punja, Z. 2020. Evaluation of disease management approaches for powdery mildew on *Cannabis sativa* L. (marijuana) plants. *Can J Plant Pathol*. 43(3):394–412.

Smith, H., et al. 2023. Emerging *Fusarium* spp. causing head blight on hemp in Kentucky. *Plant Health Prog*. 24:132–134.

Smith, H., et al. 2026. Cover crop impacts Fusarium head blight incidence and mycotoxin contamination of *Cannabis sativa*. *Plant Disease*. https://doi.org/10.1094/PDIS-02-26-0295-RE.

Smith, H., et al. 2024. Incidence and distribution of the *Fusarium sambucinum* species complex associated with *Cannabis sativa* in Kentucky. *Plant Health Prog*. 25:1535.

Stack, G. M., et al. 2024. Genetic mapping, identification, and characterization of candidate susceptibility gene for powdery mildew in *Cannabis sativa*. *Mol Plant Microbe Interact*. 37(1):51–61.

Szarka, D., et al. 2023. *Fusarium* species recovered from postharvest hemp materials from Kentucky fields. *Plant Health Prog*. 24:468–474.

Thiessen, L. D., et al. 2020. Surveying for potential diseases and abiotic disorders of industrial hemp (*Cannabis sativa*) production. *Plant Health Prog*. 21:321–332.

Weldon, W. A., et al. 2020. Cross-infectivity of powdery mildew isolates originating from hemp (*Cannabis sativa*) and Japanese hop (*Humulus japonicus*) in New York. *Plant Health Prog*. 21(1): 47–53.

Yulfo-Soto, G. E., et al. 2022. First report of *Fusarium graminearum* causing flower blight on hemp (*Cannabis sativa*) in Kentucky. *Plant Dis.* 106:334.

Chapter 11

Weed management in hemp cultivation

Panayiota Papastylianou, Ioannis Gazoulis, Angeliki Kousta and Metaxia Kokkini, Agricultural University of Athens, Greece; and Panagiotis Kanatas, University of Patras, Greece

1 Introduction

Hemp (*Cannabis sativa* L.) is a herbaceous, wind-pollinated annual species that originated in Asia and is now cultivated in more than 30 countries for its fibers, seeds, flowers and leaves (Fike, 2016). As an industrial crop, it can provide a variety of raw materials for numerous industrial applications (Crini et al., 2020). Schumacher et al. (2020) point out that hemp fibers are economically viable and have the potential to be a more environmentally friendly alternative material than cotton (*Gossypium hirsutum* L.) in the textile industry. The plant is also considered an important candidate for cultivation on poor soils and marginal lands due to its resilience and adaptability, while it can also be used for phytoremediation and bioenergy production (Blandinières and Amaducci, 2022). The suitability of hemp biomass as feedstock for bioenergy production was also recently investigated by Zhao et al. (2020). The seeds of hemp are also known for their nutritional quality and potential functionality for human

http://dx.doi.org/10.19103/AS.2025.0161.22

health and nutrition (Burton et al., 2022). Therefore, the species is reasonably considered as a dynamic multipurpose industrial crop whose economic importance is expected to increase in the coming years (Dudziec et al., 2024).

Regarding the agronomy of hemp, the effects of various agronomic practices on the growth, yield and quality of different hemp cultivars with different characteristics and traits have been studied, including irrigation, fertilization, plant population manipulations, adjustment of sowing time, etc. (Bajić et al., 2022; Cosentino et al., 2012; Deng et al., 2019; García-Tejero et al., 2019; Papastylianou et al., 2018; Yazici, 2023). In contrast to other agronomic practices, weed management remains an unexplored area of research. In fact, hemp has long been touted as a crop that competes well with weeds and requires little intervention to prevent yield loss due to its rapid growth and canopy closure (Lotz et al., 1991). There are currently few studies on the interactions between hemp and weeds and possible yield losses due to weed competition in this crop. Amaducci et al. (2015) also pointed out that this tall and fast-growing species typically outcompetes weeds once a dense crop canopy has formed that shades the ground.

However, weed interference and competition for resources up to this stage can potentially affect stand establishment and thus crop growth, yield and quality, as is the case for other non-food annual crops with similar characteristics (Gazoulis et al., 2021). Another example that supports this point is the recent study by Campiglia et al. (2017), in which the authors mentioned that the slow growth rate of hemp plants in the very early growth stages makes weed control essential to avoid the strong effects of weed competition on crop establishment and growth. Aubin et al. (2015) also reported that they manually controlled weeds throughout the growing season as needed. Flessner et al. (2020) also highlighted the importance of early-season weed control as a necessary agronomic practice in this crop, as is common in China (Amaducci et al., 2015).

Therefore, the importance of weed management in this crop should not be underestimated, as weeds, if left unchecked, can severely affect the growth and yield performance of any crop under different conditions and circumstances. As there is a research gap in terms of available weed management options in this crop, this chapter aims to propose several approaches that can be incorporated into integrated weed management (IWM) systems for the sustainable cultivation of hemp. Our research and management experiences related here are drawn on and most applicable to production in Europe.

Common terms used in the chapter:

1 Herbicide: A chemical substance used to control or kill unwanted plants.
2 Pre-emergence treatments: Application of control measures before weeds emerge from the soil.

3 Post-emergence treatments: Application of control measures after weeds have emerged from the soil.
4 Integrated weed management (IWM): A systematic approach using multiple weed control methods.
5 Mechanical weed control: The uprooting of weeds by the use of mechanical tools and machinery.
6 Competitive crop/hemp cultivar: A crop/hemp cultivar that is able to suppress weeds by reducing their emergence and growth.
7 Weed suppression: The reduction of weed density or/and biomass through the application of weed control treatments, cultural practices or hemp competition.
8 Increased plant populations: Increasing the number of hemp plants per unit area to suppress weeds.
9 Narrow row spacing: Reducing the distance between rows of hemp plants to suppress weeds.

2 General recommendations for optimal crop establishment

When it comes to weed management in hemp, one of the most important things to consider is to follow all the appropriate cultural practices that facilitate the rapid emergence of the crop and the formation of a dense crop canopy as soon as possible in the growing season (Višković et al., 2023). Therefore, it is essential to sow seeds at a shallow depth of 2–3 cm into a warm, moist, and well-prepared seed bed (Amaducci et al., 2015). Seedbed preparation includes fall or winter ploughing at a depth of 30–40 cm, followed by disk-harrowing at a depth of 20 cm before sowing in the spring. Cultipacking can also be used to improve seed-soil contact (Kanatas et al., 2021). A suitable soil for hemp growth should have pH of 6.0–7.5; it should be deep, rich in capillary and aeration, rich in nutrients and have a good water holding capacity (Amaducci et al., 2015).

Disk-harrowing to break soil clods and compaction layers is necessary regardless of the soil type to avoid the formation of a compaction layer that can limit root development and make the tap root take an L-shape that reduces nutrient and water absorption (Desanlis et al., 2013). As for sowing date selection, it mostly depends on the prevailing climate for a given agricultural land. Seed germination is optimal when the soil temperature reaches 10–12°C, resulting in maximum crop emergence (Liu et al., 2017). With this soil temperature and with sufficient soil moisture, the crop will emerge in 3 to 5 days and will soon be able to compete well with weeds (Višković et al., 2023).

3 Chemical weed control

Currently, there are no herbicides registered for use in the European Union (EU) to control weeds in hemp fields. This indicates that weed management in this 'minor crop' should first rely on non-chemical cultural practices that exploit the species' competitiveness against weeds and are alternatives to conventional synthetic herbicides (Travlos et al., 2023).

However, minor crops, such as hemp, may be particularly affected by weeds due to a lack of registered herbicides (Sandler and Gibson, 2019). Under this context, Santín-Montanyá et al. (2017) also emphasized that there is an urgent need in the EU for research and investigation into weed management, particularly in minor crops, due to the significant reduction of approved active substances. Therefore, there should be a focus on trials evaluating herbicide efficacy and selectivity in EU countries, with a special emphasis on hemp. Given the experience from the United States of America, the scientific community can get valuable feedback from such experiments.

For instance, results by Flessner et al. (2020) indicated that the crop is tolerant to *S*-metolachlor (HRAC/WSSA Group 15) soil pre-emergence treatments for the control of broadleaf and grass weeds. The same authors also observed adequate selectivity and efficacy for post-emergence treatments with sethoxydim (HRAC/WSSA Group 1) and quizalofop (HRAC/WSSA Group 1) against grass weeds and post-emergence treatments with clopyralid (HRAC/WSSA Group 4) and bromoxynil (HRAC/WSSA Group 6) applied against broadleaf weeds. Similar were the results of the herbicide screening by Ortmeier-Clarke et al. (2022) and Singh et al. (2024).

In one of the very few studies in the EU, and specifically in Romania, Puiu et al. (2023) observed excellent efficacy on grass weeds and crop selectivity of post-emergence treatments with Acetyl-CoA carboxylase (ACCase) inhibitors (HRAC/WSSA Group 1), such as clethodim and fluazifop. These researchers also recorded sufficient efficacy on broadleaf weeds and crop selectivity of clopyralid as a post-emergence treatment and aclonifen (HRAC/WSSA Group 34) as a soil pre-emergence treatment. In another very recent study in the EU and particularly in Greece by Gitsopoulos et al. (2024), soil pre-emergence treatments with aclonifen, pendimethalin (HRAC/WSSA Group 3), and *S*-metolachlor were effective on broadleaf weeds and caused some levels of phytotoxicity, but hemp plants recovered without suffering yield reduction. In the same study, sethoxydim was effective on grass weeds and absolutely safe to the crop. The herbicides recently tested on hemp are summarized below (Table 1).

Here, it should be noted that another perspective of chemical control in this crop is given through the formation of a stale seedbed (Merfield, 2023). Creating a stale seedbed means that sowing is delayed after seedbed

Table 1 Herbicides tested for weed control in recent field trials with hemp

Herbicide	HRAC/WSSA group	Target weeds	References
Sethoxydim	1	Grass	Flessner et al. (2020)
Clethodim	1	Grass weeds	Puiu et al. (2023)
Fluazifop	1	Grass weeds	Puiu et al. (2023)
Quizalofop	1	Grass weeds	Flessner et al. (2020)
Pendimethalin	3	Grass, broadleaf	Gitsopoulos et al. (2024)
Clopyralid	4	Broadleaf	Puiu et al. (2023)
Bromoxynil	6	Broadleaf	Flessner et al. (2020), Ortmeier-Clarke et al. (2022)
S-metolachlor	15	Grass, broadleaf	Flessner et al. (2020)
Aclonifen	34	Broadleaf, grass	Gitsopoulos et al. (2024)

preparation to allow weed emergence; weed emergence is further stimulated by irrigation or incidental rainfall (Travlos et al., 2020). Weed control is carried out after weed emergence and before sowing with a non-selective herbicide, such as glyphosate (HRAC/WSSA Group 9) (Kanatas et al., 2020a). Otherwise, the crop seed can be drilled, followed by glyphosate treatment prior to crop emergence (Merfield, 2023).

The choice regarding the time between initial seedbed preparation and weed control depends on the composition of the weed flora. If perennial weeds exist in the field, growers should time their treatment until weeds reach the flowering stage to enable maximum glyphosate translocation into their underground vegetative storage organs (Peerzada, 2017; Roberts and Florentine, 2022). After the control of early weed cohorts, a competitive crop, such as hemp, can grow in a less competitive environment and smother weed growth. To date, no studies have yet evaluated the potential of stale seedbed for weed management and improved establishment in hemp.

However, there are some preliminary results from Gazoulis et al. (2024), in which the stale seedbed technique was evaluated for weed control in another similar non-food annual crop, namely kenaf (*Hibiscus cannabinus* L.). This very recent study has shown that stale seedbed reduced weed density by 84% compared to the untreated control, even in a trial field infested with noxious perennial summer weeds. This was attributed to the fact that the plants in the stale seedbed plots grew without weed competition, as weeds had already been controlled with glyphosate before sowing (Fig. 1).

In summary, the experimental results to date sound encouraging and suggest that herbicide screening trials on hemp should be conducted in EU

Figure 1 Kenaf emerging in a weed-free environment in stale seedbed trial plots where weeds had already been effectively controlled by glyphosate treatment prior to sowing (Gazoulis et al., 2024).

countries to evaluate herbicide efficacy and the crop's response to herbicide treatments under a wide range of soil-climatic conditions.

4 Mechanical weed control

Mechanical weed control can be an effective measure for weed management in hemp. Inter-row mechanical weed control can be easily done in the areas between the rows of hemp with the common machinery and equipment used for this purpose, as in maize (*Zea mays* L.) and sunflower (*Helianthus annuus* L.) and cotton (Hussain et al., 2018). It should be noted here that inter-row cultivations are less likely to cause mechanical damage to hemp plants only when the crop is established with wide row spacing, as 50 cm, compared to common practices (Campiglia et al., 2017).

The number of inter-row cultivations can vary depending on the level of weed pressure to the crop; multiple passes may be required to achieve sufficient weed control in fields with high weed density per unit area (Alba et al., 2020; Kanatas and Gazoulis, 2022). As for recommendations for mechanical weed control in hemp, Campiglia et al. (2017) recommended hoeing twice, approximately two and four weeks after crop emergence. In Greece, the results of initial trials indicate that double inter-row cultivation can accelerate canopy closure, leading to weed suppression (Kousta and Papastylianou, unpublished data; Fig. 2).

In kenaf, the industrial crop with similar characteristics to hemp, as already mentioned, Alexopoulou et al. (2015) suggest a single hoeing after germination and another one, if necessary, when the young kenaf plants are 15 cm high and weeds are between the white thread to two-leaf growth stages. A drawback of inter-row cultivation is the lack of weed control in the rows of a given hemp stand. To address this challenge, intra-row mechanical weed tools, such as finger-weeders, torsion weeders and flex-tine harrows, can increase the effectiveness of mechanical weed control (Pannacci et al., 2017).

Another option for mechanical weed control is the formation of a false seedbed (Kanatas et al., 2020b). This concept is very similar to the one already described for stale seedbed preparation. For the preparation of a false seedbed, sowing is delayed after seedbed preparation to allow

Figure 2 Hemp in trial plots that were weeded twice with inter-row cultivation (left) compared to untreated control plots (right) (Kousta and Papastylianou, unpublished data).

weed seed germination and emergence (Merfield, 2023). The emergence of weeds can be improved by irrigation or incidental rainfall (Travlos et al., 2020). Weed control is carried out after weed emergence and before sowing by a shallow tillage (5–10 cm depth) operation (Kanatas et al., 2020b). False seedbed controls the early flush of weeds and alleviates the crop from weed pressure during its crucial growth stages. This weed management practice can potentially result in faster canopy closure and improved weed suppression in hemp, but no studies have investigated this specific research topic yet.

5 Competitive cultivars for weed suppression

The selection of competitive cultivars is one of the primary components for weed management in hemp. The competitive ability of different cultivars can be attributed to various traits such as plant height, early vigor, biological cycle length (earliness) and canopy architecture (Korres, 2018).

The morphology and growth habit of some hemp cultivars do not suppress weeds. Jankauskiene et al. (2014) assessed eight monoecious hemp cultivars for weed suppression in Lithuania and reported relatively high weed densities of up to 202 weeds m^{-2}. At hemp harvest, the authors reported greatest weed densities in plots with the early cultivars *USO 31* and *Beniko*, suggesting that early flowering could potentially limit canopy development and, therefore, the ability of hemp to suppress weeds (Jankauskiene et al., 2014). In the study by Vera et al. (2006), the factor of cultivar affected weed density in all experimental years. Specifically, plots with cultivar *Fasamo* had more weeds than those of Finola every year, even though *Fasamo* plants were significantly taller than those of *Finola*. These researchers attributed the difference to the fact that *Finola* established better than *Fasamo* and in higher density suppression, thus, weed growth.

Some hemp cultivars suppress weeds, but this could come at the expense of production goals. In the study by Campiglia et al. (2017) in Italy, *Epsilon 68*, *Futura 75* and *Santhica 27* resulted in the lowest weed biomass compared with other cultivars because they grew rapidly and shadowed the ground quickly. In Greece, *Fedora 17* reduced weed density by 21% compared to *USO 31* (Kousta et al., 2023). Kousta et al. (2023) attributed this difference to the faster canopy closure of *Fedora 17* compared to *USO 31*; *Fedora 17* plants were also significantly taller than those of *USO 31*. Below is a picture of the two cultivars competing with weeds (Kousta et al., 2023; Fig. 3).

The studies on the competitiveness of hemp cultivars against weeds are summarized below (Table 2).

Considering these preliminary results, more research should focus on the competitive ability of hemp cultivars against weeds and explain the difference

Figure 3 The hemp cultivars *Fedora 17* (left) and *USO 31* (right) in a field trial conducted to evaluate their competitiveness against weeds (Kousta et al., 2023).

Table 2 Studies on the competitiveness of hemp cultivars against weeds

Cultivar	Weed infestation	References
USO 31	High	Jankauskiene et al. (2014)
Beniko	High	Jankauskiene et al. (2014)
Fasamo	Low	Vera et al. (2006)
Finola	High	Vera et al. (2006)
Epsilon 68	Low	Campiglia et al. (2017)
Futura 75	Low	Campiglia et al. (2017)
Santhica 27	Low	Campiglia et al. (2017)
Fedora 17	Low	Kousta et al. (2023)
USO 31	High	Kousta et al. (2023)

regarding the weed-suppressive potential of different crop genotypes under real-field conditions and under different soil and climatic conditions. Another research topic that should be evaluated is the allelopathic effects of hemp cultivars on weeds (Pudełko et al., 2014).

6 Increased plant populations and narrow row spacing for weed suppression

Weed management in hemp should first rely on cultural practices that accelerate the development of a dense crop canopy that can shade the ground and suppress weed emergence. This can be achieved by increasing the seed rate of hemp in the row, thereby increasing the plant population per unit area. There is some evidence of the positive effects of increased plant population on weed management in this crop.

In the study by Hall et al. (2014), increasing the hemp population from 100 to 200 plants m^{-2} reduced the weed biomass by 72%. Weed biomass decreased even further (≥ 95%) in plots with higher plant populations (300–400 plants m^{-2}). Increasing the seed rate from 20 to 80 kg seed ha^{-1} increased populations of hemp, which led to a significant reduction in weed density (26–41%) in the field trials by Vera et al. (2006). In a more recent study by Campiglia et al. (2017) in Italy, increasing the seed rate of hemp from 7 kg seed ha^{-1} to 28 kg seed ha^{-1} led to an increase in crop populations from 40 to 120 plants m^{-2}. In the plots where the *Fedora 17* variety was grown, the increase in plant population led to a 53% decrease in weed biomass.

In addition to increasing the seed rate to increase plant populations, another method of suppressing weeds through crop competition is to reduce the row spacing of the crop. Narrow row spacing allows the crop to utilize more light through increased light interception, nutrients and moisture when competing with weeds and accelerates canopy closure, resulting in reduced weed infestation (Peerzada et al., 2017). The study by Vera et al. (2006) is one of the very few that investigated the effects of narrow row spacing on weed suppression in hemp. These authors reported that in the first year of their field trials, the narrow row spacing of 18 cm reduced weed density by 16% compared to the wider row spacing of 36 cm. Other researchers point out that row spacing can be as narrow as 9 cm to achieve a dense and well-developed stand that suppresses weeds (Amaducci et al., 2015).

The studies that reported or indicated the suppression of weeds in hemp by cultural practices that improve crop competitiveness, such as increased plant population and narrow row spacing, are summarized below (Table 3).

Although there are no other reports on the positive effects of narrow row spacing on weed management in this crop, there is much relevant evidence in cotton – a crop most often compared with fiber hemp. For example, Iqbal et al. (2022) found that a narrow row spacing of 50 cm suppressed weed growth by 55% compared to a double row spacing (100 cm). The results of Tursun et al. (2016) also indicate that cotton grown in a narrow row spacing (50 cm) is more competitive against weeds than cotton grown in a wide row spacing (90 cm). In addition, the study by Gazoulis et al. (2024) provides more recent evidence

Table 3 Studies that reported or indicated the suppression of weeds in hemp by cultural practices that improve crop competitiveness, such as increased plant populations and narrow row spacing

Cultural practice	References
Increased plant populations	Hall et al. (2014)
Increased plant populations	Vera et al. (2006)
Increased plant populations	Campiglia et al. (2017)
Narrow row spacing	Vera et al. (2006)
Narrow row spacing	Amaducci et al. (2015)

of the weed suppression potential of narrow row spacing in kenaf. In particular, this study found that a narrow row spacing of 30 cm reduced total weed density by 42% compared to a narrow row spacing of 60 cm. The effects of narrow row spacing on weed biomass were similar. Hemp has also been grown with different row spacings in preliminary trials by Kousta and Papastylianou in Greece, where in trial plots with narrow row spacing, hemp smothered weed emergence and growth (unpublished data; Fig. 4).

7 The role of crop rotation and cover crops

Crop rotation is one of the most valuable crop diversification practices that can be used to manage weeds in all crops, including hemp. As different crops

Figure 4 Hemp stand established at a wide row spacing (left) and a narrow row spacing (right) (Kousta and Papastylianou, unpublished data).

with different characteristics are grown in a given field over the years, different agronomic practices are applied each year, i.e. tillage, fertilization and herbicide use (Travlos et al., 2023). This 'unbalances' the weeds (Brankov et al., 2021). As weeds are unable to adapt, they become less competitive and persistent in the field (Korres, 2018).

Maize is probably one of the most suitable preceding crops for hemp in terms of weed management. This can be attributed to the fact that a wide range of herbicides with different Sites of Action (SoAs) are available for use in maize. Therefore, weed control with different herbicides SoAs in maize years can reduce the reproductive dynamics of weeds, resulting in lower weed pressure on the hemp plants in the following year. Crop rotation with maize is one of the best recommended practices for weed control in deciduous leafy crops, as also highlighted in the review by Matzrafi et al. (2023).

To suppress weeds in hemp, a cover crop grown in winter can also be included in the crop rotation, as is generally recommended for other important summer crops such as cotton, maize and soybean (*Glycine max* (L.) Merr.) (Büchi et al., 2020; Toler et al., 2019; Wallace et al., 2018). Under Mediterranean conditions, as also shown in the study by Antonopoulos et al. (2024), winter-grown cereals, legumes and species from the botanical family Brassicaceae are very promising cover crops in terms of biomass production and weed suppression (Fig. 5).

The most common practice is to plant such cover crops in early fall so that they accumulate significant amounts of biomass during the winter rainy season. In spring, the cover crops are removed using mechanical methods such as disk-harrowing, and the residues are incorporated into the soil to suppress the emergence of weeds (Travlos et al., 2023).

Although there is no evidence of the potential of cover crops for weed suppression in hemp, field trials should be conducted to investigate this area of research and facilitate agroecological weed management in this new dynamic multipurpose non-food crop. The same applies to the effects of crop rotation on weed management in hemp.

8 The role of fertilization management

Hemp is characterized as a high-yielding crop that requires low-input cultivation practices (Amaducci et al., 2015; Tedeschi et al., 2020). Nevertheless, experimental data show the crop's high dependence on the supply of nutrients and water (Amaducci et al., 2015). Among the nutritional elements, nitrogen is the most important macronutrient as it plays a crucial role in the growth and yield of hemp (Masclaux-Daubresse et al., 2010). The increased plant height and dry biomass, inflorescence length and weight, seed oil content and cannabinoid

Figure 5 Cover crops (cereals + legumes + cruciferous plants) suppressing weed growth (Antonopoulos et al., 2024).

(CBD) content due to the additional nitrogen supply make this cultivation practice essential (Kakabouki et al., 2021; Papastylianou et al., 2018; Vera et al., 2004).

Nutrients are also necessary for weeds, especially for those that are very sensitive to nitrogen (Blackshaw et al., 2003). Nitrogen applications influence competition between crops and weeds and promote weed establishment, emergence, density and development (Kaur et al., 2018; Kousta et al., 2023; Travlos et al., 2019). A precise nitrogen supply is necessary to minimize the losses in hemp yield and quality caused by weeds, as well as difficulties during harvest. Fertilizer application under the crop rows is an efficient approach as it contributes to optimal access of hemp to nutrients and facilitates rapid emergence and growth of hemp (Gazoulis et al., 2021). In contrast, the application of fertilizers promotes germination, especially of annual weeds in the upper soil layers.

Another practice is the application of fertilizer at a certain depth, which reduces the germination of weed seeds and hinders the germinated seedlings' access to the supplied nutrients (Kaur et al., 2018). Experimental data have shown that weed biomass and weed seed production are reduced by subsurface application of banded and spot-applied nitrogen compared to surface application of nitrogen (Blackshaw et al., 2004). Applying fertilizers below the crop rows and at a certain depth promotes germination and rapid growth of hemp. Once hemp reaches a height of 0.15–0.20 m, it can suppress most weed species (Visković et al., 2023), especially late-germinating weeds. Nevertheless, the effectiveness of fertilization on the interactions between hemp and weeds may vary depending on the weed species.

Particularly, fertilizer application in the soil depth of 0–15 cm seemed to have a pronounced effect on weeds with fibrous root systems, whereas it had little impact on weeds with tap roots (Blackshaw et al., 2004; Gazoulis et al., 2021). Special attention should be paid to the optimal N fertilization rate to maximize the financial gain. Nitrogen over-sufficiency reduces hemp growth and yield (Saloner and Bernstein, 2021; Yang et al., 2021) while promoting the appearance and growth of several weeds.

9 The role of irrigation management

As far as the plant's water requirements are concerned, hemp, as a short-day plant, requires a lot of moisture (Adesina et al., 2020). The addition of irrigation water is considered necessary for the cultivation of hemp, as drought is the most important factor for abiotic stress compared to other relevant stress factors (Gao et al., 2018). Short-term water stress is weighted by the hemp plants with various physiological and morphological changes, but the long-term drought causes irreversible effects on plant growth and productivity (Gill et al., 2022).

As the severity and frequency of water scarcity intensifies, the provision of irrigation becomes essential, especially in the semi-arid areas of the Mediterranean region (Di Bari et al., 2004). Studies in Europe have shown that the crop requires a total of 500–700 mm of water (Amaducci et al., 2015; Di Bari et al., 2004; Cosentino et al., 2013; Blandinières and Amaducci, 2022), of which at least 250–350 mm is required during the vegetative stage (Adesina et al., 2020; Blandinières and Amaducci, 2022; Cosentino et al., 2013; Herppich et al., 2020). The high water demand of hemp during the vegetative stage increases its vulnerability to weed infestation.

Drip irrigation and subsurface drip irrigation can lead to optimal irrigation management that provides hemp with a certain amount of water (Coolong, 2013). The installation cost of drip irrigation is higher than that of conventional sprinkler irrigation, but it seems to play a conspicuous role in the design of weed management methods in integrated agricultural systems. More specifically, surface drip irrigation retains available soil water and provides water to established plants while weed seeds in the row and near the surface remain inactive and dry (Schonbeck and Tillage, 2011). Both drip irrigation and surface irrigation therefore, improve the growth of the plants' root system and their competitiveness against weeds (Coolong, 2013).

A field trial with tomatoes (*Solanum lycopersicum* L.), a crop long used as a surrogate in *Cannabis* spp. research, confirmed this assertion and showed a 90–99% reduction in weed density after subsurface drip irrigation compared to furrow irrigation (Shrestha et al., 2007). Similar results were shown by de Freitas Souza et al. (2021) in onions (*Allium cepa* L.), where drip irrigation reduced weed infestation by 62–97% in the three following growing seasons compared to conventional methods. In addition, irrigation timing could be a useful practice to better control weed infestations in hemp. Several studies provided valuable information on the optimal irrigation rates for this crop based on environmental conditions (Amaducci et al., 2015; Blandinières and Amaducci, 2022; Cosentino et al., 2013; Di Bari et al., 2004), genotype (Cosentino et al., 2013; Tang et al., 2018) and the required product quality traits (Spano et al., 2022), while no research information on the available effects of irrigation timing can be found in the literature.

10 Conclusion and future trends

The aim of this chapter was to propose several approaches that can be integrated into IWM systems for the sustainable cultivation of hemp in European agriculture. When it comes to weed management in hemp, it is particularly important to follow all appropriate cultural practices that allow the crop to emerge quickly and form a dense crop canopy as early as possible in the growing season to suppress weed growth while at the same time not diminishing production goals. Currently, no herbicides are approved for weed control in hemp fields in the EU. Therefore, field trials are needed to evaluate the efficacy and selectivity of herbicides in EU countries and much of the world, with a particular focus on hemp. A perspective of chemical control in this crop is given by the formation of a stale seedbed, where glyphosate applications can facilitate weed control. In the absence of chemical weed control options, weed management in hemp should primarily rely on non-chemical cultivation methods that exploit the species' competitiveness against weeds. The selection of competitive cultivars is one of the most important components for weed management in hemp. Agronomic practices such as increased seed rates and plant populations, and narrow row spacing that accelerate the development of a dense crop canopy that can shade the ground and suppress weed emergence are valuable. Crop diversification through crop rotation and cover cropping is considered important, while the role of optimal fertilization and irrigation management should also play a central role in the design of IWM systems for this dynamic multipurpose crop. The IWM tools for sustainable hemp cultivation are summarized in the scheme below (Fig. 6).

Figure 6 Summary of IWM tools for sustainable hemp cultivation.

Further research is needed to fill the research gap in hemp-weed interactions and to identify the most troublesome weeds in hemp fields at different geographical latitudes. In addition, robust, spatially and temporally replicated field trials are needed to evaluate the combined effects of weed control methods and agronomic practices on weed infestation, crop yield and quality of industrial hemp. It should always be kept in mind that although hemp is a competitive crop, weeds can still be a serious obstacle to the establishment and productivity of the crop. Therefore, the development of sustainable IWM strategies for hemp is essential to increase its spread and adoption as a commercial field crop with multiple industrial applications.

11 Where to look for further information

The following links offer useful insights into weed management in hemp. They summarize the experiences with various agronomic practices from the United States and Canada, where the particular research topic was studied:

- Industrial Hemp Tolerance to Soil-applied Herbicides. University of Nebraska–Lincoln https://cropwatch.unl.edu/2020/industrial-hemp-tolerance-soil-applied-herbicides.

- Creating a weed management plan for hemp. Michigan State University Extension https://www.canr.msu.edu/news/creating-a-weed-management-plan-for-hemp.
- Weeds - Purdue Agriculture. Purdue University https://ag.purdue.edu/hemp-project/pest/weeds.html.
- Scientists Focus on Weed Control to Support Hemp Market. Greenhouse Product News. https://gpnmag.com/news/scientists-focus-on-weed-control-to-support-hemp-market/.
- Weed Control. Canadian Hemp Trade Alliance. https://www.hemptrade.ca/content.aspx?page_id=22&club_id=950211&module_id=409584n.
- Handout: 2021 Weeds and Weed Management. Cornell University. https://bpb-us-e1.wpmucdn.com/blogs.cornell.edu/dist/a/7491/files/2021/08/2021-Hemp-Field-Day-Handout-Weeds-Weed-Mgt-1.pdf.
- 2022 Weeds and Weed Management - Cornell Hemp. Cornell University. https://hemp.cals.cornell.edu/resources/reports-factsheets/2022-cornell-hemp-field-day-handouts/2022-weeds-and-weed-management/.
- In the Weeds. Cannabis Business Times. https://www.cannabisbusinesstimes.com/article/tips-identify-prevent-control-weeds-hemp/.

12 References

Adesina, I., et al. (2020), A review on the current state of knowledge of growing conditions, agronomic soil health practices and utilities of hemp in the United States, *Agriculture* 10: 129.

Alba, O. S., et al. (2020), Increased seeding rate and multiple methods of mechanical weed control reduce weed biomass in a poorly competitive organic crop, *Field Crops Res.* 245: 107648.

Alexopoulou, E., et al. (2015), How kenaf (*Hibiscus cannabinus* L.) can achieve high yields in Europe and China, *Ind. Crops Prod.* 68: 131–140.

Amaducci, S., et al. (2015), Crop yield and quality parameters of four annual fibre crops (hemp, kenaf, maize and sorghum) in the North of Italy, *Ind. Crops Prod.* 11: 179–186.

Antonopoulos, N., et al. (2024), The potential of cover crop (cc) mixtures to suppress winter and summer weeds in citrus orchards, *XX International Plant Protection Congress: Healthy Plants Support Human Welfare*, Greece, 75.

Aubin, M. P., et al. (2015), Hemp response to nitrogen, phosphorus, and potassium fertilization, *Crop Forage Turfgrass Manag.* 1: 1–10.

Bajić, I., et al. (2022), The effects of irrigation, topping, and interrow spacing on the yield and quality of hemp (*Cannabis sativa* L.) fibers in temperate climatic conditions, *Agriculture* 12: 1923.

Blackshaw, R. E., et al. (2003), Differential response of weed species to added nitrogen, *Weed Sci.* 51: 532–539.

Blackshaw, R. E., et al. (2004), Weed species response to phosphorus fertilization, *Weed Sci.* 52: 406–412.

Blandinières, H. and Amaducci, S. (2022), Adapting the cultivation of hemp (*Cannabis sativa* L.) to marginal lands: a review, *GCB Bioenergy* 14: 1004–1022.

Brankov, M., et al. (2021), The influence of maize–winter wheat rotation and pre-emergence herbicides on weeds and maize productivity, *Crop Prot.* 143: 105558.

Büchi, L., et al. (2020), Cover crops to secure weed control strategies in a maize crop with reduced tillage, *Field Crops Res.* 247: 107583.

Burton, R. A., et al. (2022), Hemp seed: From the field to value-added food ingredients, *J. Cannabis Res.* 4: 45.

Campiglia, E., et al. (2017), Plant density and nitrogen fertilization affect agronomic performance of hemp (*Cannabis sativa* L.) in Mediterranean environment, *Ind. Crops Prod.* 100: 246–254.

Coolong, T. (2013), Using irrigation to manage weeds: A focus on drip irrigation, in S. Soloneski and M. Larramendy (eds), *Weed and Pest Control-Conventional and New Challenges*, 161–179. Intech, London, UK.

Cosentino, S. L., et al. (2012), Sowing time and prediction of flowering of different hemp (*Cannabis sativa* L.) genotypes in southern Europe, *Ind. Crops Prod.* 37: 20–33.

Cosentino, S. L., et al. (2013), Evaluation of European developed fibre hemp genotypes (*Cannabis sativa* L.) in semi-arid Mediterranean environment, *Ind. Crops Prod.* 50: 312–324.

Crini, G., et al. (2020), Traditional and new applications of hemp, in G. Crini and E. Lichtfouse (eds), *Sustainable agriculture reviews 42: hemp production and applications*, 37–87. Springer, Cham, Switzerland.

de Freitas Souza, M., et al. (2021), Can irrigation systems alter the critical period for weed control in onion cropping?, *Crop Prot.* 147: 105457.

Deng, G., et al. (2019), Planting density and fertilization evidently influence the fiber yield of hemp (*Cannabis sativa* L.), *Agronomy* 9: 368.

Desanlis, F., et al. (2013), Hemp agronomics and cultivation, in P. Bouloc, et al. (eds), *Hemp: industrial production and uses*, 98–124. CAB International, Wallingford, UK.

Di Bari, V., et al. (2004), Potential productivity of fibre hemp in southern Europe, *Euphytica* 140: 25–32.

Dudziec, P., et al. (2024), Hemp as a multi-purpose crop: last achievements and research in 2018–2023, *J. Nat. Fibers* 21: 2369186.

Fike, J. (2016), Hemp: renewed opportunities for an ancient crop, *Crit. Rev. Plant Sci.* 35: 406–424.

Flessner, M. L., et al. (2020), Evaluating herbicide tolerance of hemp (*Cannabis sativa* L.), *Crop Sci.* 60: 419–427.

Gao, S., et al. (2018), Dynamic responses of tree-ring growth to multiple dimensions of drought, *Glob. Change Biol.* 24: 5380–5390.

García-Tejero, I. F., et al. (2019), Seeking suitable agronomical practices for hemp (*Cannabis sativa* L.) cultivation for biomedical applications, *Ind. Crops Prod.* 139: 111524.

Gazoulis, I., et al. (2021), Weed management practices to improve establishment of selected lignocellulosic crops, *Energies* 14: 2478.

Gazoulis, I., et al. (2024), Cultural practices for weed management in kenaf crop (*Hibiscus cannabinus* L.), *XX International Plant Protection Congress: Healthy Plants Support Human Welfare*, Greece, 75.

Gill, A. R., et al. (2022), Physiological and morphological responses of industrial hemp (*Cannabis sativa* L.) to water deficit. *Ind. Crops Prod.* 187: 115331.

Gitsopoulos, T., et al. (2024), Response of hemp (*Cannabis sativa* L.) to herbicides and weed control, *Int. J. Plant Biol.* 15: 281–292.

Hall, J., et al. (2014), Effect of hemp (*Cannabis sativa* L.) planting density on weed suppression, crop growth, physiological responses, and fibre yield in the subtropics, *Renew. Bioresour.* 2: 1–7.

Herppich, W. B., et al. (2020), Effects of drought and heat on photosynthetic performance, water use and yield of two selected fiber hemp cultivars at a poor-soil site in Brandenburg (Germany), *Agronomy* 10: 1361.

Hussain, M., et al. (2018), Mechanical weed control, in K. Jabran and B. S. Chauhan (eds), *Non-chemical weed control*, 133–155. Academic Press, Cambridge, MA, USA.

Iqbal, N., et al. (2022), Effect of narrow row-spacing and weed crop competition duration on cotton productivity, *Arch. Agron. Soil Sci.* 68: 355–367.

Jankauskienė Z., et al. (2014), Potential of industrial hemp (*Cannabis sativa* L.) genotypes to suppress weeds. *Zemdirbyste*, 101: 265–270.

Kakabouki, I., et al. (2021), Effect of fertilization with urea and inhibitors on growth, yield and cbd concentration of hemp (*Cannabis sativa* L.), *Sustainability* 13: 2157.

Kanatas, P. J. and Gazoulis, I. (2022), The integration of increased seeding rates, mechanical weed control and herbicide application for weed management in chickpea (*Cicer arietinum* L.), *Phytoparasitica*, 50: 255–267.

Kanatas, P. J., et al. (2020a), Yield, quality and weed control in soybean crop as affected by several cultural and weed management practices, *Not. Bot. Horti Agrobo. Cluj Napoca* 48: 329–341.

Kanatas, P. J., et al. (2020b), The combined effects of false seedbed technique, post-emergence chemical control and cultivar on weed management and yield of barley in Greece, *Phytoparasitica* 48: 131–143.

Kanatas, P. J., et al. (2021), Irrigation timing as a practice of effective weed management in established alfalfa (*Medicago sativa* L.) crop, *Agronomy* 11: 550.

Kaur, S., et al. (2018), Understanding crop-weed-fertilizer-water interactions and their implications for weed management in agricultural systems, *Crop Prot.* 103: 65–72.

Korres, N. E. (2018), Agronomic weed control: a trustworthy approach for sustainable weed management, in K. Jabran and B. S. Chauhan (eds), *Non-chemical weed control*, 97–114. Academic Press, Cambridge, MA, USA.

Kousta, A., et al. (2023), Effect of fertilization and weed management practices on weed diversity and hemp agronomic performance, *Agronomy* 13: 1060.

Liu, M., et al. (2017), Comparison of traditional field retting and *Phlebia radiata* Cel 26 retting of hemp fibres for fibre-reinforced composites, *AMB Express* 7: 58.

Lotz, L. A. P., et al. (1991), Reduction of growth and reproduction of Cyperus esculentus by specific crops, *Weed Res.* 31: 153–160.

Masclaux-Daubresse, C., et al. (2010), Nitrogen uptake, assimilation and remobilization in plants: challenges for sustainable and productive agriculture, *Ann. Bot.* 105: 1141–1157.

Matzrafi, M., et al. (2023). *Amaranthus palmeri* S. Watson: a new threat to agriculture in Europe and the Mediterranean region, *Weed Res.* (in press). https://doi.org/10.1111/wre.12596.

Merfield, C. N. (2023), Integrated weed management in organic farming, in S. Chandran, et al. (eds), *Woodhead Publishing series in food science, technology and nutrition, organic farming* (2nd ed.), 31–109. Woodhead Publishing, Cambridge, UK.

Ortmeier-Clarke, H. J., et al. (2022), Dose–response screening of hemp to herbicides commonly used in corn and soybean, *Weed Technol.* 36: 245–252.

Pannacci, E., et al. (2017), Non-chemical weed management strategies in minor crops: a review, *Crop Prot.* 96: 44–58.

Papastylianou, P., et al. (2018), Effect of nitrogen fertilization on growth and yield of hemp (*Cannabis sativa* L.), *Not. Bot. Horti Agrobo. Cluj Napoca* 46: 197–201.

Peerzada, A. M. (2017), Biology, agricultural impact, and management of *Cyperus rotundus* L.: the world's most tenacious weed, *Acta Physiol. Plant.* 39: 270.

Peerzada, A. M., et al. (2017), Weed management in sorghum [*Sorghum bicolor* (L.) Moench] using crop competition: a review, *Crop Prot.* 95: 74–80.

Pudełko, K., et al. (2014), Allelopathic effect of fibre hemp (*Cannabis sativa* L.) on monocot and dicot plant species, *Ind. Crops Prod.* 56: 191–199.

Puiu, I., et al. (2023), Weeds control in hemp (*Cannabis sativa* L.) by using herbicides in pre-emergency and post-emergency, *Sci. Pap. Ser. A Agron.* 66: 356–361.

Roberts, J. and Florentine, S. (2022), Biology, distribution and management of the globally invasive weed *Solanum elaeagnifolium* Cav (silverleaf nightshade): a global review of current and future management challenges, *Weed Res.* 62: 393–403.

Saloner, A. and Bernstein, N. (2021), Nitrogen supply affects cannabinoid and terpenoid profile in medical cannabis (*Cannabis sativa* L.), *Ind. Crops Prod.* 167: 113516.

Sandler, L. N. and Gibson, K. A. (2019), A call for weed research in hemp (*Cannabis sativa* L), *Weed Res.* 59: 255–259.

Santín-Montanyá, M. I., et al. (2017), Weed control in minor crops in Spain after adoption of European regulation 1107/2009/EC and directive 2009/128/EC, *Outlook Agric.* 46: 66–73.

Schonbeck, M. and Tillage, B. (2011), *Principles of sustainable weed management in organic cropping systems*, Workshop for Farmers and Agricultural Professionals on Sustainable Weed Management, Clemson, SC, USA, Vol. 3, 1–24.

Schumacher, A. G. D., et al. (2020), Hemp fiber: A sustainable and economical alternative to cotton, *J. Clean. Prod.* 268: 122180.

Shrestha, A., et al. (2007), Subsurface drip irrigation as a weed management tool for conventional and conservation tillage tomato (*Lycopersicon esculentum* Mill.) production in semi-arid agroecosystems. *J. Sustain. Agric.* 31: 91–112.

Singh, A., et al. (2024), Impact of soil-residual herbicides on hemp (*Cannabis sativa*) phytotoxicity and biomass yield in west Texas, *Weed Technol.* 38: e29.

Spano, M., et al. (2022), Hemp (*Cannabis sativa* L.) inflorescences as novel food: the effect of different agronomical practices on chemical profile, *Foods* 11: 3658.

Tang, K., et al. (2018), Water-and nitrogen-use efficiencies of hemp (*Cannabis sativa* L.) based on whole-canopy measurements and modeling, *Front. Plant Sci.* 9: 1–14.

Tedeschi, A., et al. (2020), Soil fertilization with urea has little effect on seed quality but reduces soil N_2O emissions from a hemp cultivation, *Agriculture* 10: 240.

Toler, H. D., et al. (2019). Global meta-analysis of cotton yield and weed suppression from cover crops, *Crop Sci.* 59(3): 1248–1261.

Travlos, I., et al. (2019), Changes of weed flora due to nitrogen addition in sunflower, *Not. Bot. Hort. Agrobo. Cluj Napoca* 47: 1337–1339.

Travlos, I., et al. (2020), Key factors affecting weed seeds' germination, weed emergence, and their possible role for the efficacy of false seedbed technique as weed management practice, *Front. Agron.* 2: 1.

Travlos, I., et al. (2023), The underestimated role of cultural practices in ecologically based weed management approaches, in N. E. Korres, et al. (eds), *Ecologically-based weed management: concepts, challenges, and limitations*, 75–92. John Wiley & Sons, Inc., Hoboken, NJ, USA.

Tursun, N., et al. (2016), Row spacing impacts the critical period for weed control in cotton (*Gossypium hirsutum*). *Phytoparasitica* 44: 139–149.

Vera, C. L., et al. (2004), The effect of N and P fertilization on growth, seed yield and quality of hemp in the Parkland region of Saskatchewan, *Can. J. Plant Sci.* 84: 939–947.

Vera, C. L., et al. (2006), Seeding rate and row spacing effect on weed competition, yield and quality of hemp in the Parkland region of Saskatchewan, *Can. J. Plant Sci.* 86: 911–915.

Visković, J., et al. (2023), Hemp (*Cannabis sativa* L.) agronomy and utilization: a review, *Agronomy*, 13: 931.

Wallace, J. M., et al. (2018), Integrated weed management strategies in cover crop–based, organic rotational no-till corn and soybean in the mid-Atlantic region, *Weed Sci.* 66: 94–108.

Yang, Y., et al. (2021), Effect of nitrogen supply on growth and nitrogen utilization in hemp (*Cannabis sativa* L.), *Agronomy* 11: 2310.

Yazici, L. (2023), Optimizing plant density for fiber and seed production in hemp (*Cannabis sativa* L.), *J. King Saud Univ. Sci.* 35: 102419.

Zhao, J., et al. (2020), Bioconversion of hemp biomass for bioethanol production: a review, *Fuel* 281: 118725.

Part 4

Social, business, economic and political issues

Chapter 12

Canada's regulatory stance on industrial hemp production

Richard Haigh and Eli Remington, Osgoode Hall Law School – York University, Canada

1 Introduction

Canada is a federal nation. Like all federations, legislative power is divided between national and sub-national units. In Canada, these powers are largely set out in ss. 91 and 92 of the *Constitution Act, 1867*.[1] Section 91 grants to the federal Parliament certain exclusive powers over certain subject matters, and s. 92 grants certain other legislative powers to the provinces. There is no true residuary power.[2]

Unlike many federal nations where criminal law powers are devolved to the more local level, the power to pass criminal laws in Canada is granted exclusively to the federal Parliament in s. 91(27) of the *CA, 1867*. This means that criminal laws are uniform throughout Canada – most of these are codified in the *Criminal Code*, a longstanding federal statute. Laws dealing with illicit drugs, for reasons not relevant to this chapter, are generally not found in the

1 30 & 31 Victoria, c. 3 (U.K). (Hereafter, "*CA, 1867*"). Sections 93–95 also set out legislative powers but are of lesser relevance to the discussion.

2 As noted above, ss. 91 and 92 of the *CA, 1867*, divide power amongst the federal and provincial governments. Much of this authority is distributed under enumerated 'heads of power'; for instance, s. 91(10) grants the federal legislature authority over 'Navigation and Shipping.' The non-exhaustive design of this formula was originally accounted for by the opening of s. 91, which grants legislative power over subject matters falling outside of these federal and provincial heads of power – that is, all residuary legislative powers – to the federal legislature as a power to enact laws for the 'Peace, Order and good Government of Canada'. While this arrangement historically tipped the balance of power in favour of the federal government, an increasingly expansive interpretation by the courts of the provinces' power over 'Property and Civil Rights in the Province' (s. 92(13)) has seen it transform into 'a kind of residuary power itself' (see Peter Hogg and Wade Wright, *Constitutional Law of Canada*, 5th ed. (Toronto, Canada: Carswell, 2024), s. 17:1).

http://dx.doi.org/10.19103/AS.2025.0161.25

Criminal Code but in a separate federal statute, the *Controlled Drugs and Substances Act* (*CDSA*).

Cannabis, in both marijuana and hemp form, was once illegal in Canada but lawsuits and legislation slowly changed things. Industrial hemp licenses began to be issued in the 1990s with the *Industrial Hemp Regulations* codifying much of the industry in 1998.[3] In addition, challenges to simple marijuana possession charges were fought around the same time using the *Canadian Charter of Rights and Freedoms*.[4] For example, those with medical conditions such as epilepsy were able to establish that medical marijuana could control seizures and thus, criminalizing such use was a breach of one's right to life, liberty and security of the person under s. 7 of the *Charter*.[5]

The culmination of this de-criminalizing trend took place on October 17, 2018, when the *Cannabis Act* and *Cannabis Regulations* came into force.[6] From that day onward, cannabis was no longer listed as an illegal drug under the *CDSA* and instead became regulated under its own specific law, allowing it to be used in much the same way as tobacco.[7] As part of these changes, hemp was also brought under the cannabis umbrella, regulated mainly through its own delegated legislation, the *Industrial Hemp Regulations*.

Exactly 1 year later, on October 17, 2019, a new set of *Cannabis Regulations* were introduced to include additional categories of cannabis edibles, topicals and extracts (considered colloquially as 'Cannabis 2.0').[8] This 1-year delay was due, in part, to the complexity of ensuring that regulations involving Cannabis 2.0 products addressed unique public health and public safety concerns, including the likely appeal they would have to youth, their risk of accidental consumption, overconsumption and the possibility of foodborne illness.[9]

This chapter is intended mainly to outline Canada's approach to the licensing and regulation of hemp, as a product derived from the cannabis plant. Before proceeding, it is important to provide some terminological clarification for the purposes of the discussion that follows. These terms are

3 Hemp regulations were promulgated under the old *Controlled Drugs and Substances Act*.

4 *Constitution Act, 1982* [*CA, 1982*].

5 *R v Parker* (2000) OJ No. 2787 (ONCA) [*Parker*].

6 *Cannabis Act*, SC 2018, c 16; *Cannabis Regulation*, SOR/2018-144.

7 As a side effect of this, the *Access to Cannabis for Medical Purposes Regulations* and the *Industrial Hemp Regulations* previously enabled under the *CDSA* were repealed and replaced by Part 14 of the *Cannabis Regulations* and the new *Industrial Hemp Regulations* enabled under the *Cannabis Act*. See *Access to Cannabis for Medical Purposes Regulations*, SOR/2016-230, as rep. by *Regulations Amending and Repealing Certain Regulations Made Under the Controlled Drugs and Substances Act*, SOR/2018-147, s. 33; *Cannabis Regulations*, SOR/2018-144; and *Industrial Hemp Regulations*, SOR/2018-145.

8 "Health Canada Finalizes Regulations for the Production and Sale of Edible Cannabis, Cannabis Extracts and Cannabis Topicals" *Health Canada* (June 14, 2019) online: Health Canada News Release https://www.canada.ca/en/health-canada/news/2019/06/health-canada-finalizes-regulations-for-the-production-and-sale-of-edible-cannabis-cannabis-extracts-and-cannabis-topicals.html.

9 Canada, Department of Health, *Regulations Amending the Cannabis Regulations (New Classes of Cannabis)–Regulatory Impact Analysis Statement*, Volume 152 (Ottawa: Canada Gazette, Part 1, 2018), Number 51.

Canadian definitions – based on Canadian laws or legal decisions – but they mirror closely terms used in many other jurisdictions:[10]

- *Cannabis*: the plant and plant parts that contain both 'marijuana' and 'industrial hemp'.
- *Marijuana*: although no longer used in a legal sense in Canada, cannabis plants or plant parts that contain 0.3% % weight per weight ('w/w', or 0.3 g of THC for every 100 g) or more of the intoxicant delta-9-tetrahydrocannabinol (THC) are still frequently referred to as marijuana (or marihuana). We will use this term to differentiate from hemp as it is still commonly used.
- *Industrial hemp*: a cannabis plant – or any part of that plant – in which the concentration of THC is no more than 0.3% in the flowering heads and leaves.[11]
- *De-criminalization*: the removal of an activity, substance, etc. from the purview of the criminal law. Of course, this does not mean that a substance is not otherwise regulated – think of almost all foods as subject to some form of regulation, be it labelling, grading, etc.; nor does it mean complete removal from criminalization as a substance may be dealt with by criminal laws related to trafficking but may not be criminalized for simple possession.
- *Medical marijuana*: the use of cannabis for a medical purpose. The broad definition reflects the disputed nature of this category, a point that will be returned to when examining the case law below.
- *Parliament*: the legislature responsible for passing the country's federal laws.
- *Canadian constitutional law*: there are two primary legislative components to Canada's constitutional law that matter for purposes of this chapter. The *CA, 1867* deals largely with the federal structure of Canada; courts may rely on provisions of the *CA, 1867* to strike down laws that are beyond a legislature's competence within the subject matters created by the *CA, 1867*. In contrast, the *Canadian Charter of Rights and Freedoms*, passed as a part of the *CA, 1982*, introduced a constitutional bill of rights in the country. Each right has developed its own jurisprudence, although some have received significantly more attention than others. Courts can also strike down legislative provisions as offending these rights.

10 This book is an attempt to provide some clarity to the botanical complexities of hemp. In Canadian law, for example, it is still unclear whether cannabis is one plant, three plants or four. The Supreme Court of Canada, in *Perka v R* [1984] 2 SCR 232, has held that there are three forms (cannabis sativa, cannabis indica, and cannabis ruderalis), but the case for four has been made by J.M. McPartland and E. Small, "A Classification of Endangered High-THC Cannabis (*Cannabis sativa subsp. Indica)* Domesticates and their Wild Relatives" (2020) 144 *Phytokeys*. Of course, the latter paper is related to high-THC cannabis which therefore is not inclusive of hemp varieties!

11 *Industrial Hemp Regulations*, SOR/2018-145, s. 1(2).

- *Section 1 of the Charter*: signifies that rights under the *Charter* are not absolute. Governments may encroach on *Charter* rights and a court can deem a government action or law infringing an individual's *Charter* right as constitutional if it can be 'demonstrably justified in a free and democratic society'.[12]
- *Regulations*: regulations are more particular formulations of statutory law – referred to sometimes as 'delegated' laws. They are not produced by legislatures but rather by the executive. Regulations can play a significant role in particularizing the legal governance in an area. As noted below, there are over 370 regulations in the *Cannabis Act*.

One of the problems arising out of the terms cannabis, marijuana, and hemp – and the basis for the title of our chapter – is that Canada has chosen to use the umbrella term 'cannabis' to describe the plant and regulate broadly under that basis both the 'recreational use' of the plant, the medicinal use and its use as hemp.[13] There are both benefits and pitfalls of this decision, which we will outline further.[14]

The chapter unfolds as follows: in the next section, we provided a brief historical overview of the regulation of cannabis and hemp in Canada, from the origins of the country in the late 1800s through to the present. In section 3 of this chapter, we delve deeper into the licensing requirements for cultivation, possession, and importation of industrial hemp before examining the regulatory framework for medicinal marijuana in Canada in section 4. Section 5 provides some broad conclusions. Note that as legal scholars, we use expansive footnotes, which is the legal scholars' method of providing a reader with further information.

2 History of regulation of cannabis and hemp

Canada formed as a country in 1867, 'confederating' from three distinct and separate provinces (Canada, Nova Scotia, and New Brunswick) into a single 'dominion'. Even before this event, it is known that cannabis existed and was

12 *CA, 1982*, *supra* note 4, s. 1.

13 Our use of general rather than scientific terms reflects the Canadian legal discourse that is the focus of this chapter. We will retain the use of 'marijuana' in this chapter generally to indicate recreational and medical use (albeit recognizing that there are concerns over the term; see Robert A. Mikos and Cindy D. Kam, "Has the 'M' Word been Framed? Marijuana, Cannabis, and Public Opinion" (2019) 14(10) *PLoS One* e0224289). (Note that 'adult use' is preferred by some in the hemp industry to differentiate the two. We are agnostic on this but will use recreational as it tends to be much more common and easily understood.) As should become abundantly clear, terminological variations complicate matters related to cannabis, and part of the difficulty of having a clear legal landscape for its various uses.

14 It's an imperfect analogy but bringing hemp regulation under a de-criminalization and administrative scheme related to cannabis is akin to having grapes grown for eating brought under a broader regulatory scheme focused on wine. Or, if you will, using cannabis as an umbrella term for the entirety of the plant, but then also for a particular (recreational) use – all while recognizing hemp as something different – would be akin to using 'cow' to represent the animal, 'milk' to represent the liquid commodity, and then 'cow' again (instead of 'beef') to represent the solid meat product.

used – a 1606 record noted that Samuel de Champlain's apothecary, Louis Hébert, cultivated it in Nova Scotia for its hemp fibres.[15] During the mid-eighteenth-century Seven Years' War, Britain paid a premium for hemp rope from Russia. When Russia raised its prices, Britain tried to encourage production in the colonies, including Canada.[16] As well, the plant was regularly harvested for medical purposes beginning around 1840 and continued to be prescribed into the early 1900s.[17]

Although it was a recognized industry, hemp production remained small. A number of factors contributed to this: a limited supply of consistent and viable industrial hemp seed; the burdensome task of clearing and preparing land; the attractiveness of alternative industries (wheat had the advantage of fetching high prices and providing the farmers with food; lumber was in high demand from the U.S.); a lack of available industrial hemp-specific machinery; and the government's preference for processed, rather than crude, industrial hemp.[18]

At the same time, the cannabis plant was also being cultivated for its psychogenic properties. With the lack of a robust and established hemp industry, the recreational aspect of cannabis overtook its more benign uses. And that led to criminalization. Starting in 1908, laws prohibiting some drugs became more commonplace.

The *Opium Act*, initially focusing on the criminal trafficking of opium, was Canada's first legislation dealing expressly with drugs. While originally urged by legislators as a method of crime control, more recent scholarship sees its purposes tied more to dynamic forces related to immigration, racism, economics, labour, and control. By cracking down on the main Chinese users and suppliers, criminalization became a tool to engineer social order – controlling Asian immigration – as much as it was rationalized as a policy focused on substance abuse.[19] More drugs were then added to the prohibited list. The *Opium Act* became the *Opium and Drug Act* in 1911; in 1920, the name changed again to the *Opium and Narcotic Drug Act* (*ONDA*) and in 1923, cannabis was added to the schedule of listed drugs. Why cannabis? Little to no evidence exists of its prevalent use nor of a concerned media, as with opium,

15 Canada, Commission of Inquiry into the Non-Medical Use of Drugs, *Cannabis* (Ottawa: Information Canada, 1972).
16 Norman Macdonald, "Hemp and Imperial Defence" (1936) 17(4) *Canadian Historical Review* 385–398.
17 Leah Spicer, 'Historical and Cultural Uses of Cannabis and the Canadian "Marijuana Clash"' (Report prepared for the Senate Committee on Illegal Drugs, April 12, 2002, s. 3(a)). The story of hemp is complicated somewhat by the presence of another form of hemp, known colloquially as 'Indian hemp'. Technically speaking, Indian Hemp (*Apocynum cannabinum*) is distinct from cannabis and was sometimes treated differently at law. For instance, the *Ontario Pharmacy Act* of 1871 listed a number of 'poisons' that only licensed pharmacists could trade in (on the grounds of 'familiarity' with the customer). 'Hemp Indian' was included on the list (see Dan Malleck, *When Good Drugs Go Bad: Opium, Medicine and the Origins of Canada's Drug Laws* (Vancouver: UBC Press, 2015) at 74). It is likely that Indian hemp was included on the pharmaceutical list because *A. cannabinum* is toxic to humans (and some animals) but this is speculative as clear evidence of why it was there remains elusive.
18 Macdonald, *supra* note 16 at 395–396.
19 See B. Fischer, K. Ala-Leppilampi, E. Single, and A. Robins, "Cannabis Law Reform in Canada: Is the 'Saga of Promise, Hesitation and Retreat' Coming to an End?" (2003) 45(3) *Canadian Journal of Criminology and Criminal Justice* 265 at 267. There were laws dealing with banned substances prior to the *Opium Act*, of course: legislation controlling ingredients in food has a long history in Canada beginning with *The Inland Revenue Act* in 1874; another example, the *Ontario Pharmacy Act* of 1871, is discussed below.

stirring up xenophobia by targeting certain ethnicities of cannabis users. In fact, cannabis was not originally listed in the 1923 amendments to the Act but was added as an afterthought at the last minute.[20]

On the other hand, industrial hemp remained unregulated until it was added to the *ONDA* in 1938, when amendments to that Act required anyone growing any form of cannabis to first get a licence from the Department of Health.[21] Hemp was therefore captured by the new laws, even though the legislative debates over the amendments make it clear that the hemp industry was of no commercial significance at the time. The most likely reason for including hemp was the conflation of 'Indian hemp' with hemp from cannabis. Although taxonomically distinct, 'Indian hemp' was used to refer to both *Apocynum cannabinum* and *Cannabis sativa*.[22] The former is toxic to humans, which helps explain the Parliamentary discussions of industrial hemp that preceded its strict regulation in 1938.[23]

For example, during the second reading for the *ONDA* amendments in late February of 1938, Hugh McKenzie, the Liberal MP for the Lambton-Kent riding in Ontario, related how he had been in touch with a farmer in Lambton county who grew 'Indian hemp for commercial purposes'.[24] The farmer had 'entered into this project encouraged by the Department of Agriculture and sponsored by them,' and, McKenzie added, the then-current Canadian Almanac included industrial hemp seed as free to import.[25] McKenzie believed that an overly stringent licensing scheme may discourage individuals from commercial participation in the hemp industry. Even though the farmer had been in the industry for 24 years – during all of which there had 'been no complaint from the narcotics branch nor from any other source whatever' – McKenzie suggested that, given the small size of the industry and the near insurmountable restrictions proposed for licenses, the federal government should compensate such individuals for their inconvenience.[26] McKenzie was a tiny minority voice in an overwhelming desire to include hemp in the amendments. Bruce McNevin, Liberal MP for Victoria, Ontario, echoed the majority view noting that any hardships hemp farmers faced could not compare with the 'evil effects that the continuance of the growing of the plant would have'.[27]

20 See Solomon and Green, "The First Century: The History of Non-medical Opiate use and Control Policies in Canada, 1870–1970" in J. Blackwell and P. Erikson, eds., *Illicit Drugs in Canada: A Risky Business* (Toronto: Nelson, 1988) 88 [reprinted from (1982) 20(2) *University of Western Ontario Law Review* 307].

21 Legal and Social Affairs Division, *The Legal Regulation of Marijuana in Canada and Selected Other Countries*, by Robin MacKay and Karin Phillips, Report No 2016-94-E (Ottawa: Library of Parliament, 6 September 2016) at 6.

22 Ernest Small and David Marcus, "Hemp: A New Crop with New Uses for North America" in J. Janick and A. Whipkey, eds., *Trends in New Crops and New Uses* (Alexandria: ASHS Press, 2002) at 284.

23 A 1960s example of this confusion: George Andrews and Simon Vinkenoog, *The Book of Grass: An Anthology on Indian Hemp* (New York: Grove, 1967) at 102.

24 "Opium and Narcotic Drug Act, 1929," 2nd reading, *House of Commons Debates*, 18th Parliament, 3rd Session, Volume 1 (1 March 1938) at 910, available online at: https://parl.canadiana.ca/view/oop.debates_HOC1803_01/1.

25 *Ibid*.

26 *Ibid*.

27 "Opium and Narcotic Drug Act, 1929," 2nd reading, *House of Commons Debates*, 18th Parliament, 3rd Session, Volume 1 (24 February 1938) at 775, available online at: https://parl.canadiana.ca/view/oop.debates_HOC1803_01/1.

The strict regulation of industrial hemp 15 years after the criminalization of marijuana reflects growing concerns around cannabis without recognizing hemp's different qualities.[28] While it is not entirely clear what the federal government's motivation behind the criminalization of marijuana was in 1923, it has been suggested that it followed from the belief amongst a handful of important governmental officials that cannabis would soon fall under international regulation.[29] By contrast, the 1938 Parliamentary debates indicate that many parliamentarians feared that naïve Canadians were at risk of falling under the influence of a very dangerous and, to some, a foreign threat. For example, along with the bootlegging of codeine, John Knox Blair (MP for Wellington North) thought the revised Act was necessary to combat the use of 'hemp on the sly' amongst the Asian–Canadian population.[30] Conversely, William Hayhurst (MP for Vegreville) noted that cannabis grew in many western areas of the country but that 'most of the English speaking people do not realize to what grave uses it could be put'.[31]

The social threat of cannabis was further substantiated by the U.S. approach to the problem which served as a consistent point of reference for Canada's Parliament. Given the challenges of enforcing absolute prohibition, in 1937 the U.S. had simultaneously passed a licensing scheme and taken an aggressive approach to seizing and destroying cannabis plants.[32] C G Power, the Minister of Pensions and National Health and sponsor of the bill to amend the *ONDA*, described the drug as 'a new menace to the youth of the country'.[33] This description was strikingly similar to that of U.S.' Commissioner of Narcotics H F Anslinger, and clearly indicated familiarity with the US approach.[34]

Ultimately, Parliament's choice to regulate industrial hemp through the passage of the revised *ONDA* was driven by three main factors. First, a majority of Parliamentarians saw cannabis as a threat to the nation and made no important distinctions for industrial hemp. Second, there were no strong competing interests – particularly, commercial interests – that outweighed the risks of cannabis, and by extension, industrial hemp. And third, given that 'Indian hemp' was already considered a toxic substance, it was relatively straightforward to sow confusion between the two forms.[35]

Things changed both for better and for worse in the 1960s. Cannabis use increased dramatically; so did its criminalization. The *Narcotic Control Act*

28 MacKay and Phillips, *supra* note 21 at 6.
29 Catherine Carstairs, *Jailed for Possession: Illegal Drug Use, Regulation, and Power in Canada, 1920–1961*, Studies in gender and history (Toronto: University of Toronto Press, 2006) at 31–32.
30 "Opium and Narcotic Drug Act, 1929," 2nd reading, *House of Commons Debates*, 18th Parliament, 3rd Session, Volume 1 (24 February 1938) at 777, available online at: https://parl.canadiana.ca/view/oop.debates_HOC1803_01/1.
31 *Ibid* at 775.
32 *Ibid* at 774.
33 *Ibid* at 772.
34 *Ibid* at 773.
35 One could possibly add a fourth factor, which is a tendency in much Canadian law to follow the American example, where, in the early 20th century cannabis, marijuana and hemp were conflated.

superseded the *Opium and Narcotic Drugs Act* and cannabis was listed as a Schedule 1 drug (most serious) with maximum penalties of 7 years imprisonment for simple possession and life imprisonment for trafficking.[36] Much of the legislative change arose as a result of international concern regarding illicit drugs: the Single Convention on Narcotic Drugs, 1961, provided a template for a comprehensive regulatory regime and banning of illicit substances except for medical and scientific purposes. Yet, towards the end of the decade, a proliferation of counter-cultural, anti-authoritariansim espoused 'pot' as the drug of choice for many, replacing the secrecy of opium dens and back-alley heroin scores of the 1920s. In Canada, the 'hipster' Prime Minister Pierre Trudeau began the process of reform that his son completed in 2018. In 1968, Pierre Trudeau's government established a Commission of Inquiry into the Non-Medical Use of Drugs (known as the Le Dain Commission). Its report, tabled in Parliament in May of 1972, called for the repeal of the offence of simple possession of cannabis.[37] An attempt to turn the report into actual legislation, however, was never fulfilled.[38] But the *Narcotic Control Act* was amended in 1969 to allow simple possession of marijuana to be tried either on summary conviction or indictment (similar to 'felony' and 'misdemeanor' in the U.S.), so the potential penalties could be much lower.[39]

This leniency was reversed again during the 'war on drugs' era, largely fuelled by American policies under Presidents Nixon and Reagan. As a result, reports like the Le Dain Commission became a mere footnote as little was done to advance the legalization of cannabis until the early 2000s.

The first inroads were medical. In Canada, cases arose where advocates argued that cannabis could help control life-threatening epileptic seizures in situations where conventional medications could not. These cases were successful.[40] In *R v Parker* for example, the defendant argued that s. 7 of the *Charter* – the right to life, liberty, and security of the person – guaranteed a right to use medical marijuana. Ultimately, the Court held that a complete prohibition on possession and cultivation of cannabis for personal use against

36 SC 1961, c 35. Strictly speaking, cannabis is not a narcotic. Pharmacologically, it is an endocannabinoid; for purposes of drug laws, however, it is treated as a narcotic. See Mary Bridgeman and Daniel Abazia, "Medical Cannabis: History, Pharmacology, and Implications for the Acute Care Setting" (2017) 42(3) *Pharmacy and Therapeutics* 180, available at: https://www.ncbi.nlm.nih.gov/pmc/articles/PMC5312634/.

37 *Cannabis: A Report of the Commission of Inquiry into Non-Medical Use of Drugs* (Gerard Le Dain, Chair) (Ottawa: Information Canada, 1972), available online at: http://www.druglibrary.org/schaffer/Library/studies/ledain/ldctoc.html (Le Dain Commission Report). The Report's recommendations themselves were controversial and not unanimous. Three commissioners, including Le Dain, recommended de-criminalization, one favoured retaining possession as a criminal offence and a fifth favoured adopting a regulated industry approach which is the ultimate route taken under the 2017 *Cannabis Act*.

38 Mike Bryan, a member of the Le Dain Commission, felt that the government was somewhat duplicitous in the aftermath of the Report, in claiming it would not imprison persons found guilty of simple possession: see Foreward to Cannabis Control Policy: A Discussion Paper http://cfdp.ca/canrep/.

39 *Ibid*.

40 See, for example, *Parker*, *supra* note 5. We discuss Canada's approach to medical marijuana more fully in Part 4.

epilepsy was overbroad and irrational because it deprived an individual access to a drug that could have benefits on health.[41]

These cases led the Canadian government to revise the regulations contained in the *CDSA* to allow possession of cannabis for medical purposes. Additional cases followed which held that a person's rights were further infringed where they were unable to access an adequate legal supply of cannabis where they had a medically prescribed need for it.[42] (A classic example of governments being penalized for trying to help, but not helping enough!)

A parallel set of concerns arose regarding industrial hemp at around the same time, leading to its legalization in 1998. Health Canada recognized that research had shown that hemp could be grown successfully as a 'separate entity from cannabis (marijuana),'[43] acknowledging that pressure from various sources within the agricultural, industrial, political, and popular sectors helped to sway the decision.[44]

By prying open a wedge allowing cannabis for medical purposes – where evidence was often led regarding the lack of harm caused by medical marijuana use – and by recognizing hemp as a distinctly different product, it was only a matter of time before a challenge to the very criminalization of cannabis itself was brought to the Supreme Court of Canada (SCC). In *R v Malmo-Levine*, the proponents sought de-criminalization of marijuana for simple possession. While the Court ultimately decided that marijuana use is a proper subject matter for the criminal law power of the federal Parliament, and that the provisions in the *Narcotic Control Act* did not contravene the *Charter*, it nevertheless noted how views of cannabis had changed radically since the 1920s, and research showed unequivocally that occasional use was not harmful.[45]

Thus it was in 2015 that the newly formed government of Justin Trudeau embarked on fulfilling a campaign promise to legalize marijuana.[46] The end result, as we have already described, was a comprehensive regulatory scheme contained in a number of enactments. The most crucial document is the *Cannabis Act* which contains more than 200 sections and over 370 regulations.[47] A number of consequential regulations were also made to other statutes,

41 *Parker, supra* note 5 at para 144. The Court made an interesting observation about Canada's history of cannabis regulation in Canada as being "short", "lacking a significant foundation in our legal tradition" and an "embarrassing history based upon misinformation and racism." (at para 126, citing with approval the decision of Justice McCart in *R v Clay* (1997) 9 CR (5th) 349 (Ont Gen Div).

42 *Marihuana Medical Access Regulations*, SOR/2001-227; superseded by *Marihuana for Medical Purposes Regulations*, SOR/2013-119; superseded by *Access to Cannabis for Medical Purposes Regulations*, SOR/2016-230.

43 Health Canada, "Hemp and the Hemp Industry Frequently Asked Questions", (30 August 2006), available online at: https://www.canada.ca/en/health-canada/services/drugs-medication/cannabis/producing-selling-hemp/about-hemp-canada-hemp-industry/frequently-asked-questions.html Last Modified: 2018-11-07.

44 Arthur Hanks, "Canada: Hemp Industry-In-Progress 1998–2002" (2002) 7(2) *Journal of Industrial Hemp* 87–93 at 88.

45 *R v Malmo-Levine* [2003] 3 S.C.R. 571 [*Malmo-Levine*]. See paras 192–201 for a discussion by Justice Arbour on the changing understanding of marijuana (dissenting on other grounds). As the majority stated in paragraph 5, "it is open to Parliament to decriminalize or otherwise modify any aspect of the marihuana laws that it no longer considers to be good public policy."

46 See Liberal Party platform, "Real Change: A New Plan for a Strong Middle Classe" at 55, available at: https://www.liberal.ca/wp-content/uploads/2015/10/New-plan-for-a-strong-middle-class.pdf.

47 *Cannabis Regulations* SOR/2018-144.

including the *CDSA* and the *Criminal Code*, particularly related to impaired driving.

But because Canada is a federal nation, the passing of one federal law is not enough. There is a need for jurisdictional cooperation in a federal system, and the *Cannabis Act* is only part of the picture. For one, provincial and territorial governments need to provide legislative schemes related to distribution and sales, and municipalities may also enact by-laws (as they do with tobacco smoking in certain areas) regulating local use. Another issue is whether and how Indigenous authorities will regulate cannabis in their own ways, separate and apart from other governmental authorities.

That leads us to hemp. As already noted, cannabis is an unusual plant because it can be used recreationally, pharmacologically, and industrially. Since marijuana is derived from it, Canadian legislators took a cautious approach in 2018 by regulating hemp and cannabis separately but including references to the other. Thus, the *Hemp Regulations* exclude products with too high a THC content, and the *Cannabis Regulations* exclude industrial hemp where the THC is below a threshold amount. An industrial hemp license is a license granted under the *Industrial Hemp Regulations*.[48]

In sum, while cannabis is now 'legal' in Canada, it is more accurate to think of it as legal only within the very narrow confines of the *Cannabis Act*. While possessing, distributing, cultivating, selling, and processing is allowed, importing and exporting cannabis is an offence, punishable by imprisonment or a fine, depending on the circumstances. The *Cannabis Act*, the *Cannabis Regulations* and the *Industrial Hemp Regulations*, together with other applicable provincial legislation, set out what is permitted for both the end consumers and those involved in bringing a final cannabis product to market. The next section looks at the specifics of how hemp products are licensed in Canada.

3 Current licensing and law related to hemp in Canada

3.1 Hemp statistics

Hemp is becoming more common in Canada. Table 1 shows some statistics related to its production. The active number of licenses, shown in the first row, showed an initial increase upon full de-criminalization, but since 2020 these have been decreasing, largely due to consolidation of industry players.[49] Prior to October 17, 2018, licencees were not required to have a licence for sites that limited themselves only to cultivation. These sites were called 'Registries'.

48 *Industrial Hemp Regulations*, supra note 11, s 3.
49 Health Canada, "Industrial Hemp Licensing Statistics," (18 June 2024), available online at: https://www.canada.ca/en/health-canada/services/drugs-medication/cannabis/producing-selling-hemp/about-hemp-canada-hemp-industry/statistics-reports-fact-sheets-hemp.html Last Modified: 2024-06-18.

Table 1 Overview of Canadian Hemp Statistics

Year	2018	2019	2020	2021	2022	2023	2024
Active industrial hemp licences per calendar year	542	866	1269	1050	992	737	N/A
Production of hemp (metric tonnes)	N/A	N/A	26 500	25 820	32 988	24 862	N/A
Seeded area (hectares)	16 700	36 900	21 400	25 207	31 100	22 400	14 800
Harvested area (hectares)	N/A	32, 500	20 400	22 500	28 800	22 100	N/A
Average yield (kilograms per hectare)	N/A	1003	1299	1148	1145	1125	N/A
Top three most popular cultivars based on hectarage for cultivation	1. Finola (12 687.51 ha) 2. X-59 (Hemp Nut) (6760.79 ha) 3. Katani (4225.44 ha)	1. Finola (10 032.96 ha) 2. X-59 (9680.49 ha) 3. Katani (8957.46 ha)	1. Finola (10 618.11 ha) 2. Katani (2496.63 ha) 3. CanMa (1516.47 ha)	1. Finola (7844.37 ha) 2. Granma (1865.38 ha) 3. CFX-2 (1593.97 ha)	1. Finola (5142.38 ha) 2. CFX-2 (1678.03 ha) 3. Anka (1492.86 ha)	1. Finola (4691.57 ha) 2. Vega (1153.77 ha) 3. CFX-2 (994.13 ha)	N/A

Production numbers, set out in the second row, also exhibit an initial rise then fall, somewhat similar to licensing.[50] This is attributable to the CBD extraction market not expanding as the anticipated boom in the recreational market following the 2018 *Cannabis Act* did not materialize.[51] Present efforts to have hemp approved for the use of animal feed are now the main basis for such hopes.[52] It is perhaps not surprising that some members of the Canadian hemp industry contend that the industrial hemp sector is overly regulated.[53]

Finally, as already noted, there are 91 cultivars of hemp allowed under the current regulation.[54] Three cultivars were added in 2024 and none were removed. Eight cultivars are currently under observation.[55] Early on in Canada's industrial hemp program, Small and Marcus observed that cultivars generally preserved their legal THC limits and were thus rarely removed for exceeding the threshold.[56] The final row of the table shows the top three cultivars – clearly Finola is the most favoured form and has been since statistics were kept.[57] While Finola is clearly the most popular Canadian cultivar in recent years, there is some variability in the next most favoured cultivars.

The *Cannabis Act* provides for broad regulation and administration of both marijuana and hemp. Seven separate regulations are in force under the *Cannabis Act* – for our purposes it is the *Industrial Hemp Regulations* (*IHR*) that are most pertinent. However, it is important to understand that there are implications for hemp created by the the *Cannabis Act* itself.[58] For example, while the *IHR* is clear in setting out that only those who hold a licence may cultivate it (s. 3(1)(c)), the details of how to obtain a licence are set out in the *Cannabis Act*. Section 62 of that Act grants the Minister of Health the authority to issue licences. Another example is the provision in the *Cannabis Act* allowing individuals to cultivate up to four plants in their own dwelling house.[59] Since the Act applies to all cannabis plants, this means that hemp is subject to the same law. One cannot, for example, grow more than four cannabis plants at home even if using them to make hemp rope, despite the fact that such a cannabis plant might be benign pharmacologically. Since all cannabis plants, no matter

50 Statistics Canada, "Crop Production: Visualization Tool [Data visualization tool]," (9 November 2023), available online at: https://www150.statcan.gc.ca/n1/pub/71-607-x/71-607-x2020025-eng.htm.

51 United States Department of Agriculture–Foreign Agricultural Service, *Update on Industrial Hemp Production Trade and Regulation*, Report No. CA2023-0034 (Ottawa: USDA, 2023) at 2.

52 *Ibid*.

53 Wesley Tourangeau, "Re-defining Environmental Harms: Green Criminology and the State of Canada's Hemp Industry" (2015) 54(4) *Canadian Journal of Criminology and Criminal Justice* 528 at 540–541.

54 Health Canada, "List of Approved Cultivars for the 2024 Growing Season: Industrial Hemp Varieties Approved for Commercial Production", (03 September 2024), available online at: https://www.canada.ca/en/health-canada/services/drugs-medication/cannabis/producing-selling-hemp/about-hemp-canada-hemp-industry/statistics-reports-fact-sheets-hemp.html Last modified: 2024-09-03.

55 *Ibid*.

56 Small and Marcus, *supra* note 22 at 294.

57 Finola was at one time under probation due to high THC levels – see Small and Marcus, *supra* note 22 at 294.

58 For a good overview of the interaction between the *Cannabis Act, Cannabis Regulations* and *Industrial Hemp Regulations,* see Bruce A. MacFarlane, Robert J. Frater, and Croft Michaelson, *Cannabis Law* (Toronto: Thomson Reuters, 2021) at chapter 22.

59 *Cannabis Act,* s. 12(4).

the end usage, come under the broader umbrella of cannabis, the authorizing statute is key.

De-criminalization of cannabis has other flow-on effects related to hemp. Whereas prior to 2018, anyone applying for a license to grow hemp would need to undergo a criminal background check, this is no longer the case. And while growing and selling are often treated as distinct activities, regulated separately, the *IHR* allows the holder of a license of sale to sell 'flowering heads, leaves and branches' of industrial hemp to another hemp licence holder or to a person holding a licence under the *Cannabis Regulations*.[60]

3.2 Hemp and hemp-related disputes

Although hemp use is generally less contentious than marijuana, its commercial production has been the subject of some litigation. *Carmel Pharms Corp v Oro-Medonte (Twp)* highlights some of the local concerns that have arisen since 2018.[61] In this case, the Ontario town of Oro-Medonte amended its zoning by-laws to exclude the production and processing of hemp and cannabis from the definition of acceptable agricultural activities. The two disputants – one an industrial hemp company and the other a cannabis company – argued before the Ontario Land Tribunal (OLT) that the amendment's relegation of hemp and cannabis production sites to 'Industrial Zones' undermined any economic incentive for these industries.[62] They claimed, for instance, that a requirement that hemp and cannabis farms be at least 150 metres away from 'newly defined sensitive land uses' (e.g. schools) meant it would 'be virtually impossible to grow industrial hemp which requires large acreage'.[63] The town, by contrast, maintained that limiting hemp and cannabis farming to industrial zones managed the objectionable byproducts of these practices (e.g. light, noise, odours), some of which had been the subject of complaints from community members.[64]

The OLT decided in favour of both companies, determining that the town's by-law contradicted Ontario's *Provincial Policy Statement, 2020* (PPS) that 'covers policies about managing growth, using and managing natural resources, protecting the environment, and public health and safety'.[65] This decision was overturned on review;[66] the companies appealed further to the Ontario Superior Court of Justice where it was dismissed on two counts, both more procedural than substantive.[67]

60 *IHR*, s. 28.
61 2023 CarswellOnt 1239 [*Carmel Pharms*].
62 *Ibid* at para 22.
63 *Ibid* at paras 96, 103.
64 *Ibid* at paras 40, 85.
65 Ontario, Ministry of Municipal Affairs and Housing, *Provincial Policy Statement, 2020* (Queen's Printer, 2020).
66 *Carmel Pharms Corp v Oro-Medonte (Twp)*, 2023 CarswellOnt 13055.
67 *2541005 Ontario Ltd v Oro-Medonte (Township)*, 2023 ONSC 5569. First, the Court determined that the standard of procedural fairness was not breached and second, that the matter was not one of "general public importance"

It is worth noting that the case revolved around the issue of zoning bylaws and did not discuss the *IHR*. However, the hemp licensing scheme was alluded to when Derek Witlib, an expert witness for the town, noted that the mechanism for dealing with complaints through Health Canada was 'severely lacking'.[68] More broadly, the case also suggests that localities that are inhospitable to the hemp industry may employ creative legal approaches to stonewall legitimate enterprise.

In a markedly similar case to *Carmel Pharms*, Minhal Jaffer challenged zoning by-laws that prevented a budding hemp company from building its processing facility in the town of Grimsby.[69] His appeal to the OLT was based on his being denied, by a Committee of Adjustment, a 'Minor Variance' to the zoning by-law.[70] The Committee, an Ontario quasi-judicial body, had been asked to amend the permitted use of a 'Licenced Cannabis Cultivation Facility' in order to allow for a 'Hemp Processing Facility'.[71] Prior meetings between the applicant and Town Planning had resulted in a Memorandum supporting the variance, with a condition that hemp could only be processed (with the exception of hemp leaves) and not grown on the proposed site. The Committee denied the application on four grounds: (1) it did not align with the intent of the by-law; (2) it did not align with the intent of Grimsby's Official Plan; (3) it was not 'minor' in nature; and (4) it was 'not desirable for the subject lands'.[72]

The case is particularly interesting because Jaffer was initially allowed to build a marijuana facility but was, instead, seeking to change that to a hemp facility. Given the backlash he faced at the Committee of Adjustment hearing, one wonders whether it would have been even more severe had he gone ahead with his original plan.[73] The Committee members, and the public interveners, focused on problems such as noxious odours, the impact on property values, and the variance allowing an industrial facility in an agricultural zone.[74] The public's evidence was strongly resistant to Jaffer's argument that hemp was similar to other agricultural crops. Much of the testimony indicated a clear misunderstanding of hemp and how it differed from marijuana.

Ultimately, Jaffer was successful. The OLT found that the variance met the four tests of the Planning Act and allowed the appeal. It held that the Town Planning's condition prohibiting the growing of hemp on the proposed site

(at para 82).

68 *Carmel Pharms*, *supra* note 61 at para 47.

69 *Minhal Jaffer v Grimsby (Town)*, 2023 CarswellOnt 12198 [*Jaffer*].

70 Town of Grimsby, "Minor Variances", (Town of Grimsby, 2024), online available at: https://www.grimsby.ca/build-and-invest/building-and-planning/minor-variances/.

71 *Jaffer*, supra n 69 at para 4.

72 *Planning Act*, RSO 1990, c P13 2014, s. 45(1).

73 Abby Green, "Grimsby Residents Frustrated by Hemp Facility after OLT Approval" *NiagaraThisWeek.com* (11 September 2023), available online at: https://www.niagarathisweek.com[perma.cc/news/grimsby-residents-frustrated-by-hemp-facility-after-olt-approval/article_02b45dfc-32a1-50a8-be67-ff7e7253f6a5.html]; Town of Grimsby, "Committee of Adjustment–December 20, 2022" (20 December 2022), available online (video) at: www.youtube.com[perma.cc/watch?v=m-uiqsZq7A0].

74 *Jaffer*, *supra* n 69 at para 14.

was, in the main, a reaction to public outcry and not grounded in law.[75] The OLT recognized that the proposed hemp facility represented a 'relatively new' industry, but it was one that fit within existing regulatory mechanisms and aligned with the economic objective of agricultural sector growth.[76]

The *IHR* has been the subject of a constitutional challenge, albeit an earlier version to the current incarnation. In *R v Seeman*, three accused were charged with illegally trafficking medical marijuana between August 2016 and October 2018.[77] In addition to arguments related to the required online distribution, the accused took issue with the *IHR*, particularly s. 15(2). This required that '[a] person who holds a licence to cultivate industrial hemp shall, at the time of harvesting the crop, dispose of the branches, leaves and flowering heads by retting or by otherwise rendering them into a condition such that they cannot be used for any purpose not permitted under the Act'.[78] They argued that the section was arbitrary in that it 'criminalized medical cannabis patients' by criminalizing medicalized hemp.[79] Justice Klatt rejected this argument.[80] She did not accept the evidence of a licensed hemp producer – that the disposal requirement under s. 15(2) was wasteful since growing hemp was more cost-efficient than growing cannabis and could be done outdoors – determining that this was insufficient to give rise to a s. 7 *Charter* breach.[81] For her, the clear, cohesive, and overarching objective of the laws was protecting public health and safety and the lack of evidence regarding the production of hemp meant that there were no proven breaches of s. 7 rights.

Despite this limited discussion of a previous version of the *IHR*, the case offers some insight into how a Court may construe the current *IHR*. The Crown had tried to argue that the purposes of the various drug laws were not limited to public health and safety but broader: in effect attempting to balance access to medical cannabis with risks posed by the drug.[82] Justice Klatt rejected this position, relying on a Supreme Court of Canada case, *R v Moriarity*,[83] to say that the Crown was conflating the objectives and the means of the legislation: balancing the accessibility and risks of medical cannabis was the method used to achieve public health and safety and was not an objective in itself.[84] This is important to understanding the present incarnation of the *IHR*. Given that a primary focus of today's *Cannabis Act* is the protection of public health and safety, the *IHR* that is empowered under it should also be seen to reflect the same objective.

75 *Ibid* at para 90.
76 *Ibid* at para 92.
77 *R v Seeman*, 2022 SKKB 232 at para 2 [*Seeman*].
78 *Industrial Hemp Regulations*, *supra* note 11, s 15(2).
79 *Seeman*, *supra* note 77 at para 8.
80 *Ibid* at para 124.
81 *Ibid* at paras 121, 123.
82 *Ibid* at paras 78–80.
83 2015 SCC 55 [*Moriarty*].
84 *Seeman*, *supra* note 77 at para 81.

Another potential argument surrounding hemp and the *IHR* can be drawn out of the *Charter*-based ruling in *Bennett v Canada (Attorney General).*[85] Bennett sought a ministerial exemption to consume 7 g of marijuana per day in accordance with his religious beliefs. This was refused. He then sought judicial review of the decision before the Federal Court, contending that it violated three of his *Charter* rights: s. 7's life, liberty, and security of the person; s. 2(a)'s freedom of conscience and religion; and s. 15's equality rights. At trial, Justice Shore determined there was no violation of Bennett's s. 2(a) right to freedom of conscience and religion. Bennett's position was that he had been a member of the Church of the Universe since 1990, and his marijuana consumption reflected the concomitant religious belief that cannabis is 'the tree of life'.[86] Despite the biblical origins of the phrase, Justice Shore found that Bennett's was a secular belief grounded in the benefits of marijuana consumption and not on any Christian faith. He concluded that it did not meet the standard of a 'comprehensive system of faith and worship'.[87] A unanimous Federal Court of Appeal overturned the trial decision, holding that Justice Shore had relied on 'conflicting factual findings' in reaching his conclusion that Bennett's belief was secular, not religious.[88] More specifically, the appellate court found that the lower court had inconsistently characterized Bennett's beliefs as both spiritual and secular. As a result, the case was sent back down to the Federal Court. There is no record of what happened subsequently.

What does the case have to do with hemp? Although the facts relate to medical marijuana, it raises the possibility of religious-based exemptions to existing cannabis and hemp laws. At trial, Justice Shore alluded to the *IHR* as an example of the comprehensive legal framework that is in place to regulate cannabis as a whole. Put otherwise, the *IHR* helped substantiate the Federal Court's position that the government has been dutiful in their regulation of cannabis. The case also illustrates how marijuana, hemp, and cannabis are parts of the same whole and may, for some purposes, be treated as such.

Finally, as noted in Section 1, in this chapter, cannabis has been used by Indigenous peoples in Canada and around the world for centuries. Indigenous rights have been at the forefront of many dramatic legal changes in Canadian constitutional law in the last few decades. Examples include title to land, activity-specific rights related to hunting, fishing and agriculture, duty on governments to consult, and a signature judicial creation, the 'Honour of the Crown' which is intended to govern all relations between Indigenous peoples and governments. Inevitably, with these changes, have come conflicts.

85 2011 FC 1310 [*Bennett*].
86 *Ibid* at para 9.
87 *Ibid* at para 83.
88 *Bennett v Canada (Attorney General)*, 2013 FCA 161 at para 6.

In *R v Marshall*, the Provincial Court of Nova Scotia dismissed an attempt by four accused to establish a constitutionally protected Indigenous right to the use and trade of cannabis and hemp.[89] The accused did not challenge the factual basis of the charges – they admitted to trading illegally in cannabis – but rather claimed that, as Mi'kmaq peoples, the licensing and tax requirements violated their constitutionally protected right to trade. The Court determined that a lack of oral traditional evidence and supporting expert evidence meant the case had to be dismissed. This, despite the fact that the Court recognized the need for, and importance of, 'strongly advanc[ing] the law on aboriginal and Treaty rights'.[90]

In *R v Brennan*, ten accused were criminally charged in Ontario with operating cannabis dispensaries without the required licences.[91] As in *Marshall*, the accused did not challenge the factual basis of the charges. Instead, the accused argued in defence that they belonged to an unrecognized First Nation – the Amikwa – and as such are beyond the scope of criminal laws. Two reasons were given: (1) Amikwa territory is unceded; and (2) they have a traditional right based on their Amikwa identity to trade in both marijuana and hemp.[92]

The case has become entangled in the justice system. The Crown attempted to summarily dismiss all five constitutional questions raised by the accused. The Ontario Superior Court agreed with four of the Crown's arguments, but not one related to Indigenous use of cannabis. Thus, the parties have now narrowed the issue down into a single constitutional question: '[w]hether the laws under which the charges have been brought do not apply because they infringe the accused persons' traditional use and trade of cannabis and hemp'.[93] This question remains to be answered. There have been extensive delays, partly due to the need for expert reports regarding the use of cannabis and hemp in Amikwa communities.

In contrast with *Marshall, Brennan* can be seen as giving some indication of what the judiciary is willing to accept as viable constitutional arguments for Indigenous rights regarding cannabis and hemp. While the Court did not accept that there was an economic right to trade in hemp and cannabis under s. 7 of the *Charter*, it was willing to recognize certain Indigenous rights.[94] It remains to be seen whether or not the evidentiary record will support the Amikwa claim. A recent voir dire has determined that the probative value of

89 2024 NSPC 33 [*Marshall*].
90 *Ibid* at para 51.
91 2022 ONSC 1986 [*Brennan*].
92 *Ibid* at para 5. "Unceded" means that the land was never given to the Crown and so remains as potentially Indigenous land.
93 *Ibid* at Part V.
94 It would require a volume to properly explain Canada's complicated system of individual rights and Indigenous rights. For purposes of this chapter, it is sufficient to say that the *Constitution Act, 1982* establishes both a *Charter of Rights and Freedoms* – largely individual, traditional rights, such as freedom of expression and religion, rights against discrimination on grounds such as sex, race, and many more – and also protects "Aboriginal Rights" in another part that is separate from the *Charter*.

the accused's expert evidence outweighed its prejudicial risk. If the Court ultimately decides in favour of the ten accused, it will support the notion that Indigenous groups can establish a constitutional right to the traditional use and trade of cannabis and hemp. This would mean that they may be exempted from generally applicable cannabis laws. It is likely, given the possible conflict with *Marshall*, that the Supreme Court of Canada will need to resolve this issue.

We have already described a number of situations where hemp and medical marijuana have been treated in similar ways by the courts. The final section of this chapter explores Canada's unique experience with medical marijuana.

4 Health Canada and legalization of medicinal marijuana

4.1 Medical marijuana prior to the Cannabis Act

Medical access to cannabis in Canada is presently regulated through the *Cannabis Regulations*, having replaced the *Access to Cannabis for Medical Purposes Regulations* (*ACMPR*) in 2018.[95]

Internationally, Canada was an early adopter in allowing for regulated medical cannabis use, introducing its first program in 1999 by allowing for special exemptions from the criminal law.[96] The exemption process, however, was flawed. It suffered from excessive ministerial discretion, lack of adequate supply, and a regulatory regime that maintained a criminalized approach to marijuana. Thus, the necessity for proponents to turn to the *Charter* and courts for relief – as we have already seen, an approach that is all-too-common, and one might say, not always the most efficacious since it frequently requires multiple cases to force a more comprehensive ruling from the Supreme Court of Canada.

Well before Canada's de-criminalization of cannabis, the Ontario Court of Appeal (ONCA) held in *R v Parker* that the *Charter* protected a right to medical marijuana.[97] As with cases already discussed, it took a criminal charge and a defence lawyer mounting a constitutional challenge to resolve. Parker was charged with cultivating and possessing marijuana that he claimed substantially improved his severe epilepsy. He maintained that growing and smoking marijuana was the only reasonable avenue available to him, given his condition. Justice Rosenberg, writing for the Court, confirmed the trial judge's factual finding that medical cannabis provided Parker with relief that he had

95 Health Canada, "Understanding the Access to Cannabis for Medical Purposes Regulations", (August 2016), available at: https://www.canada.ca/en/health-canada/services/publications/drugs-health-products/understanding-new-access-to-cannabis-for-medical-purposes-regulations.html Last Modified: 2016-09-30.

96 Minsup Shim, Hai Nguyen and Paul Grootendorst, "Lessons from 20 years of Medical Cannabis use in Canada" (2023) 18(3) *PLoS One* e0271079.

97 *Parker*, *supra*, note 5.

not obtained from conventional treatments.[98] In recognizing the therapeutic value of marijuana, Justice Rosenberg determined that forcing Parker to choose between imprisonment and 'effective medical treatment'[99] violated his s. 7 rights on two counts. First, the threat of a criminal sanction infringed on his right to liberty via the potential of imprisonment and by limiting his ability to make decisions of 'fundamental personal importance'.[100] Second, the threat of a criminal sanction interfered with his physical and psychological integrity and thus imposed upon his security of the person.[101] Rosenberg determined that the infringements on Parker's s. 7 rights were not in accordance with the principles of fundamental justice and were not saved by s. 1 of the *Charter*.

Following *Parker*, the federal government created the *Marihuana Medical Access Regulations* (*MMAR*) to regulate medical access to cannabis in Canada.[102] Again, the regulations were challenged under the *Charter*. The new regulations limited medical access to a 'dried' form of cannabis. In *R v Smith*, the accused challenged this ban.[103] The SCC upheld the trial court's finding that the ban on medical access to non-dried forms of marijuana violated s. 7 of the *Charter*. The Court mirrored the reasoning relied on in *Parker*, determining that the threat of imprisonment deprived medical marijuana users of their right to liberty.[104] Moreover, the criminal sanctions deprived medical marijuana users of their right to make a 'reasonable' choice of how they wished to administer such legal medication.[105]

A year later, a further challenge occurred. In *Allard v Canada,* the Plaintiffs challenged the *Marihuana for Medical Purposes Regulations* (*MMPR*), which had replaced the previous *MMAR*.[106] The revised *MMPR* limited medical cannabis patients to purchasing their medicine from a defined group (deemed 'Licensed Producers') and limited possession to a maximum of 150 g of dried marijuana. Justice Phelan at the Federal Court found the government lacked evidence to support much of the policy underlying the *MMPR*.[107] He found little evidence that supported the prohibition on patients personally cultivating cannabis; on the contrary, '[t]he evidence establishes that the great majority of patients were able to produce their own cannabis as medicine without any threat to their own health or safety or that of the public'.[108] He determined that the *MMPR* violated s. 7 and was not saved by s. 1. In his opinion, *Allard* belonged to a 'line of cases

98 *Ibid* at para 100; 103–104.
99 *Ibid* at para 107.
100 *Ibid* at paras 101–102.
101 *Ibid* at paras 93–97, 110.
102 Health Canada, *supra* note 95.
103 *R v Smith*, [2015] SCC 34 at para 3 [*Smith*]. Although the *MMAR* at this point had been replaced by the *Marihuana for Medical Purposes Regulations* (*MMPR*), this did not impact the focal point of the challenge: the limitation to dried cannabis in the *CDSA*. Consequently, the Court largely discussed the case in relation to the *MMAR*.
104 *Ibid* at para 17.
105 *Ibid* at para 18.
106 *Allard v Canada*, [2016] FC 236 [*Allard*].
107 *Ibid* at para 12.
108 *Ibid* at para 260.

starting with *R v Parker* and culminating in *R v Smith* that have examined, often with a critical eye, the efforts of government to regulate the use of marihuana for medical purposes and the various barriers and impediments to accessing this necessary drug'.[109]

The end result meant the federal government once again had to revise the medical cannabis regime. This time they passed the *Access to Cannabis for Medical Purposes Regulations* (*ACMPR*) which came into force in 2016, just before the main overhaul decriminalizing cannabis in 2018.

4.2 Medical marijuana after the 2018 cannabis reforms

With the changes brought about in 2018, one would expect judges to find it difficult, if not impossible, to ignore the fundamental philosophical turnaround on recreational marijuana use and the concomitant impact that would have on medical marijuana use.

There were some cases that had entered the system before the changes and so were assessed under the old *ACMPR* and *CDSA*. For example, in the 2020 Alberta case of *R v Howell*, Shaun Howell was charged with the unlawful production of cannabis and the unlawful possession of cannabis for the purpose of trafficking. Howell owned a cannabis production facility that he used to provide medical marijuana to individuals outside the regulatory regime provided for in the *CDSA* and the *ACMPR*. The main reason for the charge was that Howell was not licensed to provide medical marijuana. Like others before him, he challenged the constitutionality of the provisions on the grounds that his s. 7 rights were infringed.

Justice Graesser declared that certain sections of the *ACMPR* were invalid, but found the same was not true of the *CDSA* trafficking provisions.[110] The court agreed with Howell that the *ACMPR*'s prohibition of cannabis oil and extracts with THC levels exceeding 30 mg/100mL violated his s. 7 rights. Echoing the *Parker* court, Graesser concluded that it infringed the rights of medical cannabis users 'to make reasonable medical choices' in relation to their own treatment.[111] Furthermore, the lack of medical evidence linking the purpose of the THC limit to the reduction of health risks meant the infringement was arbitrary.[112]

Howell also alleged that the regulatory regime outlined in the *ACMPR*, requiring mail-based home or medical office delivery, undermined reasonable access to medical marijuana. The court agreed that these requirements arbitrarily violated the s. 7 rights of many medical cannabis users, particularly houseless individuals who had no mailing address.[113] Without steady and

109 *Ibid* at para 3.
110 *R v Howell*, 2020 ABQB 385 at para 4 [*Howell*].
111 *Ibid* at para 276.
112 *Ibid* at paras 300–317.
113 *Ibid* at para 346.

reliable access to licit medical marijuana, individuals could be forced to use the 'black market' and risk criminal sanction.[114] Graesser rejected the Crown's alternative argument that s. 56 of the *CDSA* saved any s. 7 *Charter* violations, agreeing that medical exemptions should not be left to ministerial discretion.[115]

In the decision, Graesser recounts a line of case law in which the Canadian state struggles to strike an appropriate balance between ensuring access to medical marijuana while simultaneously prohibiting recreational marijuana use (e.g. *Parker*, *Smith*, *Allard*). Noting that medical cannabis 'was essentially decriminalized' in *Parker*, he observed that the federal Crown continues to drag its feet, appearing to respond to medical marijuana issues only when the courts force it to do so.[116] As we've noted a number of times already, while litigation is not perhaps the best way to deal with difficult social issues, sometimes it is the only way when governments continue to equivocate, or only partially deal with problems.

Even where a government is acting progressively in one realm, it may fail to do so in a related realm. This is illustrated in the Ontario case of *R v Ferkul*. Here, the progressive idea – the 2018 decriminalization of recreational marijuana – had little to no effect in the realm of medical marijuana. The applicants challenged the constitutionality of the same *ACMPR* regulations as in *Howell*. Justice Rondinelli of the Ontario Court of Justice determined that the government's approach was reasonable and therefore constitutional.[117] This time, issues regarding access did not reach the threshold required for a s. 7 infringement: 'the Applicants are seeking a system that provides perfect access (which is not constitutionally required), rather than one that provides reasonable access (which is constitutionally protected)'.[118] According to Rondinelli, the case was distinguishable from previous SCC cases as the evidence fell short of establishing that any serious long-lasting harm or life-threatening conditions arose due to delays with accessing the product.[119]

The applicants also argued that s. 15 of the *Charter* – the equality right – was infringed as certain groups using medical cannabis, namely the disabled and the unhoused, were disadvantaged by the existing medical marijuana regime. Rondinelli rejected the arguments of both. For the disabled, he found that rather than exacerbate any discriminatory disadvantage, the *ACMPR* have improved historical access to medical marijuana.[120] For the homeless group, he found that it was not a group that could be clearly delineated enough to warrant

114 *Ibid* at para 277. Graesser noted *Hitzig v. R.* (2003), where the Ontario Court of Appeal rejected the Crown's argument that the lack of licit medical cannabis was acceptable because medical cannabis users are able to safely navigate the black market.
115 *Howell*, *supra* note 110 at paras 354, 359.
116 *Ibid* at paras 22, 270.
117 *R v Ferkul* [2019] ONCJ 893 [*Ferkul*].
118 *Ibid* at para 17.
119 *Ibid* at para 20.
120 *Ibid* at paras 31–33.

a *Charter* right. A final argument related to the prohibition of storefront access to medical marijuana was based on freedom of expression rights under s. 2(b) of the *Charter*. Rondinelli determined that neither the *CDSA* nor the *ACMPR* prevents businesses from setting up informative storefronts that buttress the mail-order system, and therefore did not give rise to a unconstitutional limit on free expression.[121]

The Nova Scotia case of *R v Daniels* suggests that criminal sanctions related to cannabis offences will differ within a medical context.[122] Daniels was selling marijuana from his dispensary to those with 'medical authorizations'. The dispensary was not authorized under s. 10 of the *Cannabis Act*; as a result Daniels was charged with selling marijuana illegally. Justice Bodurtha distinguished medical dispensary cases from other trafficking cases on the grounds that the evidence supported Daniels' belief that offering marijuana for medicinal purposes provided a net benefit to society.[123] As Bodurtha declared, 'due to the apparent condonation by police, and the legalization of authorized marijuana sale and possession, Courts have recognized that a dispensary operating in the open is less morally blameworthy than a grow-op or drug smuggling racket'.[124] This leniency has not been granted where accused holding medical marijuana licences that allow for the production of personal-use plants use such licences as 'shields' to hide much larger operations suggestive of old-fashioned trafficking.[125]

Through its recent decision in *Murray-Hall v Quebec (Attorney General)*, the Supreme Court has also clarified that the criminal nature of Canada's *Cannabis Act* means the legislation merely provides minimal thresholds that provinces can enhance with further regulation.[126] In upholding ss. 5 and 10 of Quebec's *Cannabis Regulation Act* – provisions that prohibit individuals in the province from possessing or cultivating cannabis plants for personal purposes – the Court indicated that the *Cannabis Act* does not necessarily grant positive cannabis rights in the realm of personal use.

Canadian federalism cases like *Murray-Hall* are often concerned with whether or not a government was acting within its constitutionally-delegated powers in passing a particular law. These 'division of powers' cases require courts to determine the subject matter of the contentious law, and then to classify it as falling under federal or provincial authority.[127] In this case, the Applicant alleged that the fine-backed ss. 5 and 10 were invalid in one of two ways. First, the Quebec government had passed laws that were criminal in nature and therefore of exclusive federal domain; in this instance, the law is

121 *Ibid* at paras 41–42.
122 2021 NSSC 103 [*Daniels*].
123 *Ibid* at para 54.
124 *Ibid* at para 2.
125 See *R v Coffey*, 2018 BCPC 386 at paras 40, 7.
126 2023 SCC 10 [*Murray-Hall*].
127 *CA, 1867*, *supra* note 1, ss. 91–92.

constitutionally invalid. Or second, the Quebec government passed laws that conflicted with the operation of paramount federal law; in this instance, the law is constitutionally inoperative. The Court rejected both positions.

In the first instance, the Court determined that Quebec had passed constitutionally valid law under its powers over 'Property and Civil Rights in the Provinces',[128] and its authority to deal with matters of a 'local or private nature'.[129] Central to the Court's reasoning was that the challenged provisions had to be read in the context of the entire *Cannabis Regulation Act*. In so doing, the Court concluded that prohibiting the possession and cultivation of personal cannabis plants was not an objective in itself, but rather one means of ensuring public health and security through the province's regulated cannabis market.

In the second instance, the Court determined that Quebec had passed constitutionally operational laws because they did not conflict with the purpose or operation of the federal government's *Cannabis Act*, which allowed individuals to personally possess and cultivate as many as four cannabis plants. Quebec had simply expanded the prohibitory nature and crime-deterring aspirations of the existing federal law.

Needless to say, s. 7 of the *Charter* and the standard of 'reasonable' access continue to dominate medical cannabis case law. More generally, the impact of the *Cannabis Act* on the medical cannabis regime is not yet entirely clear. Still in its infancy, the transition appears to be in a grey zone where cannabis legislation and decisions of the past continue to remain relevant for the time being. While societal views around the perils of cannabis have certainly lessened in Canada, the above cases suggest that the legal regulation of the cannabis industry post-2018 to some extent legitimizes the prosecution of unregulated cannabis activities.

4.3 Current concerns over efficacy

While it is safe to say that litigation over the accessibility and use of medical marijuana has been favourable to promoters of its efficacy, it would be remiss of us not to make note of some of more recent concerns regarding its value. There are five points to consider: uncertainty, efficacy, potential for abuse, negative psychoactive effects, and lax retail standards.

First, the critical literature on medical marijuana is full of uncertainty. While there is a reasonable degree of support for the therapeutic use of cannabis in treating conditions such as chronic pain, nausea from chemotherapy, epilepsy, and multiple sclerosis, there are enough concerns regarding risks and potential negative consequences to raise alarms. Part of this is due to the inherent

128 *Ibid*, s. 92(13).
129 *Ibid*, s. 92(16).

variability and lack of standardization in the products. Cannabis contains hundreds of chemical compounds, including THC and CBD, which interact with the body's endocannabinoid system in complex ways. The percentages of these compounds vary widely between strains and products. This inconsistency makes it difficult to conduct reliable clinical trials and provide precise medical prescriptions.[130]

The US Food and Drug Administration (FDA) has approved only a few cannabinoid-based medications (such as Epidiolex for epilepsy and Marinol for nausea), but these are isolated compounds rather than whole-plant extracts. The whole-plant approach, advocated by many medical marijuana proponents, introduces greater variability in dosing, potency and efficacy, making it harder to measure outcomes in clinical trials. This lack of rigorous, large-scale studies poses a challenge to fully understanding the risks and benefits of medical marijuana, leading to unease over its medical legitimacy.[131]

Second, despite medical marijuana being touted as a cure-all for various ailments, the evidence supporting its effectiveness is mixed. For instance, the National Academies of Sciences, Engineering, and Medicine (NASEM) report from 2017 concluded that while there is substantial evidence that cannabis is effective for chronic pain, chemotherapy-induced nausea, and spasticity in multiple sclerosis, the evidence is far less convincing for other conditions like anxiety, PTSD, and sleep disorders. It has also been shown to cause negative effects on the cardiovascular system,[132] the respiratory system,[133] brain functioning,[134] and psychological health.[135] Moreover, for many conditions,

130 National Academies of Sciences, Engineering, and Medicine (2017), "The Health Effects of Cannabis and Cannabinoids: The Current State of Evidence and Recommendations for Research." This report highlights the complexity of cannabis as a plant with hundreds of compounds and the difficulty in conducting standardized research. Available at: https://www.nap.edu/catalog/24625/the-health-effects-of-cannabis-and-cannabinoids-the-current-state. Health Canada also suggests caution to users in its "Information for Health Care Professionals: Cannabis (marihuana, marijuana) and the Cannabinoids." This comprehensive guide reviews the current evidence on medical cannabis, emphasizing the need for standardized products and reliable dosing. Available at: https://www.canada.ca/en/health-canada/services/drugs-medication/cannabis/information-medical-practitioners.html. Health Canada also plays a significant role in regulating medical marijuana in Canada. The agency has repeatedly emphasized the importance of standardization and the need for more robust clinical research on cannabis – see https://healthycanadians.gc.ca/task-force-marijuana-groupe-etude/framework-cadre/alt/framework-cadre-eng.pdf

131 See "FDA and Cannabis: Research and Drug Approval Process." The FDA has only approved a few cannabinoid-based medications, such as Epidiolex (CBD) for epilepsy and Marinol (THC) for nausea. These drugs isolate specific cannabinoids rather than using the whole plant, which raises issues of consistency and standardization. Available at: https://www.fda.gov/news-events/public-health-focus/fda-and-cannabis-research-and-drug-approval-process.

132 See Ravi et al., "Associations Between Marijuana Use and Cardiovascular Risk Factors and Outcomes" (2018) 168(3) *Annals of Internal Medicine*, available at: https://www.acpjournals.org/doi/full/10.7326/M17-1548 and Singh, et al., "Marijuana and Related Substances: A Review" (2018) 7 *Cardiology and Therapy* 45, available at: https://link.springer.com/article/10.1007/s40119-017-0102-x (indecisive findings, "more research needed"); Pasha, et al., "Cardiovascular Effects of Medical Marijuana: A Systemic Review" (2021) 134(2) *American Journal of Medicine* 182, available at: https://www.sciencedirect.com/science/article/pii/S0002934320308913 (possible risks.)

133 Vásconez-González, et al., "Effects of Smoking Marijuana on the Respiratory System: A Systemic Review" (2023) 44(3) *Substance Use & Addiction Journal*, available at: https://journals.sagepub.com/doi/full/10.1177/08897077231186228.

134 Testai, et al., "Use of Marijuana: Effect on Brain Health" (2022) 53(4) *Stroke*, available at: https://www.ahajournals.org/doi/full/10.1161/STR.0000000000000396.

135 Sheila Gupta, Sanjay Gupta, and Tiffany Phelan, "Medical Marijuana: Do the Benefits Outweigh the Risks?" (2018) *Current Psychiatry*, available at: https://www.researchgate.net/profile/Sheila-Gupta/publication/322603065_Medical_marijuana_Do_the_benefits_outweigh_the_risks/links/5a6243fc0f7e9b6b8fd6158f/Medical-marijuana-Do-the-benefits-outweigh-the-risks.pdf.

the studies that do exist are often small, poorly designed, or rely on subjective self-reported outcomes rather than objective measurements. Critics argue that the enthusiasm surrounding medical marijuana has outpaced the evidence, leading to its use for conditions where it may not provide meaningful relief, or worse, may exacerbate symptoms.[136]

Third, the potential for abuse and dependence on medical marijuana is not insignificant. While cannabis is often perceived as benign, research suggests that about 10% of users may develop cannabis use disorder which can lead to physical and psychological impairment. The problem is made worse by the increased availability of medical marijuana, as has happened in Canada. The risk is that patients might turn to marijuana as a 'quick fix' for chronic conditions, particularly in pain management, where opioid addiction remains a significant public health crisis. That Canadian federal law allows individuals to grow their own cannabis plants, makes it even harder to control and monitor usage.[137]

Fourth, the psychoactive effects of THC are not always positive. THC can impair cognitive function, memory, and attention, especially in young people, whose brains are still developing. Studies have shown that regular cannabis use in adolescents can lead to long-term cognitive deficits and increase the risk of developing psychiatric disorders, including depression, anxiety, and schizophrenia, particularly in individuals with a genetic predisposition to these conditions. Given that medical marijuana is commonly taken by patients with mental health disorders such as PTSD and anxiety, overuse is a concern. In addition, the relationship between cannabis use and mental health is complex and not fully understood. Some evidence suggests that while cannabis may provide temporary relief from anxiety or stress, long-term use can worsen psychiatric symptoms. Thus, there is now a body of research that cautions against the use of medical marijuana in populations vulnerable to mental health issues, arguing it may do more harm than good.[138]

Finally, medical marijuana products in Canada are typically available in dispensaries not subject to the same standards as ordinary pharmacies. This can lead to problems with contamination, inconsistent dosing, and inaccurate labeling. Critics argue that the current regulatory system fails to protect patients from these risks, particularly in locations with poorly regulated medical marijuana programs.[139]

136 Whiting et al., "Cannabinoids for Medical Use: A Systematic Review and Meta-analysis" (2015) 313(24) *JAMA* 2456, available at: https://jamanetwork.com/journals/jama/fullarticle/2338251.

137 Hasin et al., "Prevalence of Marijuana Use Disorders in the United States Between 2001–2002 and 2012–2013" (2015) *JAMA* 1235–prevalence of cannabis use disorder increased as marijuana became more widely available and legalized, available at: https://jamanetwork.com/journals/jamapsychiatry/fullarticle/2464591.

138 Volkow et al., "Adverse Health Effects of Marijuana Use" (2014) 370 *NEJM* 2219, available at: https://www.nejm.org/doi/full/10.1056/NEJMra1402309.

139 See Health Canada, "Compliance and Enforcement Report: Cannabis Inspection Data Summary 2023 to 2024" noting the issuing of 21 non-compliant reports and almost 50% of medical inspections were scored 2 or below (out of 3), available at: https://www.canada.ca/en/health-canada/services/drugs-medication/cannabis/research-data/compliance-enforcement-report-cannabis-inspection-data-summary/2023-2024.html.

In sum, while medical marijuana has shown promise in treating certain conditions, the critical literature raises valid concerns about its efficacy, safety, and societal impact. The lack of standardized research, potential for abuse, cognitive risks, and regulatory shortcomings suggest that it is not a panacea. More rigorous, large-scale studies and better regulatory frameworks are needed to ensure that patients can benefit from cannabis therapy without facing undue risks. Until these issues are addressed, critics argue, medical marijuana should be approached with caution, both by patients and policymakers.

Cannabis certainly presents taxonomic complexities in botanical circles,[140] but one of the major themes of this chapter is to illustrate that classificatory complications are needlessly exacerbated in the legal context. It would be naïve to assume that the long and illicit history of cannabis plays no role in its present-day governance, even if there were no current fears over its potential deleterious effects (at least as they relate to marijuana). Today, the legal regulation of cannabis in Canada reflects two fundamental concerns. First, at their core, cannabis classifications are aimed at managing the potential intoxicating effects of various forms of the plant. Second, the law remains preoccupied with the particular context in which cannabis arises (e.g. personal or medical use).

In such a milieux, the potential for hemp suffers. Its strict regulation clearly reflects its historical association with marijuana, despite indications that the legal threshold for THC in Canadian hemp is a 'conservative' drop-off from intoxicating levels (expected to occur around 1%).[141] To help alleviate this, we propose one preliminary and simple fix. Clear terms, especially in the legal context, are both important and clarifying. Why not alter the legally defined terms and adopt the following: 'marijuana' for recreational use; 'medical marijuana' for medical use; 'hemp' for fibrous applications; and 'cannabis' for the category/plant as a whole? Although modest, the aim is to clarify existing distinctions to reflect the reality that what wine is to grapevine, marijuana is to cannabis. And what grape juice is to grapevine, hemp is to cannabis. Or, in answer to our own question, hemp is and is not cannabis.

5 Conclusion

Canada's regulatory approach to industrial hemp illustrates the complexities that arise when diverse cannabis-derived products are governed under a single legislative framework. While the Cannabis Act represents a progressive approach to drug policy reform, its umbrella structure has created unintended consequences for the hemp industry. The conflation of hemp with marijuana

140 Ernest Small, "Evolution and Classification of *Cannabis Sativa* (Marijuana, Hemp) in Relation to Human Utilization" (2015) 81(3) *Botanical Review* 189.
141 Small and Marcus, *supra* note 22 at 292.

– both linguistically and legally – perpetuates historical misconceptions that date back to the 1938 amendments to the Opium and Narcotic Drug Act.

Hemp regulation in Canada suffers from its association with psychoactive cannabis despite containing negligible THC levels. Municipal zoning disputes, declining industry participation, and unnecessary regulatory burdens reflect the practical challenges of this approach. Courts have struggled to balance public health objectives with economic realities and constitutional rights, particularly regarding Indigenous treaty claims and medical access.

Canada would benefit from adopting clearer terminological distinctions: 'marijuana' for recreational use, 'medical marijuana' for therapeutic applications, 'hemp' for industrial purposes, and 'cannabis' as the overarching plant category. This modest reform could reduce regulatory confusion, support industry development, and better reflect the functional differences between these products. Such clarification would honour both the plant's diverse applications and the legitimate regulatory concerns that initially drove cannabis prohibition and recognize that hemp's non-intoxicating nature warrants differentiated treatment in law.

6 Where to look for further information

Several comprehensive resources provide accessible overviews related to hemp regulation. Small and Marcus's "Hemp: A New Crop with New Uses for North America" (ASHS Press, 2002) offers an excellent technical introduction to industrial hemp cultivation and regulation. MacFarlane, Frater, and Michaelson's *Cannabis Law* (Thomson Reuters, 2021) provides detailed analysis of Canadian cannabis legislation including hemp provisions. For U.S. perspectives, the National Agricultural Law Center maintains comprehensive hemp regulatory resources, while the European Industrial Hemp Association offers comparative international regulatory frameworks.

The Canadian Hemp Trade Alliance (CHTA) serves as Canada's national industry association, promoting Canadian hemp products globally while disseminating information and developing standards (https://www.hemptrade.ca). In the United States, the Hemp Industries Association (HIA) has advanced the use and acceptance of hemp for almost 30 years, while the Federation of International Hemp Organizations (FIHO) provides a global platform promoting research, development, and standards harmonization. The National Industrial Hemp Council of America focuses on consumer safety and regulatory oversight (https://nationalhempassociation.org).

For linkages connecting medical cannabis and hemp, Thomas Jefferson University's Lambert Center for the Study of Medicinal Cannabis & Hemp is an excellent source, being the first major U.S. health sciences university to provide comprehensive academic resources. Health Canada's hemp regulatory website provides current licensing requirements, approved cultivar lists, and industry statistics. The USDA's hemp regulatory framework offers comparative U.S. policy analysis, while provincial agricultural departments maintain region-specific guidance for Canadian producers.

For information on academic journals and conferences, the *Journal of Industrial Hemp* remains the primary academic publication for hemp research. The Canadian Journal of Criminology and Criminal Justice frequently publishes cannabis policy analysis. Legal scholarship appears regularly in constitutional law journals and agricultural law reviews.

Chapter 13

Consumers and hemp: demand, trust and sustainability

Jane Kolodinsky, University of Vermont and Colorado State University, USA; Hannah Lacasse, University of Vermont, USA; Tyler Mark, University of Kentucky, USA; Rebecca Hill, Colorado State University, USA; Heather Darby, University of Vermont, USA; and William Snell and Yuqing Zheng, University of Kentucky, USA

1 Introduction

This chapter focuses on an aspect previously ignored in the limited consumer-related hemp literature, trust in sources of information. Information is a precursor to understanding the market for hemp-based products. How and to whom to communicate messages about hemp can effectively drive the hemp product market to meet consumer needs. Using a nationally representative sample in the United States, we analyzed trust in eight different sources of information and identified three classes of consumers: untrusting, neither trusting nor untrusting, and trusting. A multivariate model identified that positive perceptions about the sustainability of hemp are associated with trust. Given the current unsettled provision of information landscape in terms of THC levels in hemp foods, our findings reinforce the necessity of transparent and accurate information if consumers are to make informed decisions about the hemp-based food and CBD products they consume.

Key concepts used in this chapter:

- Hemp – Industrial hemp is from the plant species *Cannabis sativa* and has been used worldwide to produce a variety of industrial and consumer

http://dx.doi.org/10.19103/AS.2025.0161.26

products. By definition, industrial hemp is high in fiber and low in active tetrahydrocannabinol (THC), the psychoactive ingredient in marijuana.
- Sustainability–The practice of meeting current needs without compromising the ability of future generations to meet their own needs through the balancing of environmental, social, and economic considerations.
- Consumer – An individual or entity that purchases or uses goods and services to satisfy needs or wants.
- Hemp information – Data on hemp that is processed, organized, or structured to provide meaning or context that enables understanding and aids decision-making.
- Trust – The confident belief in the reliability, integrity, or ability of a person, organization, or system.

Hemp has a thousands of purported uses and a history going back to ancient times (Figure 1) (Robinson, 1996). Hemp was removed (*Cannabis sativa L.*) from the definition of marijuana in the Controlled Substances Act (CSA) in the Agriculture Improvement Act of 2018 if it contains levels of the psychoactive compound delta-9-tetrahydrocannabinol no more than 0.3% delta-9 THC on a dry-weight basis (USDA, 2019). The resulting expectation was that hemp products would have an almost immediate demand by American consumers (Johnson, 2019). Yet, rebuilding the hemp industry after an 80-year ban has proven difficult. There are many challenges facing the hemp industry. This chapter focuses on the role of consumer demand.

Historically, hemp was used across the globe for paper and textiles (Bouloc & van der Werf, 2013; Naeem et al., 2023); sailing canvas and rope (J. H. Fike et al., 2020; Naeem et al., 2023); medicine, therapy, and healing (Chen et al., 2023; Fike, 2016); and food (Adesina et al., 2020; Pihlanto et al., 2017). Despite hemp's utility, globalization, and technological progress introduced unprecedented competition. The steam engine replaced sailboats, reducing demand for sailing cloth and rope (Fike et al., 2020; Fortenbery & Bennett, 2004). The cotton gin and enslavement made cotton the dominant material for clothing and textiles (Choi et al., 2018), and jute imports provided a more cost-effective rope alternative (Fortenbery & Bennett, 2004).

There is some evidence that both synthetic fiber development and competition of hemp paper with tree pulp dampened the growth of the hemp fiber industry going back to the early 1900s. In 1917, American George W. Schlichten patented a new machine for separating bast fiber from the internal woody core (hurds), reducing labor costs by a factor of 100 and increasing fiber yield from hemp significantly (Figure 2) (Schlichten, 1919). This invention allegedly threatened paper pulp from trees (Beach, 2021).

The historical account suggests that dominant companies in American industry (including Scripts, Hearst, and Dupont) were instrumental in both the

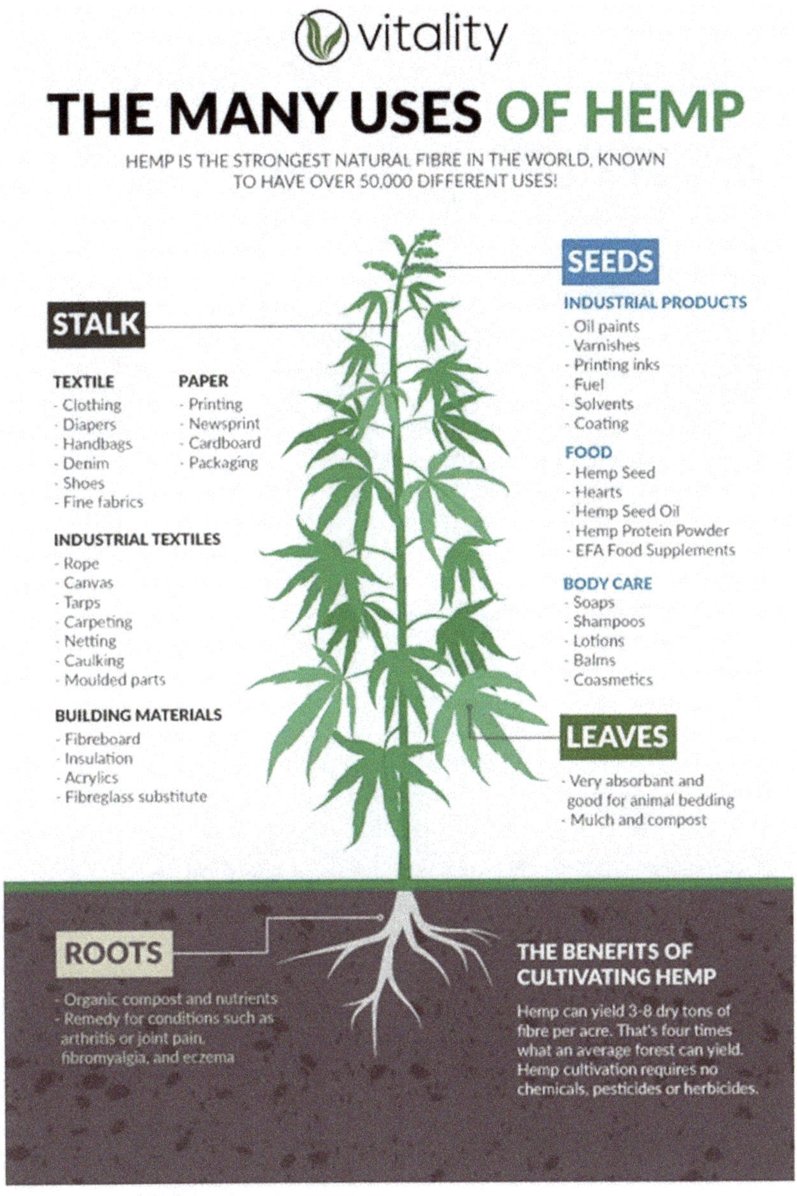

Figure 1 The many end uses of Hemp (*The Many Uses of Hemp*, 2020).

development of taxes and the legal criminalization of hemp, as they supported the US Marijuana Tax Act of 1937 (French & Manzanárez, 1984). Adding to the pressure on hemp was a growing perception of the crop as a drug and its relationship to marijuana, as well as racist and xenophobic tensions associated

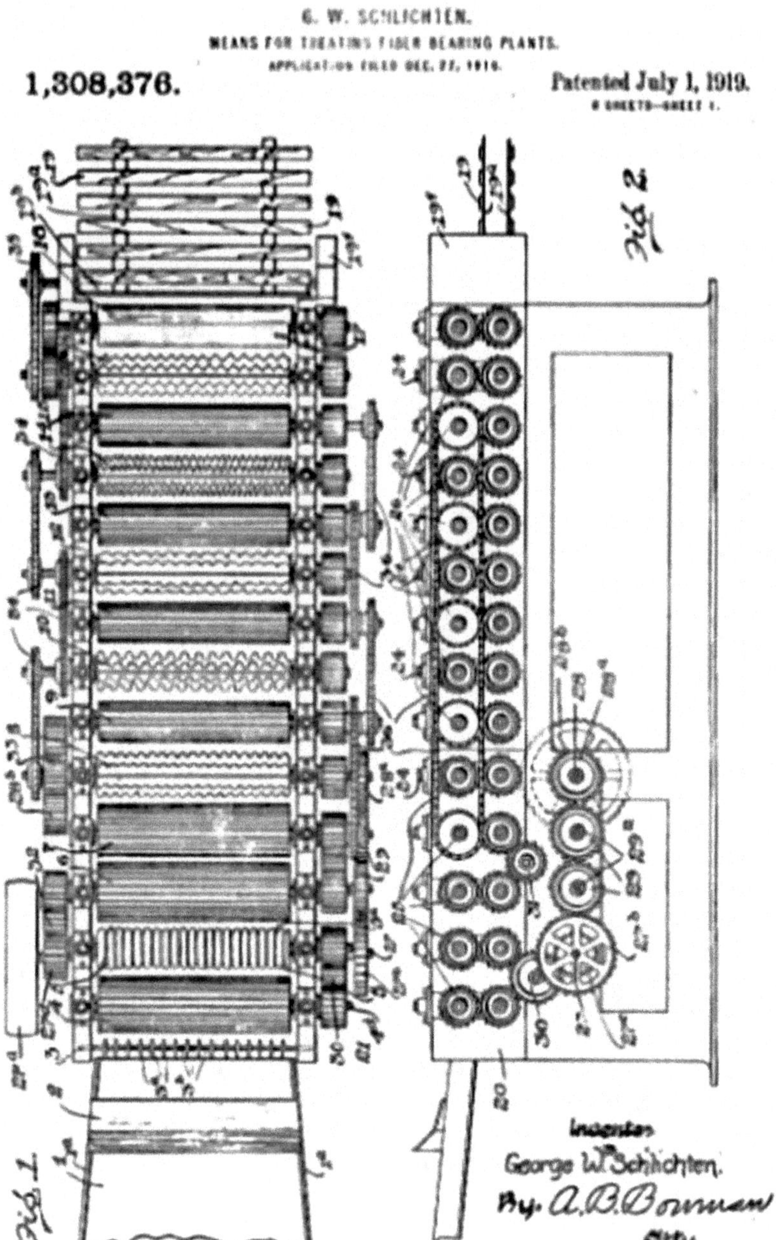

Figure 2 Bast fiber hemp decorticator patent drawing (Schlichten, 1919).

with its use (Bewley-Taylor et al., 2014; Choi et al., 2018; Gray, 2000; Yano & Fu, 2023). In the name of public health and safety, many countries prohibited hemp, including Canada, China, the European Union, and the United States (Cherney & Small, 2016a; McGrath, 2020; Putnam, 2022; Yano & Fu, 2023). Many countries also legalized hemp, up to decades before the United States, or following the 2018 US Farm Bill (Law Library of Congress, 2022). While we focus on the US experience, the global history of hemp is illustrative.

Prohibition was lifted across many parts of the world as the distinction between hemp and marijuana was clarified and interest in hemp's environmental and economic potential rose (Ahmed et al., 2022a; Mark et al., 2020a; World Health Organization, 2018). Industrial hemp is defined as *Cannabis sativa* L. with a tetrahydrocannabinol (THC) concentration not exceeding a certain percentage, which varies from country to country (DeCarlo & Weaver, 2023). A naturally occurring cannabinoid produced by the *Cannabis sativa* plant, THC is the psychoactive compound that distinguishes hemp from marijuana (World Health Organization, 2018). The most common thresholds for THC in defining hemp across the globe are 0.3% and 0.1% on a dry-weight basis (DeCarlo & Weaver, 2023).

2 Understanding consumer demand for hemp-based products

Rising demand for plant-based and environmentally friendly food, textile, and plastic alternatives, coupled with a realization of the crop's potential to contribute to modern economic and sustainability goals, set the stage for hemp's legalization during 2018 (Ahmed et al., 2022b; Crini et al., 2020; Ely et al., 2022b; Musto, 1972). Hemp is perceived to be an environmentally and biodiversity friendly crop that can remediate soil, complement food production, and grow quickly (Fortenbery & Bennett, 2004; Montford & Small, 1999). This perception of sustainable hemp production extends to the crop's resulting products, such as hemp grain for plant-based milk substitutes or textiles made with hemp fiber rather than cotton. Efforts to reduce dependence on petroleum-based products (Babu et al., 2013) stimulated technological progress for the conversion of hemp fibers to plastic, construction material, and polymer alternatives (Ahmed et al., 2022b; Filimonova et al., 2022). Particular attention was also being paid to another emerging use of hemp: cannabidiol (CBD), a naturally occurring, non-psychoactive compound in the hemp flower used for therapeutic and recreational purposes (Adesina et al., 2020; Mark et al., 2020a; World Health Organization, 2018). High profit estimations piqued the interest of farmers and entrepreneurs (Mark et al., 2020a CBD was added to edibles, tinctures, and common food and beverage products (Malyshev & McDonough, 2019).

With lucrative prospects for CBD oil and hemp's general untapped potential in mind, national acreage for hemp increased nearly 350% from 2018 to 2019 and largely focused on cultivation for floral biomass for CBD production (Mark et al., 2020b). Though the applications of this crop are many, hemp is not without limitations. It competes with the cost efficiency of conventional alternatives and new technologies targeting sustainable outcomes, such as flax and other plant-based bioplastics and building materials (Cherney & Small, 2016b; Fike, 2016; Notaro et al., 2022; Papadopoulou et al., 2015). The production efficiencies championed by cotton and petroleum-based products helped to push hemp out of production in the nineteenth and twentieth centuries (Fortenbery & Bennett, 2004; Meijer et al., 1995). Current processing constraints include a lack of textile infrastructure in general making hemp inputs a relatively more expensive alternative (Mark et al., 2020b; Mark & Will, 2019; Sterns, 2019). Production constraints can also influence the ability of products to make it to store shelves, restricting availability for consumer purchase. All along the supply chain, constraints in obtaining appropriate varieties of hemp seed and clones, harvesting and processing equipment, and supply transport impact profitable production. Though CBD drove the initial ascent of hemp production, an unknown market and overinflated profit estimates led to CBD's oversupply and price crash in 2019 and 2020 (Mark et al., 2020b; Quinton, 2021). Exacerbating these barriers to the hemp sector is limited research across the supply chain for informed decision-making (Ellison, 2021; Ely et al., 2022a, 2022b; Mark & Will, 2019; Schluttenhofer & Yuan, 2017).

Interest in hemp production legalization was partially prompted by the potential environmental and sustainable production practices associated with its cultivation (Adesina et al., 2020; Ely et al., 2022b). Perceived benefits to hemp production include crop rotation (Fike et al., 2020); low pesticide and water use (Yano & Fu, 2023); high biomass yield (Ely et al., 2022a); carbon sequestration (Adesina et al., 2020); soil remediation (Guo et al., 2024); and weed suppression (Adesina et al., 2020). In addition, hemp has the potential to contribute to circular economies through the production of plant-based, renewable alternatives that can decrease fossil fuel dependence (Ahmed et al., 2022a; Ely et al., 2022a). Due to the relatively short period since hemp's legalization, the industry's ability and willingness to contribute to these sustainability goals remain to be seen. Statements regarding production feasibility before legalization were largely based on conjecture. It is too early to tell whether these environmentally friendly and sustainable outcomes will be achieved, and more research is needed (Adesina et al., 2020; Fike et al., 2020). That said, these characteristics of hemp may help drive the marketing for hemp-based products.

If hemp's promise as a sustainable crop rings true, then there are segments of consumers who could be brought into the hemp marketplace based on demographic and attitudinal criteria, including sustainability characteristics. Research on consumer demand and market opportunities for hemp has been cited as a need (Mark et al., 2020a; USDA, 2024), with markets, economics, and consumer perceptions identified as the second most important area of research to surveyed US hemp stakeholders (Ellison, 2021).

Despite an expressed need for peer-reviewed research on consumer demand for hemp-based products made by stakeholders (Ellison, 2021; Mark et al., 2020a), available consumer demand research on hemp in the US is largely limited to industry reports on CBD products that focus on demographic characteristics such as gender and age (New Frontier Data, 2020). Much of the literature on consumer behavior has also focused on hemp-derived CBD products and how consumers perceive and use them (Colclasure et al., 2021; Goodman et al., 2022; Kolodinsky et al., 2020; McFadden & Malone, 2021; Nguyen et al., 2023; Schilling et al., 2021; Staples et al., 2022; Wheeler et al., 2020; Wysota et al., 2022). These studies frequently sample young people or students and find that knowledge about CBD is low (Colclasure et al., 2021; Wheeler et al., 2020; Wysota et al., 2022). Findings indicate that consumers of all ages perceive CBD to be beneficial for health and effective for pain, anxiety, and sleep (Goodman et al., 2022; McFadden & Malone, 2021; Nguyen et al., 2023; Schilling et al., 2021; Staples et al., 2022; Wheeler et al., 2020; Wysota et al., 2022), despite the prohibition of health claims on CBD. The marketing messages about CBD products appear to have influenced public opinion, with health benefits being the dominant factor for its use.

Market estimates (Grandview Research, n.d.). and existing peer-reviewed research (Kim & Mark, 2023; Kolodinsky & Lacasse, 2021) demonstrate that other, non-CBD, hemp products are of interest to consumers and that demographic data alone provide an incomplete picture of hemp demand (Kolodinsky et al., 2020). Studies following the 2018 Farm Bill are characterized by an examination of awareness, attitudes, and policy support for hemp, which was relatively low during the early years of hemp legalization (Borkowska & Bialkowska, 2019; Colclasure et al., 2021; Kolodinsky et al., 2020; Rampold et al., 2021; Ruth et al., 2022), though a study of Vermont consumers found that awareness may increase over time (Kolodinsky & Lacasse, 2021). A review of the existing literature finds a handful of consumer studies on hemp products across the globe. Metcalf et al. (2021) model early adopters of newly legal hemp-based food products in Australia and found that consumer associations with hemp and marijuana do not significantly influence adoption, while consumers do value health and ethical sourcing of hemp food products. Willingness to purchase a food product (meat loaf) with hemp ingredients significantly increased when consumers in Poland were provided information about hemp's positive health impacts (Zając et al., 2019). Though consumer studies on hemp products in other countries can offer some insight, the unique context of hemp in the US necessitates specific exploration of its consumers.

An application of the diffusion of innovations to a sample of Vermont consumers highlights that hemp product choice is related to their perceived advantages compared to alternatives (Lacasse et al., 2023). This includes attributes related to health benefits, durability, taste, and quality. Velasco et al. (2022) finds that consumers may perceive hemp to be natural, which positively influences attitudes. Using the theory of planned behavior, and the only currently available representative sample of hemp consumers, Lacasse et al. (Lacasse et al., 2024) found a consistent relationship between attitudes toward hemp and intent to purchase hemp foods, clothing, personal care products, CBD, and pet products. Attitude included overall support of hemp, the idea that hemp is 'good' and the perception that hemp is beneficial. Increases in age were negatively associated with intent to purchase across all product categories and are the only demographic with consistent results, leading to the idea that consumer demand may increase over time as we 'age out' of a demand constraint.

Some studies focus on the use and purchase of hemp foods and offer implications for health and taste. Consumers appear more interested in hemp food if it has health benefits (Goel et al., 2023; Metcalf et al., 2021; Zając et al., 2019), though they may find hemp alternatives less palatable than conventional counterparts (Hayward & McSweeney, 2020; Zając et al., 2019). Kim and Mark (2023) find that young people are more likely to purchase hemp food products. Overall, existing studies highlight that acceptability of hemp foods may be low for taste but high for perceived health benefits.

There is a growing segment of consumers who are interested in sustainable products and a large literature that has investigated demographic characteristics, attitudes, knowledge, and beliefs about sustainable product adoption in general (de Carvalho et al., 2015; Farla et al., 2012; Fischer & Newig, 2016; Gallo et al., 2023; Huttunen et al., 2021). However, this has not yet translated into a substantial literature focused on the consumption of hemp products. One pilot study using two national consumer panels in the US and Canada examined willingness to pay (WTP) for hemp textiles, including hemp jeans and bedlinens (Kolodinsky, 2023; Kolodinsky et al., 2023). A major finding of this pilot was that the perception that hemp textiles were more sustainable led to the highest WTP premium for both hemp jeans and bedlinens compared to other product characteristics, controlling for demographic characteristics.

During 2024, the initial excitement following hemp's legalization appears to have tempered and the sector profitability has been thwarted by a CBD price decrease. Though challenges to the sector remain, a new equilibrium is emerging within the industry. Production and prices are stabilizing, the implementation of the Hemp Final Rule solidified some policy uncertainties, financing opportunities are increasing, and the market for fiber and grain is rising (Grand View Research, 2023; Mark et al., 2020a; Markets & Markets, 2022). There is potential for the industry to learn, adapt, and move forward with a more informed understanding of the consumer market landscape. To do so, however, requires research and data, which have been a limiting factor for hemp stakeholders.

Changing policy, processing bottlenecks, and price uncertainty (Mark et al., 2020b) make knowledge of hemp demand essential to inform producer strategies in a historically turbulent sector. Indeed, hemp's success is conditional on demand for its products (Ely et al., 2022a). Otherwise, investments made by hemp producers could be fruitless, a reality already experienced by many early growers of CBD (Quinton, 2021).

The case study for this chapter focuses on an aspect previously ignored in the limited consumer-related hemp literature, trust in sources of information about hemp. Information is a precursor to understanding the market for hemp-based products and how and to whom to communicate messages about hemp that can effectively drive the hemp product market in a way that meets consumer needs. In the presentation of the study below, we focus on two major market categories of hemp produced from floral material and grain: hemp foods and CBD.

3 Case study: trust, sustainability and consumer markets for hemp foods and cannabidiol (CBD)

Hemp foods and CBD are mostly consumed (although there are also CBD topicals). This differentiates them from other hemp-based products which might be used as topically applied (health and beauty products), in the animal and pet sector, or in other utilitarian ways (clothing, paper, rope, plastics, or construction materials). CBD products are made from hemp floral material. Hemp foods are made from hemp grain (seeds) which can be pressed into oils, made into meal, or hulled (hemp hearts), and may also contain cannabinoids.

Trust in information sources is a key element in using information to make choices. For new and novel products in general, literature has focused on social networks, self-confidence, general trust, and risk aversion (Buskens, 2020; Siegrist, 2021). Recent general literature discusses trust across the supply chain including interpersonal, organizational (general and specific), food chain, and product (Benson et al., 2020). Price Waterhouse Coopers (Price Waterhouse Coopers, 2015) specifically points to scandals, empowered consumers, changing food demand, and certifications as areas to build consumer trust. Specifically regarding hemp foods and CBD products, some have called the hemp marketing landscape the 'wild wild west' (Drewery & Hustvedt, 2024; Fair, 2020; Gelski, 2019; Kastel, 2020; Nieves, 2021). Reference to this unsettled landscape has come from industry, government, and industry groups, all of which have been cited in the food system literature on trust (Wu et al., 2021).

The analysis presented here aims to answer the following research questions:

- R1. To what extent do consumers trust sources of information that originate from the hemp value chain (producers, processors, sellers); from advocacy and industry groups, friends and family; from the news media, social media, and the internet; and from government sources and academic research?
- R2. Does trust in information differ by whether a respondent is aware of hemp foods or hemp CBD compared to other hemp products?
- R3. Is trust in hemp information sources associated with demographic characteristics and/or perceived of hemp sustainability characteristics?
- R4. What are the implications of answers to research questions 1–3 for the emerging hemp marketplace?

We utilize a representative sample of US consumers (n = 2000). We include only those who are at least aware of one hemp product group (clothing, personal care products, rope, paper, and bioplastics, CBD, CBG, and foods) (n = 1959). Data were collected by Qualtrics during the first quarter of 2022. We categorize hemp-aware consumers into four mutually exclusive awareness groups: CBD only (33.2%), hemp foods only (13.6%), CBD and hemp foods only (22.1%), and any hemp product except CBD or food (31.1%). We also collected data on the level of respondent trust for each of the following sources: hemp processors, sellers, farmers, other consumers, friends/family, state departments of agriculture, the United States Department of Agriculture (USDA), non-governmental agencies associated with hemp, university outreach (Extension), news media, social media, and other internet sources. These data were collected on a five-point scale and collapsed into three ordinal categories of trust: not trusted, neither trusted nor non-trusted, and trusted, as this is most appropriate for our classification method, latent class analysis (LCA). Demographic information was collected from respondents, as well as their perceptions of hemp as a sustainable crop and input into value-added products. Demographics include age, education, children in the household, gender, political affiliation, race, and income.

We use latent class analysis (LCA) to classify respondents based on their trust in each of the 12 providers of information groups listed above. LCA is a statistical procedure used to identify qualitatively different subgroups within populations who share certain characteristics. LCA models work on the assumption that the observed distribution of the variables is the result of a finite latent (unobserved) mixture of underlying distributions and membership in unobserved groups (or classes) that can be explained by patterns of scores across survey questions, assessment indicators, or scales (Sinha et al., 2021; Weller et al., 2020). Latent class analysis is appropriate when data are collected as categories and the algorithm assigns probability scores that help identify which classification a respondent belongs to on each dimension. Utilizing the Delta method, we identified a three-class structure. Table 1 provides details. The appropriate latent class cut-off is at the point where the probabilities decrease. For our sample, this was at three classes.

The three latent classes are described as: respondents who have an overall trust in every source, respondents who have an overall distrust in every source, and respondents who are in the middle across all sources of information. Individual respondents were placed in a class if the probability of being in that class was >0.51. All but six respondents were classified (n = 1953).

Figure 3 shows size of the three retained classes. Sixty percent of respondents were classified as trusting (2.60; $\delta = .186$; $\eta = 2.68$), 21% as neither trusting nor untrusting (2.04; $\delta = .082$; $\eta = 2.05$), and 19% as untrusting (1.57; $\delta = .229$; $\eta = 1.51$). Figure 4 shows detail of the average trust level for each source of information and depicts visually the three latent classes. ANOVA tests of the null hypotheses that each class is not different from another class are rejected for every source of trust (Critical value of F ranges from 280.428 to 933.756; $P < .01$). One notable result is that for the least trusting class, trust in both government (United States Department of Agriculture, USDA and University information, Extension) is higher (approaching neutral) compared to the other information sources.

Next, we use Chi-square analysis to test the null hypothesis that there is no difference in latent class assignment by hemp awareness group. We reject the null hypothesis that each hemp awareness group is assigned similarly to the three latent classes (=57.87, $P <= .01$) (Figure 3). More respondents in the two food aware groups are in the trusting latent class (67.60% and 72.90%). More respondents in the CBD only group are in the neither trusting nor untrusting and untrusting latent classes (23.80% and 21.80%) (Figure 5).

We use ANOVA analysis to determine whether to reject the null hypothesis that levels of trust in each source do not vary by awareness group. Because of

Table 1 Latent class analysis fit statistics and class structure

Fit statistic	Value	Description		n = 1959
Likelihood ratio chi2_ms(1594215)	14005.86 (*P* <= .01)	Model vs. saturated		
Latent class marginal probabilities				
Delta-method classes	Margin	Std. err.	[95% conf.	interval]
1	0.2388277	0.0126247	0.2149687	0.2644427
2	0.1203782	0.0081467	0.1053025	0.137281
3	0.1549192	0.0085348	0.1389186	0.1723938
4	0.4858749	0.013752	0.4589886	0.5128433

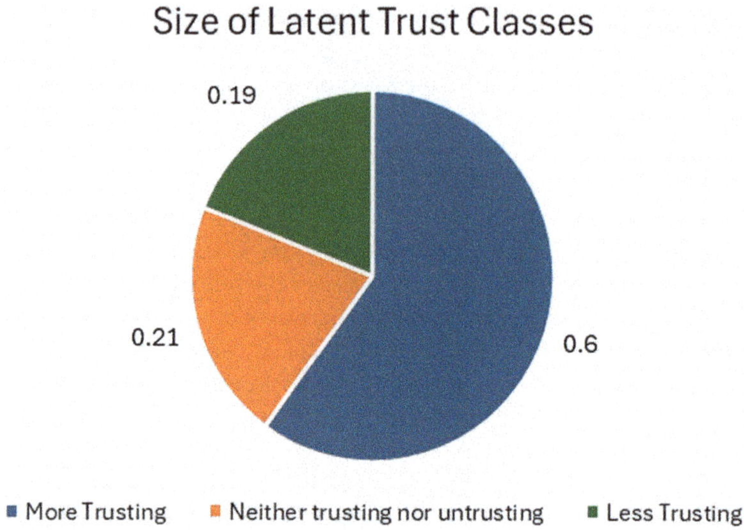

Figure 3 Size of assigned latent trust classes (n = 1953).

the ordinal nature of the trust variable, we also conducted Chi-square analysis. The results were identical to ANOVA. For ease of visualization, we use the ANOVA results. We reject the null hypothesis that there is no difference in trust for every source except from USDA (Figure 6). Overall, respondents aware of food are more trusting of sources including USDA, University Extension, and the Farmer. They are least trusting of news and social media. Sellers are not the most nor least trusted.

We expand on these general results by investigating whether demographic characteristics and perceptions of the sustainability of hemp are associated with the three classes of consumer trust. Results are shown in Table 2.

Demographic characteristics including minor children residing in a household and age are significant. More households with minor children are in the trusting classification. Consumers in the untrusting classification are older (ANOVA was used as age is a continuous variable). All three sustainability measures were significant. Almost three times as many respondents who agree with each sustainability statement are in the trusting group, a majority of respondents.

We also ran an ordinal regression analysis with trust class as the dependent variable, and the demographic and sustainability variables as independent variables. All independent variables were measured as 0/1 dummy variables, with the exception of age, which was continuous. While we do not present full results, none of the demographic variables were significant ($P > .05$). All three sustainability variables were significantly associated with the trusting class compared to the untrusting class. Respondents who agreed that hemp

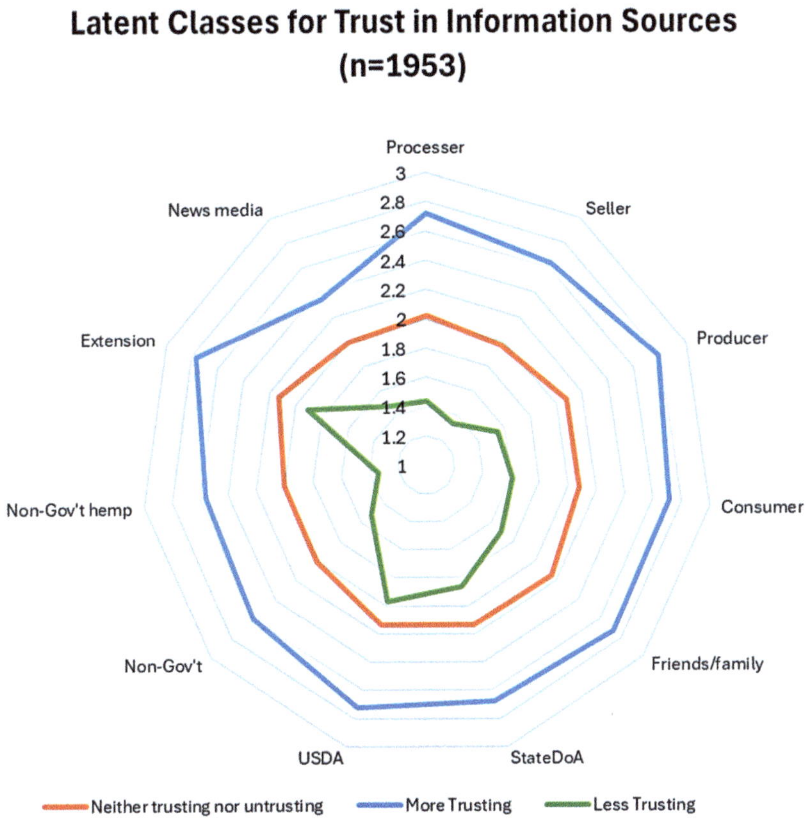

Figure 4 Latent trust classes.

products local were two times ($P < .001$) more likely to be in the trusting class compared to the untrusting class. Respondents who agreed that hemp products are sustainable were 2.2 times ($P < .001$) more likely to be in the trusting class and respondents who agreed that hemp products are environmentally friendly werethree times ($P < .001$) more likely to be in the trusting class.

What does this mean in the information marketplace? With regard to R1, using LCA, the 'salsa effect' is at play with regard to information source trust. Respondent trust from all sources of information presented was classified as distrust, neither trust nor distrust, and trust (mild, medium, and hot salsa) (Sinha et al., 2021). The majority of respondents are more trusting of all sources of information (60%). With reference to R2 and R3, more respondents who are aware of hemp foods, either with or without also being aware of CBD, are classified as trusting. But, respondents aware of food have a higher trust in more evidence-based information sources, including USDA and Extension, and are more trusting of farmers, state departments of agriculture, and in the case

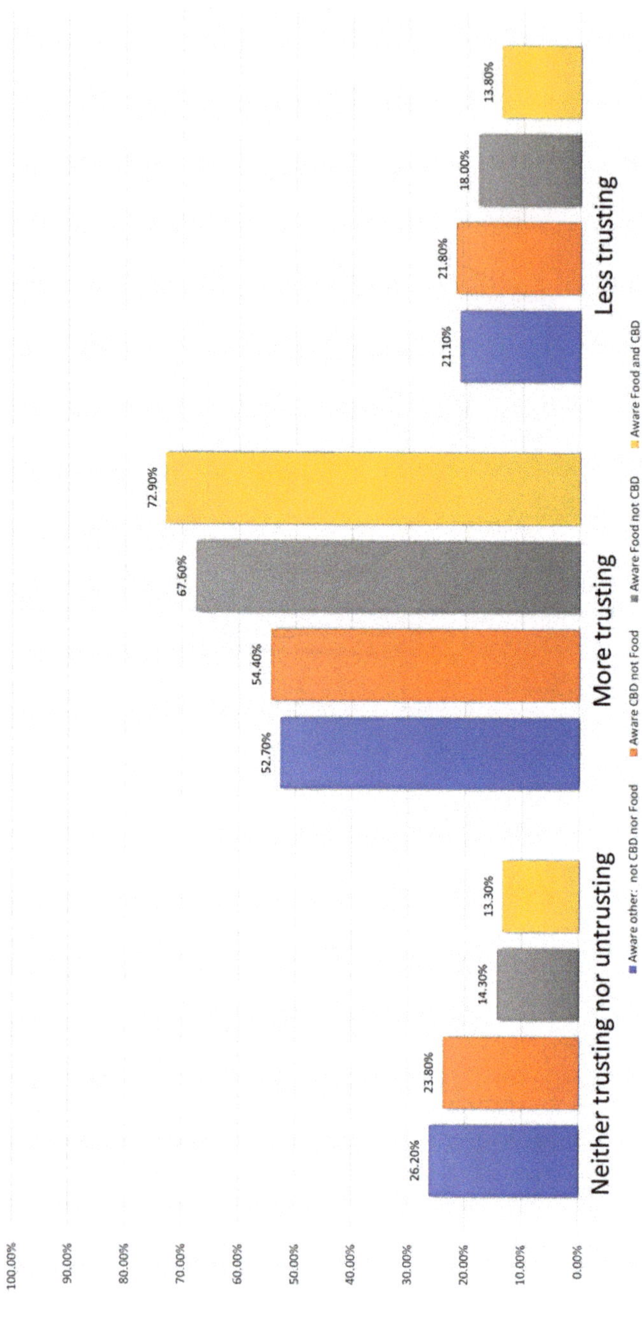

Figure 5 Percentage of aware group by latent trust class (n = 1853; $\chi^2 = 57.87$, P £ 0.1).

Source of Information	Aware Other	Aware CBD Only	Aware Food Only	Aware Both
USDA	2.42	2.44	2.51	2.51
Extension	2.42	2.43	2.58	2.55
State agency	2.37	2.37	2.49	2.48
Consumer	2.31	2.31	2.43	2.51
Farmer	2.3	2.35	2.46	2.56
Friends	2.3	2.34	2.49	2.55
Non-governmental Agricultural Organization	2.22	2.2	2.28	2.38
Processer	2.2	2.27	2.43	2.5
Seller	2.16	2.15	2.33	2.45
Non-governmental Hemp Organization	2.12	2.14	2.28	2.35
News Media	2.12	2.05	2.19	2.14
Other internet	2.09	1.97	2.16	2.19
Social Media	1.91	1.85	2.08	2.1

Trust 1= less trusting; 2= neither trusting nor untrusting; 3= more trusting.

Figure 6 Differences in mean trust by awareness categories (n = 1953). Trust 1 = less trusting; 2 = neither trusting nor untrusting; 3 = more trusting.

Table 2 Bivariate statistical results: trust and demographic/sustainability perceptions

Variable	Trusting	Neither trusting nor untrusting	Untrusting	Total	N	Chi-sq
White	0.7649	0.7486	0.7574	0.7600	1671	0.393
Female	0.7650	0.7490	0.7570	0.7600	1671	0.748
College+	0.1430	0.1270	0.1330	0.1380	1671	0.739
Republican	0.2680	0.2510	0.2810	0.2680	1671	0.778
Children <18 in household	0.3540	0.2540	0.2450	0.3120	1953	23.87***
Age (in years)[1]	43.1003	49.3184	51.6946	46.0615	1918	41.91***
Values Local	0.5550	0.1560	0.2210	0.4080	1953	265.82***
Values Sustainable	0.6790	0.2170	0.2770	0.5060	1953	354.83***
Values Environmentally Friendly	0.7140	0.2400	0.2850	0.5330	1953	387.92***

[1]ANOVA was used for the age variable because it is measured on a continuous scale.
P<=.001

of food only, the seller. Across all groups, social media is less trusted, as is the news media and 'other' internet sources.

Few demographics are significantly related to trust in information sources. Perceptions about the sustainability of hemp in terms of the environment are significantly associated with trust, with people who believe hemp has a positive contribution to environmental sustainability being more trustful. Older consumers are less trusting while households residing with children under the age of 18 are more trusting. Demographic targeting is not likely to be useful in the sense of reaching consumers who trust information or who distrust information. The key is to provide truthful information. Marketers may not always adhere to this. 'Puffing' is permissible in advertising and for the most part advertisers self-regulate unless something is brought to their attention by another agency, person, or organization. Some sellers *have* been warned by the FTC concerning false and misleading advertising. Because of the health dangers of children and unsuspecting adults ingesting Delta-8 containing products, both the FDA and FTC are surveilling the situation. During July 2024 the FDA jointly with the FTC issued warning letters to five companies (Rhodes, 2024). The news media is often used to disperse information about this type of information and about any recalls that may occur due to foods having higher levels of psychotropic THC than are allowable by law. As the US promulgates appropriate policy for hemp foods and current loopholes with regard to Delta-8, caveat emptor, or buyer beware, is the recommendation. However, lack of labeling and the selling of dangerous products is only one currently small, but noteworthy component of information dissemination about hemp food and CBD products. Consumers need to vet information sources. If the hemp industry is to thrive in the US and provide consumers with products that meet their preferences, industry, and policymakers continue to have to 'find their way.' This requires both the accurate conveyance of information by industry and the diligence of consumers to determine the trustworthiness of information sources.

With regard to R4, the implications of this study reveals that we have a way to go in order to make concrete market recommendations. That said, sellers must be honest and not deceptive; they must follow all current (and developing regulations), and they should be wary of conflating hemp foods with CBD and CBD-containing foods, especially when the food product contains Delta-8 psychotropic THC produced from legal (<.3% THC hemp). This is increasingly being seen as a slippery slope by regulators and market watchdogs, including the Federal Trade Commission.

The strengths of this study are the use of a national database of US consumers. That said, there is much research to be conducted. We studied hemp-aware consumers. We have not yet included consumers who have purchased or intend to purchase hemp products (that information is in the data set). We have not included information search patterns of consumers (that information is in the data set). And we have not included any demographic or attitudinal questions or a multivariate analysis. All these are aims for future studies.

4 Conclusion

There are very few representative studies about the hemp consumer and in the United States, most of them are based on the data set used in this chapter (Kolodinsky, 2023; Kolodinsky et al., 2020, 2023; Kolodinsky & Lacasse, 2021; Lacasse et al., 2023, 2024). Yet, knowledge of the demand for hemp products is critical if the market is to develop in a way that is profitable across the supply chain, from producer and processor to manufacturer and seller. In addition, the promise of hemp's sustainability and superior performance must be exemplified if consumer demand is to increase.

5 Where to look for further information

The following articles provide information on consumer demand for hemp products in the United States:

Ellison, S. (2021). Hemp (Cannabis sativa L.) research priorities: opinions from United States hemp stakeholders. *GCB Bioenergy*, *13*(4), 562–569. https://doi.org/10.1111/gcbb.12794.

Kim, G. & Mark, T. (2023). What factors make consumers in the USA buy hemp products? Evidence from Nielsen consumer panel data. *Agricultural and Food Economics*, *11*(1), 5. https://doi.org/10.1186/s40100-023-00245-y.

Kolodinsky, J. & Lacasse, H. (2021). Consumer response to hemp: a case study of Vermont residents from 2019 to 2020. *GCB Bioenergy*, *13*(4), 537–545. https://doi.org/10.1111/gcbb.12786.

Kolodinsky, J., et al. (2020). Making Hemp Choices: Evidence from Vermont. *Sustainability*, *12*(15), 6287. https://doi.org/10.3390/su12156287.

Lacasse, H., et al. (2024). Determinants of behavior towards hemp-based products: an application of the theory of planned behavior. *British Food Journal*, *126*(13), 394–414. https://doi.org/10.1108/BFJ-10-2023-0910.

Lacasse, H., et al. (2023). Modeling hemp as an innovative input: an application of the diffusion of innovations in a sample of hemp aware consumers. *Agriculture and Human Values*. https://doi.org/10.1007/s10460-023-10481-z.

6 References

Adesina, I., et al. (2020). A review on the current state of knowledge of growing conditions, agronomic soil health practices and utilities of hemp in the United States. *Agriculture*, *10*(4), 1–15. https://doi.org/10.3390/agriculture10040129.

Ahmed, A. T. M. F., et al. (2022a). Hemp as a potential raw material toward a sustainable world: a review. *Heliyon*, *8*(1), e08753. https://doi.org/10.1016/j.heliyon.2022.e08753.

Ahmed, A. T. M. F., et al. (2022b). Hemp as a potential raw material toward a sustainable world: a review. *Heliyon*, *8*(1), e08753. https://doi.org/10.1016/j.heliyon.2022.e08753.

Babu, R. P., et al. (2013). Current progress on bio-based polymers and their future trends. *Prog Biomater*, *2*, 8.

Beach, B. (2021). The tragedy of george schlichten; or reconsidering cannabis conspiracies. *Points*. https://www.pointshistory.org/post/the-tragedy-of-george-schlichten-or-reconsidering-cannabis-conspiracies.

Benson, T., et al. (2020). The development and validation of a toolkit to measure consumer trust in food. *Food Control*, *110*, 106988. https://doi.org/10.1016/j.foodcont.2019.106988.

Bewley-Taylor, D., et al. (2014). *The rise and decline of cannabis prohibition* (pp. 1–38). Transnational Institute. https://doi.org/10.1086/470120.

Borkowska, B. & Bialkowska, P. (2019). Evaluation of consumer awareness of hemp and its applications in different industries. *Scientific Journal of Gdynia Maritime University*, *110*, 7–16. https://doi.org/10.26408/110.01.

Bouloc, P. & van der Werf, H. M. G. (2013). The role of hemp in sustainable development. In P. Bouloc, et al. (Eds.), *Hemp: industrial production and uses* (pp. 278–289). CABI. https://doi.org/10.1079/9781845937935.0278.

Buskens, V. (2020). Spreading information and developing trust in social networks to accelerate diffusion of innovations. *Trends in Food Science & Technology*, *106*, 485–488. https://doi.org/10.1016/j.tifs.2020.10.040.

Chen, H., et al. (2023). Emerging natural hemp seed proteins and their functions for nutraceutical applications. *Food Science and Human Wellness, 12*(4), 929–941. https://doi.org/10.1016/j.fshw.2022.10.016.

Cherney, J. H. & Small, E. (2016a). Industrial Hemp in North America: production, politics and potential. *Agronomy, 6*(4), 58. https://doi.org/10.3390/agronomy6040058.

Cherney, J. H. & Small, E. (2016b). Industrial hemp in North America: production, politics and potential. *Agronomy, 6*(4), 1–24. https://doi.org/10.3390/agronomy6040058.

Choi, E., et al. (2018). *A review of hemp as a sustainable agricultural commodity: tools and recommendations for Winona LaDuke's Hemp Farm and Sovereign Native American Tribes* (Task Force Report 2018). The Henry M. Jackson School of International Studies at the University of Washington.

Colclasure, B. C., et al. (2021). Hemp, hemp, hooray: the impact of a hemp educational campaign on college students' attitudes and knowledge of industrial hemp. *Journal of Agricultural Education, 62*(1), 246–259. https://doi.org/10.5032/jae.2021.01246

Crini, G., et al. (2020). Traditional and new applications of hemp. *Sustainable Agriculture Reviews, 42*, 37–87.

de Carvalho, B. L., et al. (2015). Consumer sustainability consciousness: a five dimensional construct. *Ecological Indicators, 58*, 402–410. https://doi.org/10.1016/j.ecolind.2015.05.053.

DeCarlo, S. & Weaver, M. (2023). *Keeping the high out of hemp: global THC standards* (pp. 1–2) [Executive Briefings on Trade (EBOT)]. US International Trade Commission.

Drewery, M. & Hustvedt, G. (2024). Consumer support for hemp by-products as food and feed. *Journal of Natural Fibers, 21*(1), 2309893. https://doi.org/10.1080/15440478.2024.2309893.

Ellison, S. (2021). Hemp (Cannabis sativa L.) research priorities: opinions from United States hemp stakeholders. *GCB Bioenergy, 13*(4), 562–569. https://doi.org/10.1111/gcbb.12794.

Ely, K., et al. (2022a). Cannabis/Hemp: sustainable uses, opportunities, and current limitations. In D. C. Agrawal, et al. (Eds.), *Cannabis/Hemp for sustainable agriculture and materials* (pp. 59–87). Springer Singapore. https://link.springer.com/10.1007/978-981-16-8778-5_3.

Ely, K., et al. (2022b). Industrial hemp as a crop for a sustainable agriculture. In D. C. Agrawal, et al. (Eds.), *Cannabis/Hemp for sustainable agriculture and materials* (pp. 1–28). Springer Singapore. https://link.springer.com/10.1007/978-981-16-8778-5_1.

Fair, L. (2020, December 17). *One thing marketers of CBD products need to know right now* [Federal Trade Commission]. https://www.ftc.gov/business-guidance/blog/2020/12/one-thing-marketers-cbd-products-need-know-right-now.

Farla, J., et al. (2012). Sustainability transitions in the making: a closer look at actors, strategies and resources. *Technological Forecasting and Social Change, 79*(6), 991–998. https://doi.org/10.1016/j.techfore.2012.02.001.

Fike, J. (2016). Industrial hemp: renewed opportunities for an ancient crop. *Critical Reviews in Plant Sciences, 35*(5–6), 406–424. https://doi.org/10.1080/07352689.2016.1257842.

Fike, J., et al. (2020). Industrial hemp in the USA: a brief synopsis. In G. Crini & E. Lichtfouse (Eds.), *Sustainable agriculture reviews 42* (Vol. 42, pp. 89–109). Springer International Publishing. http://link.springer.com/10.1007/978-3-030-41384-2_3.

Filimonova, L., et al. (2022). Finding ways to solve problems of waste recycling: biodegradable hemp materials. *IOP Conference Series: Earth and Environmental Science*, *988*(3), 032040. https://doi.org/10.1088/1755-1315/988/3/032040.

Fischer, L.-B. & Newig, J. (2016). Importance of actors and agency in sustainability transitions: a systematic exploration of the literature. *Sustainability*, *8*(5), 5. https://doi.org/10.3390/su8050476.

Fortenbery, T. R. & Bennett, M. (2004). Opportunities for commercial hemp production. *Applied Economic Perspectives and Policy*, *26*(1), 97–117. https://doi.org/10.1111/j.1467-9353.2003.00164.x.

French, L. & Manzanárez, M. (1984). *NAFTA and neocolonialism: comparative criminal, human, and social justice*. University Press of America.

Gallo, T., et al. (2023). Do sustainable consumers have sustainable behaviors? An empirical study to understand the purchase of food products. *Sustainability*, *15*(5), 4462. https://doi.org/10.3390/su15054462.

Gelski, J. (2019). Consumers have trust issues with cannabis ingredients. *Food Business News*. https://www.foodbusinessnews.net/articles/13162-consumers-have-trust-issues-with-cannabis-ingredients.

Goel, R., et al. (2023). 'Hemp and Marijuana are both pots?' Young consumer awareness and perception of hemp food in India. *Journal of Agriculture and Food Research*, *14*, 100809. https://doi.org/10.1016/j.jafr.2023.100809.

Goodman, S., et al. (2022). Use and perceptions of cannabidiol products in Canada and in the United States. *Cannabis and Cannabinoid Research*, *7*(3), 355–364. https://doi.org/10.1089/can.2020.0093.

Grand View Research. (2023). *Industrial hemp market size, share, and trends analysis report*. Grand View Research. https://www.grandviewresearch.com/industry-analysis/industrial-hemp-market.

Grandview Research. (n.d.). *Industrial hemp market size, share & trends analysis report by product (Seeds, Fiber, Shivs), By Application (Animal Care, Textiles, Personal Care), By Region (North America, APAC), And Segment Forecasts, 2022–2030*. https://www.grandviewresearch.com/industry-analysis/industrial-hemp-market.

Gray, M. (2000). The devil weed and Harry Anslinger. Common Sense for Drug Policy, Real Reporting Foundation, Lancaster, PA. https://www.csdp.org/psa/anslinger.

Guo, Y., et al. (2024). Industrial hemp (Cannabis sativa L.) can utilize and remediate soil strongly contaminated with Cu, As, Cd, and Pb by phytoattenuation. *Chemosphere*, *358*, 142199. https://doi.org/10.1016/j.chemosphere.2024.142199.

Hayward, L. & McSweeney, M. B. (2020). Acceptability of bread made with hemp (Cannabis sativa subsp. Sativa) flour evaluated fresh and following a partial bake method. *Journal of Food Science*, *85*(9), 2915–2922. https://doi.org/10.1111/1750-3841.15372.

Huttunen, S., et al. (2021). Pluralising agency to understand behaviour change in sustainability transitions. *Energy Research & Social Science*, *76*, 102067. https://doi.org/10.1016/j.erss.2021.102067.

Johnson, R. (2019). *2018 farm bill primer: hemp cultivation and processing*. U.S. Congressional. Research Service.

Kastel, M. (2020, August 21). Case study: CBD products and the "wild west" of fraudulent labels. *Organic Eye*. https://organiceye.org/lawsuit-filed-the-wild-west-of-organic-cbd-marketing-targeted/.

Kim, G. & Mark, T. (2023). What factors make consumers in the USA buy hemp products? Evidence from Nielsen consumer panel data. *Agricultural and Food Economics, 11*(1), 5. https://doi.org/10.1186/s40100-023-00245-y.

Kolodinsky, J. (2023, July). *Perceptions of hemp based produces*. Agricultural and Applied Economics Annual Meeting, Washington, DC.

Kolodinsky, J. & Lacasse, H. (2021). Consumer response to hemp: a case study of Vermont residents from 2019 to 2020. *GCB Bioenergy, 13*(4), 537–545. https://doi.org/10.1111/gcbb.12786.

Kolodinsky, J., et al. (2020). Making hemp choices: evidence from vermont. *Sustainability, 12*(15), 6287. https://doi.org/10.3390/su12156287.

Kolodinsky, J., et al. (2023, June). *Willingness to pay for hemp textiles: a pilot projec*. Annual Hemp Team Conference, Burlington, Vermont.

Lacasse, H., et al. (2023). Modeling hemp as an innovative input: an application of the diffusion of innovations in a sample of hemp aware consumers. *Agriculture and Human Values, 41*, 239. https://doi.org/10.1007/s10460-023-10481-z.

Lacasse, H., et al. (2024). Determinants of behavior towards hemp-based products: an application of the theory of planned behavior. *British Food Journal, 126*(13), 394–414. https://doi.org/10.1108/BFJ-10-2023-0910.

Law Library of Congress. (2022). *Regulation of hemp* (LL File No. 2022-021696). Law Library of Congress. https://tile.loc.gov/storage-services/service/ll/llglrd/2022666115/2022666115.pdf.

Malyshev, A. & McDonough, T. (2019). The marketing and sale of products containing hemp and CBD over the internet. *Journal of Internet Law, 23*(1), 19–25.

Mark, T. B. & Will, S. (2019). Economic issues and perspectives for industrial hemp. In D. W. Williams (Ed.), *Industrial hemp as a modern commodity crop* (pp. 107–118). ASA, CSSA and SSSA. https://doi.org/10.2134/industrialhemp.c7.

Mark, T., et al. (2020a). *Economic viability of industrial hemp in the United States: a review of state pilot programs* (Economic Information Bulletin EIB-217; pp. 1–83). U.S. Department of Agriculture, Economic Research Service.

Mark, T., et al. (2020b). Economic viability of industrial hemp in the United States: a review of state pilot programs. In *United States department of agriculture* (217; p. 11). USDA. https://www.ers.usda.gov/webdocs/publications/95930/eib-217.pdf?v=6894.

Markets and Markets. (2022). *Industrial hemp market by type, application, and region*. MarketsandMarkets. https://www.marketsandmarkets.com/Market-Reports/industrial-hemp-market-84188417.html.

McFadden, B. R. & Malone, T. (2021). Homegrown perceptions about the medical use and potential abuse of CBD and THC. *Addictive Behaviors, 115*, 106799. https://doi.org/10.1016/j.addbeh.2020.106799.

McGrath, C. (2020). *2019 hemp annual report* (Voluntary Report CH2020-0018; pp. 1–12). USDA Foreign Agricultural Service.

Meijer, W. J., et al. (1995). Constraints to dry matter production in fibre hemp (Cannabis sativa L.). *European Journal of Agronomy, 4*(1), 109–117. https://doi.org/10.1016/S1161-0301(14)80022-1.

Metcalf, D. A., et al. (2021). Comparing early hemp food consumers to non-hemp food consumers to determine attributes of early adopters of a novel food using the Food Choice Questionnaire (FCQ) and the Food Neophobia Scale (FNS). *Future Foods, 3*, 100031. https://doi.org/10.1016/j.fufo.2021.100031.

Montford, S. & Small, E. (1999). Measuring harm and benefit: the biodiversity friendliness of Cannabis sativa. *Global Biodiversity*, *8*(4), 2–13.

Musto, D. F. (1972). The Marihuana Tax Act of 1937. *Archives of General Psychiatry*, *26*(2), 101–108. https://doi.org/10.1001/archpsyc.1972.01750200005002.

Naeem, M. Y., et al. (2023). Hemp: an alternative source for various industries and an emerging tool for functional food and pharmaceutical sectors. *Processes*, *11*(3), 718. https://doi.org/10.3390/pr11030718.

New Frontier Data. (2020, May). *Frequency of use among U.S. CBD consumers*. New Frontier Data. https://newfrontierdata.com/cannabis-insights/frequency-of-use-among-u-s-cbd-consumers/.

Nguyen, C., et al. (2023). Consumer perception, knowledge, and uses of cannabidiol. *Mental Health Clinician*, *13*(5), 217–224. https://doi.org/10.9740/mhc.2023.10.217.

Nieves, A. (2021, October 1). The wild west of CBD products could end soon in California. *Politico*.

Notaro, S., et al. (2022). Consumers' preferences for bioplastic products: a discrete choice experiment with a focus on purchase drivers. *Journal of Cleaner Production*, *330*, 129870. https://doi.org/10.1016/j.jclepro.2021.129870.

Papadopoulou, E., et al. (2015). Value-added industrial products from bast fiber crops. *Industrial Crops and Products*, *68*, 116–125. https://doi.org/10.1016/j.indcrop.2014.10.028.

Pihlanto, A., et al. (2017). Bioactivities of alternative protein sources and their potential health benefits. *Food and Function*, *8*(10), 3443–3458.

Price Waterhouse Coopers. (2015). *Food trust: giving consumers confidence in your food*. PWC. https://www.pwc.nl/nl/assets/documents/pwc-food-supply-and-integrity-services.pdf.

Putnam, A. (2022). *Market trends and regulations for industrial hemp in the European Union* (Voluntary Report E42022-0052; pp. 1–9). USDA Foreign Agricultural Service.

Quinton, S. (2021). The hemp boom is over. What now? In *Pew: stateline*. https://www.pewtrusts.org/en/research-and-analysis/blogs/stateline/2021/07/09/the-hemp-boom-is-over-what-now.

Rampold, S., et al. (2021). Hemp there it is: examining consumers' attitudes toward the revitalization of hemp as an agricultural commodity. *Journal of Applied Communications*, *105*(4), Article 2. https://doi.org/10.4148/1051-0834.2385.

Rhodes, C. (2024, July 16). *FDA, FTC continue joint effort to protect consumers against companies illegally selling copycat Delta-8 THC food products*. US Food and Drug Administration. https://www.fda.gov/news-events/press-announcements/fda-ftc-continue-joint-effort-protect-consumers-against-companies-illegally-selling-copycat-delta-8.

Robinson, R. (1996). *The great book of hemp: the complete guide to the environmental, commercial, and medicinal uses of the world's most extraordinary plant*. Park Street Press.

Ruth, T. K., et al. (2022). Hemp on the horizon: understanding the influences on industrial hemp purchases. *Advancements in Agricultural Development*, *3*(3), 1–13. https://doi.org/10.37433/aad.v3i3.189.

Schilling, J. M., et al. (2021). Cannabidiol as a treatment for chronic pain: a survey of patients' perspectives and attitudes. *Journal of Pain Research*, *14*, 1241–1250. https://doi.org/10.2147/JPR.S278718.

Schlichten, G. W. (1919). *Means for treating fiber bearing plants* (US1308376A). https://patents.google.com/patent/US1308376.

Schluttenhofer, C. & Yuan, L. (2017). Challenges towards revitalizing hemp: a multifaceted crop. *Trends in Plant Science*, *22*(11), 917–929. https://doi.org/10.1016/j.tplants.2017.08.004.

Siegrist, M. (2021). Trust and risk perception: a critical review of the literature. *Risk Analysis*, *41*(3), 480–490. https://doi.org/10.1111/risa.13325.

Sinha, P., et al. (2021). Practitioner's guide to latent class analysis: methodological considerations and common pitfalls. *Critical Care Medicine*, *49*(1), e63–e79. https://doi.org/10.1097/CCM.0000000000004710.

Staples, A. J., et al. (2022). CBD and THC: who buys it, and why? *Journal of Food Distribution Research*, *53*(3), 67–85.

Sterns, J. (2019). Is the emerging U.S. hemp industry yet another boom–bust market for U.S. farmers? *Choices*, *34*(3), 1–8.

The Many Uses of Hemp. (2020). VitalityCBD. https://vitalitycbd.com/hemps-many-uses/

USDA. (2019). *Farm bill*. https://www.usda.gov/farmbill.

USDA. (2024). *Hemp research needs roadmap*. U.S. Department of Agriculture, Office of the Chief Scientist.

Velasco, F., et al. (2022). Key factors that influence consumer preferences for natural-hyped products. *Psychology & Marketing*, *40*(2), 373–390. https://doi.org/10.1002/mar.21730.

Weller, B. E., et al. (2020). Latent class analysis: a guide to best practice. *Journal of Black Psychology*, *46*(4), 287–311. https://doi.org/10.1177/0095798420930932.

Wheeler, M., et al. (2020). CBD (Cannabidiol) product attitudes, knowledge, and use among young adults. *Substance Use & Misuse*, *55*(7), 1138–1145. https://doi.org/10.1080/10826084.2020.1729201.

World Health Organization. (2018). *Cannabidiol (CBD): critical review report*. Expert Committee on Drug Dependence.

Wu, W., et al. (2021). Consumer trust in food and the food system: a critical review. *Foods*, *10*(10), 2490. https://doi.org/10.3390/foods10102490.

Wysota, C. N., et al. (2022). Cannabidiol knowledge, perceptions, and use among young adults in 6 U.S. metropolitan areas. *Cannabis and Cannabinoid Research*, *8*(6), 1140–1149. https://doi.org/10.1089/can.2022.0029.

Yano, H. & Fu, W. (2023). Hemp: a sustainable plant with high industrial value in food processing. *Foods*, *12*(3), 651. https://doi.org/10.3390/foods12030651.

Zając, M., et al. (2019). The quality of pork loaves with the addition of hemp seeds, de-hulled hemp seeds, hemp protein and hemp flour. *LWT*, *105*, 190–199. https://doi.org/10.1016/j.lwt.2019.02.013.

Chapter 14

Building sustainable biobased manufacturing supply chains to help hemp reach its global economic potential

Zhaohui Wu and Jeffrey J. Steiner, Global Hemp Innovation Center, Oregon State University, USA

1 Introduction: a brief history of hemp supply chains

We start with defining a few common terms used throughout this chapter:

- Technology roadmap, a document that helps guide a long-term strategy for adopting technology to increase process efficiencies and help avoid costly mistakes.
- Commercialization, the process of bringing new products or services to market that entails activities including production, distribution, marketing, sales, customer support, and other key functions critical to achieving the commercial success of the new product or service.
- Biobased manufacturing, the processes used to convert renewable chemicals, materials, and other biobased inputs through biorefineries into final products on a commercial scale.

http://dx.doi.org/10.19103/AS.2025.0161.28

- Supply chain, the network of people, organizations, resources, activities, information and technologies involved in the creation and sale of a product or service from the initial design of a product to its final delivery to a customer.

Hemp, or *Cannabis sativa* L., is believed to have originated in Central Asia in the regions that are now modern-day China and Afghanistan. Archaeological evidence suggests that hemp fiber was being used as early as 8000 BCE. The Chinese were among the first to cultivate hemp for making textiles, rope, paper, and its grain as food. By 2000 BCE, hemp had spread to the Middle East where it was used by ancient Egyptians and the Hebrew civilizations. It made its way to Greece and Rome by 800 BCE (Bloom and Blair, 2009; Clarke and Merlin, 2013; Bourguignon, 1993).

During the Middle Ages, hemp cultivation spread from Asia to across Europe where it became an essential crop for its versatile use in producing many products. Local supply chains emerged where hemp was grown, processed, and used within communities. Hemp fibers became crucial for making ropes, paper, and clothing. In parallel to other technological innovations, hemp paper was used in the printing of many early books including the Gutenberg Bible. Hemp was introduced to North America by early European settlers in the 1600s and quickly became one of the colonial crops. In the seventeenth century, a Virginia law passed in 1690 required farmers to grow hemp, and Massachusetts and Connecticut laws similarly encouraged farmers to grow hemp, sometimes under penalty for non-compliance. Since the colonies had to be relatively self-sufficient in producing materials for shipbuilding and other maritime industries, hemp was highly valued and considered a strategic crop for the colonies because of its use in supporting commerce. As evidenced by diaries and farm records, both George Washington and Thomas Jefferson grew hemp on their plantations, recognizing its value in regional commerce and as being beneficial in diversifying the kinds of crops grown in rotations on their farms.

In the nineteenth century, hemp continued as an important industrial crop but required great laborforce to produce, harvest, and process into fiber. In California, securing labor to produce hemp fiber products led to conflicts between early California hemp farmers and Indigenous labor at the San Jose Mission dating back to the early nineteenth century, during the Spanish colonial period and into the Mexican era (Phillips, 2017). In the Eastern U.S., states such Kentucky, slave labor produced hemp fiber for the manufacturing of wrapping for baled ginned cotton produced by slaves in the southeastern cotton-producing states (KHS, 2024). Hemp production peaked between the 1820s and 1860s, as Kentucky's hemp industry flourished to meet the demand for cotton bale wrapping, driven by the growth of cotton plantations in the Deep South. The decline of Kentucky's hemp industry began around the time of the Civil War (1861–1865), as shifts in the economy, the end of slavery, and the rise of alternative materials like jute and later synthetics changed the industry.

Domestic demand for hemp products did not necessarily assure purchase of domestically produced hemp. The U.S. Navy had sourced its hemp rope during the eighteenth and nineteenth centuries based on performance and reliability, leading them to purchase Russian hemp rope despite the existence of domestic production. Russian rope, made from hemp grown in the Baltic region and Ukraine, was stronger, more durable, and long-lasting than hemp produced in other parts of the world, giving Eastern European producers an edge in the global market. This period demonstrated the establishment of more complex and global supply chains, integrating local producers in one part of the world with markets for manufactured products elsewhere. Eventually, with the advent of steamships the demand for hemp rope and sails declined, but hemp was still utilized in textile and paper product manufacturing.

The continued decline in hemp usage by various industries was influenced by many factors including innovations and advancements in manufacturing technologies, regulatory policies, and negative societal attitudes toward cannabis. While competing economic interests and lobbying by industrialists may have played a role in declined use of hemp, the story is likely more complex and not due to collusion among industry titans to supplant hemp in the marketplace with other natural materials such as wood fiber. In the earlier half of the twentieth century, hemp production declined due to the ease of manufacturing textiles from cotton, followed by emergent competition from synthetic fibers, as well as its association with intoxicating marijuana (Herer, 1985).

Domestic policies can also come into play and drive market demand, regardless of product quality and other market forces. The Marihuana Tax Act of 1937 effectively priced the cultivated hemp out of U.S. markets (Table 1) (U.S. Congress, 1937). During World War II, there was a brief resurgence of hemp production as a domestically produced strategic material. The U.S. government launched the *Hemp for Victory* campaign at that time to encourage farmers to grow hemp for the war effort, primarily for making ropes and other military products (USDA, 1942).

Even with the ban of hemp production in the U.S. caused by the Controlled Substances Act of 1970 and in other western countries by international agreements under the United Nations framework that placed restrictions on hemp production and use because of the association with intoxicating varieties of cannabis, there was continued hemp production and commerce in Russia, Eastern Europe, and China. In the late twentieth and early twenty-first centuries, many countries began to ease former restrictions on hemp cultivation for the manufacture of industrial products. In the U.S., renewed interest in hemp arose because of its historic versatility of uses and believed environmental benefits

Table 1 History summary of the banning of hemp production and use in the United States (1937–2014).

Reasons	Explanation
Industrial Interests	In the early twentieth century, paper pulp interests led by figures like William Randolph Hearst and the emerging synthetic fiber industry, particularly the DuPont company, had competing other sources of fiber than hemp. DuPont had recently patented nylon, a synthetic fiber that competed directly with not only hemp and other natural fibers such as cotton.
Association with Marijuana	Anti-drug campaigns in the early twentieth century focused on marijuana, which is from the same *Cannabis sativa* plant species as hemp. The term 'marijuana' was popularized and associated with African, Mexican, and Caribbean immigrants, linking these minority groups with criminal behavior and moral decline. These actions by the U.S. government helped fuel public fear and opposition to cannabis of all forms (USDA, 1911; Velasquez-Manoff, 2019).
Adverse Legislative Actions	Marihuana Tax Act of 1937, spearheaded by Harry Anslinger, Head of the Federal Bureau of Narcotics, effectively outlawed the cultivation of hemp by imposing heavy taxes and strict regulations on the production and sale of all Cannabis products. This came at a time when repeal of prohibition allowed Congressionally appropriated federal resources to be directed at Cannabis regulatory enforcement.The Controlled Substances Act of 1970 further solidified the ban on hemp by classifying all forms of the plant, including non-intoxicating hemp, as a Schedule I controlled substance, the same as for heroin and lysergic acid diethylamide (LSD).

that would result from its cultivation (Small and Marcus, 2002). A pivotal piece of legislation in the U.S. was the Agricultural Improvement Act of 2014 (2014 Farm Bill) that included a provision allowing the cultivation of hemp for research purposes under state-regulated pilot programs. The Agricultural Improvement Act of 2018 (2018 Farm Bill) went further and removed hemp from the list of controlled substances and legalized its production nationwide and brought its regulation under the authority of the U.S. Department of Agriculture (USDA) rather than the U.S. Drug Enforcement Agency. These policies helped foster a surge in interest with farmers, businesses, and investors exploring new opportunities across many sectors, particularly for non-intoxicating cannabinoids such as cannabidiol (CBD), as well as products manufactured from the plant stalks and seeds including textiles, foods and beverages, building materials, and cosmetics and other personal care products.

The passage of the 2018 Farm Bill triggered great enthusiasm in the U.S., especially for hemp flower production to produce CBD and other non-intoxicating cannabinoids as well as essential oils such as terpenes used as flavor and aroma ingredients. Not only did established farmers who grew widely produced commodities such as corn, soybean, and cotton begin producing hemp, but because of the relative ease of growing hemp, particularly at small scales of production, many beginning farmers did as well. With early success and high prices for biomass and extracted cannabinoids, production expanded rapidly. Farmers across the U.S. and other countries planted large areas of hemp, anticipating high returns. However, with inadequate available labor for harvest and limited handling and processing infrastructure in place to support expanded production, incomplete supply chains consolidated as a result of the lack of marketing planning and the absence of sales contracts, over-production led to a significant drop in hemp biomass prices and ultimately the sudden crash of the CBD market in 2020. The CBD isolate market has just now begun to recover in 2024. However, the number of acres of floral hemp planted and the value of extracts in 2024 is a fraction of levels reached in 2019.

With the 2018 Farm Bill legalizing U.S. hemp production and product sales, the regulatory environment for CBD and other non-intoxicating cannabinoid uses continues to remain uncertain. The U.S. Food and Drug Administration (FDA) has not established the rules regarding CBD use as an ingredient in food, beverages, and dietary supplements, particularly due to a lack of research data demonstrating efficacy and consumer safety of consumption. Regulatory uncertainty continues to create confusion for both producers and manufacturers and has limited the ability of companies to market and sell CBD products, much less expand markets and the needed supporting supply chains. As recently as November 2025, the U.S. Congress enacted a resolution for continuing U.S. Government funding package that included a rider substantially narrowing the federal definition of hemp for regulatory purposes. The amendment closed a so-called *2018 Farm Bill hemp loophole* by shifting from a delta-9 THC-only criterion to a total THC standard that includes other intoxicating cannabinoids, excludes synthetic or manufactured cannabinoids from compliantly produced hemp, and imposed a strict low-milligram cap on allowable THC per finished product container.

The crash of the CBD market in 2019 compelled many growers and processors to exit the market and for many to seek other commercial opportunities and business models, with a shift to production and manufacture of products made from other parts of the hemp plant than flowers. As a result, imaginations have turned to industrial uses of hemp such as grain for human food and beverages and feed for livestock, as well as manufacturing biobased products made from stem materials for industrial use in construction, biobased polymer composites, paper and packaging, biochar carbon as an agricultural soil amendment, and activated carbon for water and air filters and electrical storage devices. However, with the additional uncertainty injected into the market by the November 2015 enacted continuing appropriations and funding package, great reductions in hemp production are predicted for 2026 until regulatory certainty is established.

1.1 Building sustainable biobased manufacturing supply chains

Establishment of a significant business sector utilizing hemp for industrial products manufacturing requires a concerted effort of industry pioneers, collaboration with established biobased product manufacturing sectors, researchers, and policymakers to articulate a new value proposition beyond that of the chemical hemp market class sector that emerged following the 2018 Farm Bill action. For expanded production and use of industrial hemp, a strategy needs to further codify the disassociation of hemp from marijuana and the still persistent regulatory climate that is applied to industrial hemp, which hinders its ability to scale-up production. Hemp will also need to be viewed by the public, policymakers, and investors as an agricultural commodity that is grown, processed, and commercialized like other established agricultural commodities such as corn, soybean, or cotton, and not as a variety of marijuana disguised as hemp.

1.2 World landscape for natural fibers

To understand the opportunities and challenges for establishing a significant industrial hemp manufacturing sector, it is necessary to look at the present world business landscape for natural fibers. The leading hemp fiber producing countries in 2024 are France and China, followed distantly by the United States, North Korea, Poland, and the Netherlands. In addition to fiber for textiles, there are other promising and emergent product categories to be considered for building-out and scaling-up commercial production. This potential is built upon years of experience and persistent efforts by European countries that faced many of the same challenges as the U.S. to advance the hemp economy. All countries now can consider which are the most promising product sectors to expand including use for manufacturing hemplime as an insulation and non-structural construction material with a lowered carbon footprint than cement; hemp fibers as a substitute for fossil-based plastic materials in automotive door panels and consumer single-use products such as cutlery, cups, and packaging; and as a biomass source for producing biochar and activated carbon, in addition to natural textile fibers that can be used in the manufacture of premium apparel products.

With the many and diverse uses for hemp, a systems approach must be used to validate the performance and feasibility of using hemp across the wide range of already established manufactured products, not only made from natural materials but also the majority of which are synthetics made from extracted and mined hydrocarbons. To be viable, like other industrial materials, multiple factors need to be considered at the same time. A commercially viable product requires either a higher price premium or a lower production-processing cost than those for products in already established markets. With so many possible product applications for hemp, different produced materials must be characterized for their physical and chemical properties to be matched to meet manufacturers' specifications for making their products as well as find those optimized supply chain configurations to compete with products already in consumer markets.

To take advantage of the possible performance benefits of hemp-based materials and products, it is necessary to identify existing and needed resources and infrastructure for creating efficient supply chains; characterize and direct production and produced materials to their highest valued markets; document the performance of materials and products that meet manufacturers' specifications and create standards describing these; and identify financial and policy instruments that can be used to reduce financial risks and support investments for expanded biobased manufacturing. Because of evolving value propositions for consumers and the manufacturers of their purchased products, we assume good stewardship must be designed into any emergent hemp-based manufacturing ecosystems by estimating ahead of time the potential impacts different supply chain entities have on natural resources and environmental quality, job creation, and worker safety, as well as consider approaches to broadly enhance equity for all supply chain participants. Under this approach, decision-making must be grounded in integrated scientific and business data to rigorously evaluate the technical feasibility, economic viability, environmental performance, and social impacts of the manufacturing sector's potential contribution to establishing a sustainable hemp-based manufacturing sector.

Given the relative newness of hemp as a commodity and to take advantage of its unique physical and chemical properties that may differentiate it from other natural materials, there is a great opportunity to create new product markets and develop strategies to build-out and scale-up efficient supply chains in ways that address shortcomings in already established globalized supply chains that have evolved over the last 30 or more years. Domestic U.S. manufacturing capacities that were offshored in the last decade have had their vulnerabilities exposed during recent crises including the COVID-19 pandemic, trade war with China, and the war in Ukraine. These factors provide opportunities to create new domestically centered and efficient Manufacturing 4.0 (NIST, 2024) infrastructure and doing so in equitable ways so that all supply chain participants involved can benefit for prosperity more widely shared. In this sense, we are at a juncture where many stakeholders, domestic and international, can collaborate to create new product categories that utilize this new-old crop and do so in ways designed to compete with those manufactured

Table 2 History summary of hemp and the United States 2014 and 2018 Farm Bill legislation.

Key Provisions of 2014 Farm Bill

- Pilot Research Programs. States were authorized to oversee hemp research, primarily to study the growth, cultivation, and marketing of industrial hemp.
- Hemp Definition. The bill defined industrial hemp as cannabis with a delta-9 tetrahydrocannabinol (THC) concentration not exceeding 0.3% on a dry weight basis.
- Limited Cultivation. Help cultivation was limited to research institutions such as universities and state departments of agriculture, focusing on learning more about hemp's potential uses and economic viability.

Key Provisions of 2018 Farm Bill

- Decriminalization of Hemp. The bill legalized the cultivation and sale of hemp and hemp-derived products that contained no more than 0.3% THC.
- Regulation Framework. A regulatory framework for production was established that required states and Native American tribes to submit hemp production plans to the U.S. Department of Agriculture (USDA) for approval. Plans had to include procedures for testing THC levels, disposal of non-compliant plants, and licensing requirements for growers.
- Interstate Commerce. The bill provided clarity and regulatory security for businesses involved in the interstate commerce of hemp and hemp products.
- Federal Crop Insurance. Hemp farmers became eligible for federal crop insurance, providing a similar safety net similarly available for other U.S. crop commodities.
- Research and Development. The bill promoted research funding opportunities for hemp, encouraging innovation and industry sector development through federal and state grant programs, just as with any other commodity crops and agricultural products.

from non-renewable materials and to utilize more efficient infrastructure to complement other natural products in existing markets (Table 2).

2 Resilient supply chain development

In this section, we present a systems framework that integrates multiple supply chain components to be addressed in an integrated fashion, from field-to-harvest to primary processing-to-manufactured biobased materials and products in a way to identify new opportunities to increase overall system efficiencies toward improved triple-bottom-line financial, environmental, and social performances (Kaplan and McMillan, 2020). To enhance the competitiveness of hemp with established natural and synthetic biobased product markets (Statista, 2020; Kadolph, 2010; Van derWerf and Turunen, 2008), high-throughput and integrated production, harvest, processing, and manufacturing systems are needed, building on advances in earlier manufacturing sectors. It is a premise of this approach that for a domestic industrial hemp industry to achieve scale and be on par with or exceed that of established international competitors, high-throughput biobased manufacturing systems must be developed that produce multiple coproduct streams with no resulting produced wastes; target defined value-added end-uses; and be utilized by highest valued markets, as is done by other commodities such as cotton, corn, and soybean.

Multiple supply chain components need to be considered in total when scaling biobased manufacturing that utilizes hemp. We believe that such supply chains must be built with the manufactured end-products in mind and then developed backwards, linking manufacturing, materials processing, and production of raw materials in sustainable agricultural production systems that use custom-designed genetic materials for specific value-added end-uses and not as a non-descript, undifferentiated bulk commodity.

New discoveries and technical interventions must be applied to increase system mechanical process efficiencies across the multiple supply chain components. Such connection will need to be through smart logistics networks and characterizing produced materials, byproducts, and manufactured products for their physical and chemical properties that are defined by the manufacturer's product specifications. Strategies must be developed that utilize feedback loops to improve end-product quality and bioprocessing system efficiencies, including genomic and process information through the Internet of Things (IoT) and emerging AI capability for the creation of globally competitive manufacturing capabilities.

2.1 Supply chain components

Here we present prevalent materials and products manufactured from hemp that comprise the three general market classes of grain, fiber, and chemical. Figure 1 shows the basic processes from the field to the finished products from agricultural production through primary and secondary processing to products with their co-products and byproducts. Even though hemp is one of the oldest cultivated plants, it must now compete with products in the twenty-first century. Enthusiasts still often cite the 1938 popular magazine article that highlighted the 25 000 uses for hemp (Popular Mechanics, 1938), but much has changed for the kinds of products that are in demand today compared to the pre-World War II period. Just as the navies around the world progressed from sails for wind to boilers generating steam from burned coal to nuclear fission propulsion (Cutler, 2020), so it is that leaps in technology that are needed to drive new kinds of product choices . It was during the 80-year period from the Marijuana Tax Act to the U.S. Farm Bills that innovation in technology, marketing, and banking that made world agricultural output for other commodities such as corn, soybeans, and cotton, increase at an average annual rate of 2.3%, faster than the 1.2% average growth rate for product input use. This means world agricultural total factor productivity grew 1.1% annually, but hemp missed out on the benefits of these advancements during this period. Application of these advancements now needs to be applied to hemp to accelerate its progress in

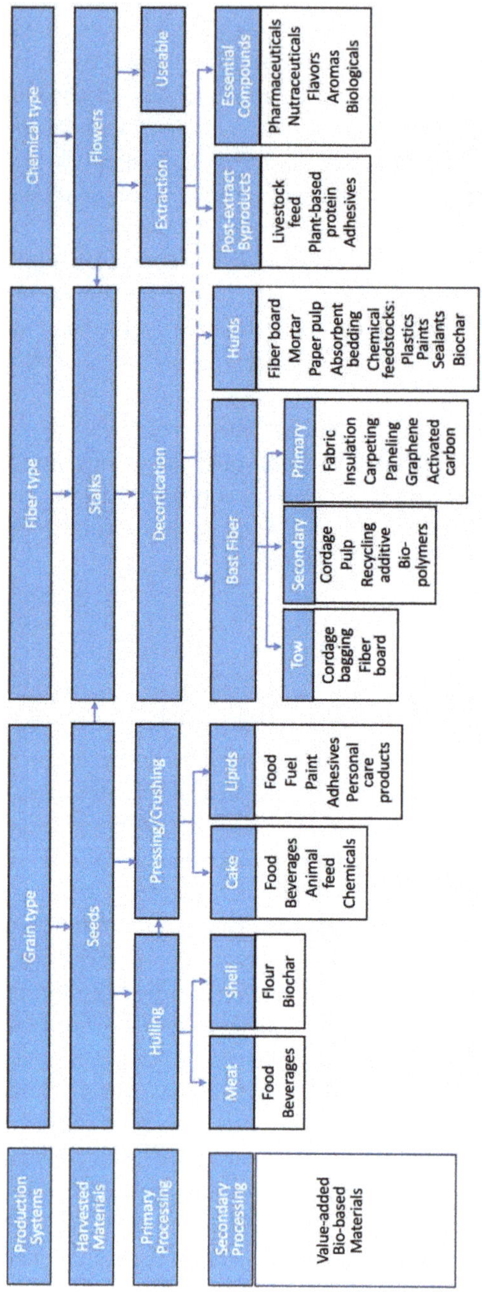

Figure 1 The general flows of harvested materials for the three hemp market classes from agricultural production systems, through materials processing, and to the manufacture of biobased products. Adapted from a Congressional Research Service report (Johnson, 2014).

reaching twenty-first century markets. Doing so provides great opportunities for new innovations in creating new materials and bringing new products to market. Concurrently, this will help create new businesses, jobs, and economic growth.

We focus on four primary products that hold strategic materials importance and can be market leaders providing opportunities for hemp industries to be built out and scaled up. These include 1) building materials, primarily as a concrete aggregate substitute and as a component in engineered wood products; 2) biopolymers and composite materials, primarily for use in consumer single-use bioplastics and automotive body parts; 3) wood fiber substitutes for paper and packaging; and 4) biomass for conversion into carbon for use in biochar, and electronics and electrical storage devices (Table 3). To achieve price competitiveness in existing markets, these products require a great volume of biomass materials to produce and so even though commanding a lower unit price at the materials supply site, they can generate great value-added sales based on volume. In addition, a price premium may be offered in the consumer markets based on real sustainability or other positive performance attribute claims. Other significant uses of hemp are also presented for making comparisons to these four materials of focus here. These include textiles and fabrics; grain for food, beverages, and livestock feed; and advanced biofuels and munitions.

2.2 Materials production

The production systems for the three general market classes of grain, fiber, and chemical hemp share common general practices including field preparation, sourcing of custom genetics directed at end-product uses, methods to establish plant stands depending on market class end-product, production culture during the growing season that includes practices required to control insect pests, application of adequate nutrients and water for crop growth, and to control weeds (Figure 1). At maturity, fields are harvested by methods depending on end-use and market class type, and then the harvested materials are handled in ways dependent on the primary processing methods used. The sequence and specifics of these activities can vary because farmers have different capital investments in equipment, value propositions, soil, field, and weather conditions, and end-product needs from their customers. Some hemp biomass producers can carry out their post-harvest operations on the farm or at close-to-farm locations. Others may require their produced materials to be shipped to primary or secondary processing facilities distant from their production fields.

The early stage of fiber hemp processing involves swathing crop plants into rows where the stems are laid on the field to dry. During the period of

Table 3 The wide range of commercial hemp-based materials and products manufactured from hemp and their uses. The potential of hemp-based products must be determined by comparing the feasibility of manufacture with those products in already established markets

Building materials:

- Hemplime. The inner woody core of the hemp plant named hurd is mixed with a lime or other binder materials and used as a non-structural building material known for its fiber strength, thermal insultation properties, and breathability.
- Hempwool. A thermal insulation product manufactured from fibers into batt-like panels used in housing construction.
- Fiberboard. A type of engineered product made of hemp fibers and bound by adhesives.
- Hempwood. A manufactured hardwood substitute made from hemp stalks into a manufactured flooring or paneling bound by biobased adhesives.

Wood fiber substitutes:

- Printing and specialty papers.
- Packaging.

Biopolymers and biobased composites:

- Biodegradable plastics incorporating hemp microfibers that are used in packaging, containers, or other disposable consumer products.
- Hemp fibers combined in composite materials to create strong, lightweight materials for use in interior automotive parts, aerospace components, and construction building and road paving products.

Soil remediation, biochar for soil amendments, and carbon-based products:

- Source of biomass for conversion to carbon through pyrolysis with utility in the manufacture of biochar and activated carbon used in air and water filtration; electronics, and energy storage devices; and munitions.
- Crop rotation benefits when used in a multiple-year production system with other crops to improve soil health and reduce the need for chemical fertilizers, herbicides, and pesticides.
- Phytoremediation. Plants are grown to remove environmental contaminants from the soil by its roots.

Grain for food, beverages, and animal feed:

- Plant-based dietary supplements made from ground hemp seeds with essential fatty acids, vitamins, and minerals.
- Seed vegetable oil used in cooking and as a dietary supplement with high content of omega-3 and omega-6 fatty acids.
- Animal feed from hemp seeds and their byproducts. Has neared approval as a feed for egg-laying poultry.

Textiles and fabrics:

- Clothing. Hemp fibers from the outer stem parts are breathable, absorbent, and demonstrate putative anti-microbial and ultraviolet light exposure resistance properties.
- Industrial textiles. Non-woven fabrics and upholstery fabrics.

(*Continued*)

Table 3 (*Continued*)

Industrial absorbent materials and animal bedding: • Post-extracted biomass as a supplemental feed additive mixed with other grains and herbage. • Hurds and post-extracted biomass used as absorbent and biodegradable bedding material for confined animals.
Advanced biofuels: • Hemp seed vegetable oil can be processed into as alternative biobased fuels that are fungible and can replace extracted fossil crude oil.

stalks laying on the field, the natural process of retting occurs involving natural microbes that break down ligneous compounds that hold bast fibers together in bundles. After retting, the rows of dried and retted stalks are typically baled and transported to a central storage area readied for further processing. The stalks are then mechanically processed at a decortication facility through machinery that breaks the bast fibers from the cellulosic hurd portion of the stalks.

For grain production, the hemp plants flower and seeds in the inflorescences are allowed to mature and dry before being combined by harvest machinery like that used for other small grains such as wheat. Due to relatively wide ranges of maturity of hemp grains in the inflorescences, harvested grains often need to be dried to relative moisture contents low enough to prevent mold growth on the seeds and to prevent further degradation of seed quality. A byproduct of hemp grain production is the stalks that remain in the field post-harvest of grain harvest. After grain harvest, the stalks can be handled similarly as stalks grown for solely for fiber and processed for bast fiber and hurd. Adequate amounts of stalk residues must remain in the field to reduce soil erosion from precipitation and to help maintain sequestered carbon levels in the soil.

Chemical-type hemp plants are grown very differently than fiber and grain hemp crops. Chemical-type plants are grown from feminized seeds that were produced from only female sex plants. To produce feminized seeds, female plants are grown and then meristem buds are treated with a chemical agent such as silver thiosulfate that, when applied to buds, induces the flowers to produce pollen. The produced pollen is genetically the same as the mother plants, and in theory, only female plants will be produced from these seeds. Feminized plant flowers produce great percentages of cannabinoids and other plant secondary aromatic and flavor compounds such as terpenes and flavonoids.

Flowers of chemical-type hemp plants are first processed after harvest by drying plants or flowers in a drying facility, and then processing by hand to prepare the usable flowers for sale. The scale of production of useable flower hemp crops may be from areas containing a few to thousands of plants that were established from individual plants grown from seeds in greenhouses and then transplanted into the field. There are also large-scale, highly mechanized flower hemp production systems of hundreds of acres or more where feminized seeds are directly planted using equipment similar to that used for planting other field crops such as vegetables and similarly produced as field corn. The plants are grown under irrigation and mechanically harvested using equipment like that used for corn silage harvest, but the biomass is then transported to forced-air drying facilities.

After drying, the friable flower and leaf materials are mechanically separated from the chopped stalk materials that consist of bast fiber and hurd. The flower material is then made into pellets for storage or is ready for further processing to extract cannabinoids. A byproduct of hemp flower production is plant stalks; the bast fiber and hurd biomass fractions. After the initial processing, the resulting bast and hurd products will be transported to facilities for further processing to continue to carry out value-added activities.

2.3 Materials utilization and processing systems

Continuous system improvement can happen by utilizing an information feedback network support system for decision-making to increase system efficiencies for the three market sectors and along each of their multiple supply chain components from custom-designed genetic materials to plants grown in fields to the manufactured products that reach consumer markets (Figure 2). Based on physical and chemical property specifications of materials used to make their products, manufacturers use information feedback loops to identify

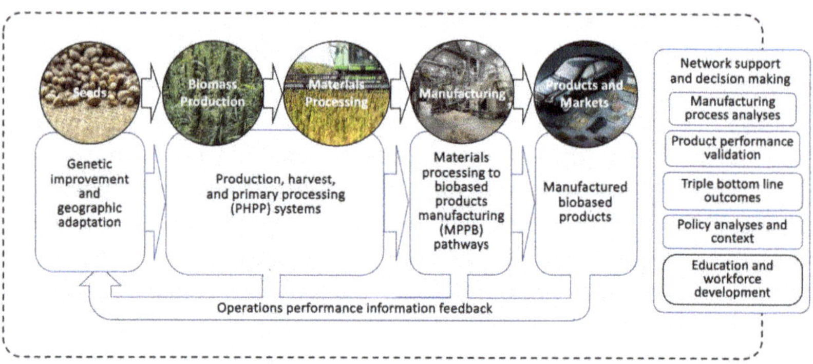

Figure 2 The flow of hemp materials and information through a supply chain from seeds to manufactured biobased products reaching consumer markets. PHPP systems provide specified materials for specific MPPB pathways that lead to the manufacture of biobased products. Shown is the feedback loop from each supply chain component used to evaluate and inform whole-systems operations performance and identify points for improvement. A network support and decision-making framework provides the criteria for evaluating triple-bottom-line performance.

ways to improve system triple-bottom-line performance. Doing so provides each supply chain component player, financial decision-makers, policy makers, federal agency service providers, regulators, and technology providers insights based on the science- and business-based information needed to evaluate the technical, economic, environmental, and social outcomes that result from incorporating a hemp-based bioeconomy across a region.

With this approach, hemp varieties must also be adapted to environmental growing conditions that meet the materials delivery needs of localized processing facilities to optimize processing operations and meet needed production output. Such customized genetics are still in the early stages of development and seed availability of desired varieties may be limited and are often at a great cost compared to the costs of seeds of other common agricultural commodities. It will be important to have ongoing improvements while meeting hopefully the growing demand for needed hemp materials. Such a continuous loop of genetic improvement and evaluation of whole-systems operations could support the development of an expanding hemp-based bioeconomy, enabling stakeholders across the supply chain to make informed decisions based on scientific and business-based evidence.

The genetics of hemp influence important traits like how it grows in different environments, the quality of its fibers and other materials, and the expression of its cannabinoid content. By adjusting hemp's genetic makeup, producers aim to improve factors like fiber or seed yield, oil composition, and the plant's ability to thrive in specific production conditions. Breeding hemp involves selecting plants with desirable traits and enhancing these traits over generations. Advanced breeding methods include techniques like genomic sequencing, marker-assisted selection, genetic modification, and CRISPR gene editing. However, genetic editing faces regulatory challenges and concerns from stakeholders about genetically modified organisms, genetic diversity, intellectual property, and intellectual property protection. As with other agricultural commodity crops, hemp can be bred to perform well in various environments and to fit into different kinds of farming systems, while also meeting the needs of manufacturers and consumers. By optimizing traits like fiber strength, biomass yield, ratios of bast fiber to hurd, and adaptability to diverse climates, the seed industry supports the development of new materials that meet manufacturer standards for these different product sectors.

The availability of superior-performing genetics adapted to specific growing environments depends on the seed industry's ability to invest in advanced breeding technologies and to deliver sales of seeds to farmers through retail distribution networks. This ensures that crops are tailored not only for agronomic performance but also to meet manufacturing facility demand for biomass supporting product schedules. Collaborative efforts with producers, scientists, and manufacturers will enable the seed industry to deliver crop seeds that enhance the production profitability, manufacturing efficiency, and quality of the final industrial products that will be produced.

3 Technology roadmap

A technology roadmap is a strategic planning process carried out by firms, industry sectors, and policymakers to identify the readiness of state-of-the-art technologies by assessing the value chain leading to product creation, gaps in information, technology development, and needed infrastructure. The technology roadmap is used as a tool to coordinate and harness the existing manufacturing infrastructure, partners, and required investments to accelerate progress in bridging any gaps and bring new products to market. As can be done for any market opportunity, technology roadmap analysis can be used to develop and advance biobased manufacturing capacities utilizing hemp as the primary feedstock material. Connecting private and public sector experts is crucial to discovering how hemp could best fit into existing manufacturing landscapes that have already been established for wood and other natural fiber sectors such as cotton. We see hemp as a complementary contributor, not a competitor, to already established biobased manufacturing sectors that utilize agricultural and forest-based feedstocks.

It is critical that hemp entrepreneurs be matched with seasoned biobased materials commodity users and other potential allies across the value chain components that connect hemp biomass producers to manufacturers. Alliances forged in such a strategic partnering process are expected to mobilize supply chain participants who together can develop realistic implementation targets for materials quantities and qualities, establish the needed linkages for creating complete supply chains, and build out scaled manufacturing facilities that produce products to reach defined consumer markets with understood price points. As the hemp industry expands, many in the sector are understandably eager to promote its potential. However, promotion that uses poorly sourced data and unsubstantiated comparisons and claims to other crops and products may alienate potentially important partners and undermine the credibility of the hemp sector if sustainable approaches that are truly productive, profitable, and done with good steward of natural resources and human capital are not established. A technology road mapping exercise can reduce this risk and help establish baseline information needed to make appropriate comparisons.

A technology road mapping exercise identifyies values and opportunities starting backwards from the end-product and applications. In Figure 3, we use the two-by-two matrix to evaluate a product's business opportunity and associated supply chain challenges. The matrix considers (1) Commercial Opportunities associated with profitability, market size, levels of consumer acceptance, and profit margin; and (2) Supply Chain and Institutional

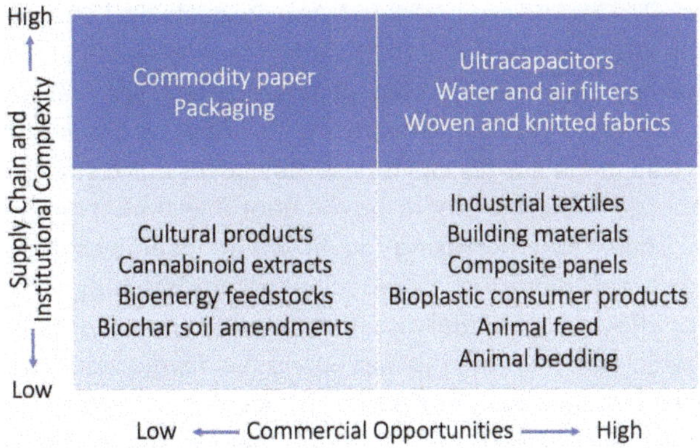

Figure 3 A technology classification for hemp-based products based on a range from low-to-high commercial opportunities associated with profitability, market size, and levels of consumer acceptance and supply chain and institutional complexity reflecting supply chain management challenges across the product life cycle. Within quadrants, hemp-based products listed are in order of low-to-high ease of manufacturing.

Complexity reflecting supply chain difficulty associated with the product life cycle from research and development to production and product recycling, and shows institutional barriers including regulation and industry standards, or the lack there of, which affect buyer-purchaser confidence that what is purchased is delivered. Such an analysis is built on industry benchmarking assessment and is an interactive process of market-supply chain research and validation by industry practitioners.

4 Value chain analysis

In supply chain and production operations, firms carry out value-added activities including behind the farm gate production, materials processing, product manufacturing, transportation logistics, wholesale, distribution, and retailing. These firms collaborate in a supply network of a complex, adaptive system with material flows that are non-linear and multi-directional, and which may often be spread across regions, countries, continents, or around the globe. We often use a more abstract framework of 'value chain' to understand and communicate the value-added activities associated with manufacturing commerce. For a given product or industry, companies assess what value-added activities they will control and which to outsource, based on their competitive advantages and abilities to move up the value chain technology ladder. Competitiveness is relevant to hemp production for two reasons. First, a value chain perspective allows all industry leaders to understand and communicate value propositions in a succinct way. Second, both business leaders and policymakers need to determine the economic development opportunities of domestic production and the risks of outsourcing, particularly for products and strategic materials needed in the national interest.

Policymakers' interests in value chains analysis are less about which firm carries out what task but in wanting to know about where the value opportunities lie in the supply chain or supply network for strategic decision-making purposes including where govenment investments may affect their consituents. There are increasing government incentives to build more regionalized domestic production capacities to mitigate supply chain risks not only for those affected by natural disasters or public unrest in sourcing nations but also as vehicles to increase job and economic development opportunities by placing production and manufacturing close to consumption. Value chain analysis helps policymakers identify weak links in production and crucial technology that may include intellectual property and determine what is adequate policy support.

As an example, much of the U.S. textile fiber manufacturing industry ceased during 1990–2010 and was replaced by production in China, Southern and Southeast Asian, Caribbean and Central American, and some African countries. Major U.S. consumer apparel brand companies only retained two ends of the value chains: high value-added activities of apparel design on one end and marketing and retailing on the other. With growing concerns about climate change, worker welfare, fair trade, and supply chain risks, there was political and economic stimuli to re-establish domestic manufacturing capacity, and so industry leaders are examining in-shoring and near-shoring manufacturing. Among the strategies currently being approached is moving certain parts of the supply chain back to the U.S. or to nearby countries such as Mexico or soveriegn Native American nations reservations. It is against this backdrop that the question is raised: How to create the U.S. domestic supply network that can utilize not only textiles but other hemp-based industrial materials in ways that are economically viable complements to other natural fibers and replace a portion of products in the market that are manufactured with non-renewable hydrocarbons?

Secondary and further value-added processing activities require great investment in new equipment to process hemp at a competitive product price compared to other natural materials, and in the case of natural fibers, in particular cotton. To increase returns on investments, great materials throughput is needed to justify high capital expenditures or there needs to be utilization of existing equipment used with other commodities that can also function as is needed for hemp. As more farms produce hemp biomass, overall processing costs will go down as more materials can be locally sourced and provide a dependable materials stream to support a facility. This follows what has been done with other established commodities such as cotton, corn, and wheat.

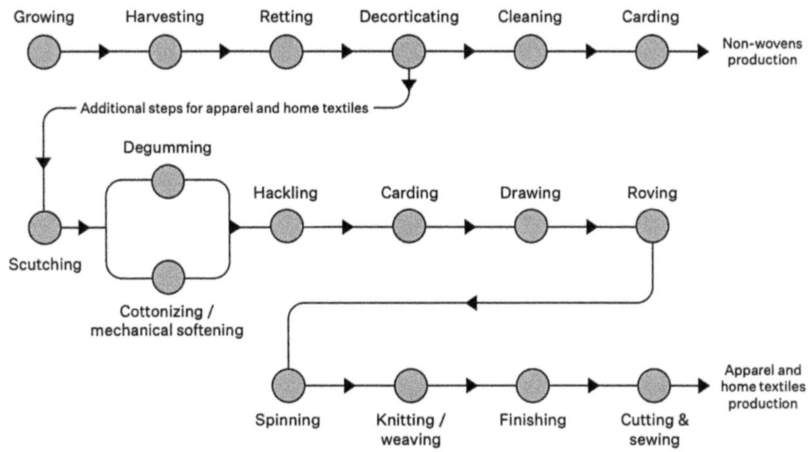

Figure 4 An industrial hemp fiber process map showing the major steps from growing hemp to making a finished textile product. Such a map is needed to assess present costs of processing plant materials and identify where technological interventions and improved processes can be applied to increase production efficiencies and lower costs of manufactured textile products. Textile Exchange, 2023.

As a new commodity, the production infrastructure for hemp needs to be identified and created to scale-up production. Often cited is the chicken and egg problem where processors are unwilling to invest until there is sufficient and dependable supplies of raw materials throughput and market outlets for their manufactured products. And growers are unwilling to grow hemp because both (1) processing infrastructure is not in place to process their crop, and (2) contracts are not in place at a price to assure the sale of their produced raw material will cover their costs. To overcome such situations, policymakers need to consider making judicious investments and provide incentives that reduce first-mover risks and promote local economic development, as has been done for other relatively novel systems such as biomass crops for biofuels production.

An example may be for a company that maps out the key processes of hemp textile fiber production (Figure 4). It starts from raw materials from a farm and ends at textile apparel production. In this case, value chain analysis is carried out by the company to understand existing production processes and identify opportunities to move and change certain supply chain processes and the needed infrastructure to enable domestic production. Once this process is mapped out, the company, in collaboration with other supply chain members, can consider which processes can be moved to different locations and how to improve the production processes. This is where the introduction of automation technology and other *Industry 4.0* approaches reduce production costs to a cost point that enables certain levels of near-shoring or domestic production to be established. The present state of hemp processing technologies has not advanced in 90 years, and significant equipment redesign and systems engineering are needed to increase manufacturing efficiencies and produce biobased materials and products in a no-waste system designed for twenty-first century bioeconomy markets.

Members of the supply network need to work across institutional boundaries to develop industry ecosystems that can optimize utilization of entire hemp plants and utilize any byproducts generated by raw material processing and utilization in their manufacturing streams. Eventually, the re-engineered value chain would be differently configured than in its original design, especially was technology innovations are introducted. For the hemp-based textile value chain, it would not necessarily be a matter of rebuilding or replicating the U.S. cotton textile industry or even hemp textile manufacturing in Europe or Asia. The reconfigured production network could entail linkages of different firms (nodes), connections (links, relationships), technologies, and institutional participants than with the existing value chain.

5 Promoting biobased manufacturing

The biobased economy encompasses a 'global industrial transition that utilizes biobased manufacturing] technology in creating renewable terrestrial and aquatic resources in energy, intermediates, and final products to the benefit of economic, environmental, and social concerns' (Golden and Handfield, 2014). The biobased economy is a form of development based on the need for societies to break from their dependence on fossil-based resources through a logic of substitution with biomass that places agriculture, forestry, and working lands-based industries at the heart of the development for many value-added industries manufacturing such products as automobiles, textiles, and pharmaceuticals (Grouiez et al., 2023; The White House, 2023). A no-waste, circular bioeconomy is a pillar of an ecological transition that can contribute to decarbonizing industry, reducing solid waste disposal, decreasing the use of non-renewable resources, and maximizing the efficiency and sustainable use of renewable materials.

Hemp has great potential as a player in a biobased economy transition because of its many potentials for use in manufactured products made from a relatively efficient growing source of biomass. It also has likely significant potential further environmental benefits such as its use in phytoremediation of contaminated soils, an additional crop choice option to further diversify producer income and crop production systems resilience, improve soil quality by incearing retained organic matter, and possibly reduce the need for certain purchased inputs including pesticides. Such attributes, though not exclusive to hemp, contribute to the agriculture industry aligning with the United Nations

17 Sustainable Development Goals (SDGs), the U.S. federal government and USDA's bioeconomy initiatives (Tables 4 and 5), and supporting the *European Green Deal* by providing materials for use in manufactured biobased products for export (USDA, 2023).

Opportunities are available to find ways to incorporate hemp into existing agricultural, business, and natural landscapes in ways that complement and not compete with existing commodity and other business enterprises. An initiative led by Oregon State University's Global Hemp Innovation Center and supported by three USDA competitive grant investments are supporting hemp production discovery and the development of biobased manufacturing

Table 4 United States government programs and initiatives to support biobased manufacturing

United States government policy instruments	Activity
Agriculture Improvement Act of 2018, The Farm Bill	Covers areas including conservation, rural development, energy, and accelerating the expansion of biobased fuels and renewable energy.
Inflation Reduction Act 2022	A $750 billion act that provided nearly $40 billion for agriculture, forestry, and rural development including $14 billion for renewable energy and biofuels.
Executive Order of the President, 2022, Advancing Biotechnology and Biomanufacturing Innovation for a Sustainable, Safe, and Secure American Bioeconomy	Places a large emphasis on expanding market opportunities for biobased products. It instructs the Secretary of Agriculture to develop a plan supporting and encouraging domestic biomanufacturing to advance biomass supply chains.
USDA BioPreferred Program	Creates a federal purchasing mandate and voluntary labeling initiative to drive biobased products in the marketplace.
Executive Order of the President, 2025, Preparing Americans for High-Paying Skilled Trade Jobs of the Future	Directs a whole-of-government effort to modernize, integrate, and realign federal workforce development programs to better match skills training with labor market demand, especially in skilled trades, advanced manufacturing, and emerging industries.
Workforce development training programs: USDA Rural Development, Rural Innovation Stronger Economy Grants and Rural Business Development Grants U.S. Department of Energy, Workforce Development for Teachers and Scientists and Industrial Assessment Centers U.S. Department of Labor, H-1B Skills Training Grants	These programs offer funding, technical assistance, and workforce training to support the growth of biobased manufacturing industries in the U.S.

Table 5 United State Department of Agriculture (USDA) programs to support establishing industrial hemp industry biobased products manufacturing supply chain infrastructure

USDA Agency/Program	Program description
Farm Service Agency/Biomass Crop Assistance Program (BCAP)	Provides incentives to help farmers grow advanced bioenergy feedstocks (crops that are well-suited to be turned into energy) and connect with relevant refineries in their area.
Risk Management Agency/Crop Insurance	Makes available market-based Federal crop insurance as a risk management tool to strengthen the economic stability of agricultural producers and rural communities.
Rural Development/Rural Cooperative Development Grant Program	Improves the economic condition of rural areas by helping individuals and businesses start, expand, or improve rural cooperatives and other mutually owned businesses through Cooperative Development Centers.
Rural Development/Socially Disadvantaged Groups Grants	Provides technical assistance to socially disadvantaged groups through cooperatives and Cooperative Development Centers.
Rural Development/Value-Added Producer Grants	Helps agricultural producers enter value-added activities related to the processing and/or marketing of new products.
Rural Development/Rural Business Development Grant program	Provides technical assistance and training for small rural businesses. Small means that the business has fewer than 50 new workers and less than $1 million in gross revenue.
National Institute of Food and Agriculture/Small Business Innovation Research and Small Business Technology Transfer grant programs	Provides competitively awarded grants to qualified small businesses to support high quality research to advance important scientific problems and technology opportunities leading to significant public benefits.
Commodity Credit Corporation (CCC)	The CCC can support such programs as the Farm Service Agency BCAP for providing incentives to farmers, ranchers, and forest landowners to establish, cultivate, and harvest eligible biomass sourced for heat, power, biobased products, research, and advanced biofuels.

capacities for economic development in Pacific Northwest region rural communities including Native American nation reservations.

In this undertaking, producers, processors, and business development organizations collaborate with one another to develop new place-based supply networks that support investments in manufacturing infrastructure to produce value-added biobased products. One such commercial opportunity is the formation of an intertribal trading network that links industrial hemp biomass production to processing facilities to the manufacture of hemplime for use in housing construction on reservation lands. Another is providing marketing incentive payments to help move produced hemp stalks to processors and then processed materials to product manufacturers, and thus help to scale other new manufacturing capacities as first investors and adopters and investors in re-engineered value chains.

Several U.S. government programs support workforce development training in biobased manufacturing, particularly through initiatives that focus on strengthening agriculture systems resilience, renewable energy production, and advanced manufacturing. Examples of programs including the USDA Rural Development programs focus on enhancing rural economies, which often include workforce development in bioenergy, bioproducts, and biobased industries; U.S. Department of Energy's Workforce Development for Teachers and Scientists (WDTS) and Industrial Assessment Centers (IACs) programs that support workforce training in bioenergy and biobased product manufacturing; and U.S. Department of Labor H-1B Skills Training Grants that fund workforce development initiatives in high-growth industries, including biobased manufacturing and advanced manufacturing sectors. These programs collectively offer funding, technical assistance, and workforce training to support the growth of biobased manufacturing industries in the U.S. and can be integrated with the design and scaling for building out the needed biobased manufacturing supply chains that utilize hemp and which can be adapted for any region in North America based on local and regional needs and available resources (Figure 2).

6 Creating supply chain resilience

The mandate for supply chain security and resiliency compels companies and policymakers to help reshoring and near-shoring supply chains to build domestic manufacturing capacity. This creates opportunities for building out regionalized production networks that can be scaled. Regionalization does not mean replicating existing supply chain networks in multiple regional locations, nor replicating the same production processes and methods of compatible products that it intends to complement (e.g. cotton, home insulation materials) or compete with same historic manufacturing approaches that were off-shored and which are already established overseas. Hemp, as a new-old commodity, presents a clean slate for all to innovate, taking advantage of existing technology and infrastructure when it makes economic sense and drastically reconfiguring the production network to create and introduce new innovations and practices and technologies. Table 6 presents several examples to capture the emerging innovation opportunities and practices.

Table 6 Hemp production and production operations network innovation opportunities

Products	Innovation opportunities	Technology and policy implications
Customized genetics delivered in seeds	Integration of gene to plant to field to factory specifications for supplying materials designed to optimize materials supply and processing operations to meet needed production outputs.	Cross-sector stakeholder collaboration. Seed supply multiplication and quality certification programs.
Hemp textiles	Collaboration with and leverage flax-based and other biofiber textile material production infrastructure. Position hemp textile as premium category in apparel, fashion sector. Super-local production and cultural products. Tax and tariff-driven production network re-engineering. Develop new raw material sources, location within and outside the U.S. Leveraging virtual reality for design and artificial intelligence for customization. Production network reconfiguration through the adoption of different supplier and technology innovators.	Cross-sector stakeholder collaboration and standards creation. Disruptive technology such as AI and automation leads to supply chain vertical integration. Infrastructure investment and workforce development training by local government. Super local production infuses craft-based production and automation.
Hemplime and hempwood	Prefabricated housing and additive manufacturing housing disrupt existing real estate development model. Regionalized production-processing networks, circular economy, and industrial ecology.	Building standards and code changes.Green building policy incentives. Infrastructure investment and workforce development training by local government. Economic development and social sustainability objectives.
Biopolymers, bioplastics, and composites	Natural materials applications. New scaled production processes development.	Federal level policy and incentives for new industry creation. Cross-sector stakeholder collaboration and standards creation.
Biochar and activated carbon	Processing biomass into carbonaceous substances with a high surface area, pore size distribution, and sorptive and catalytic properties suited for use as soil amendments; air and water filters; electronics and electrical storage devices; and munitions	Federal level policy and subsidies incentives for new industry creation. Cross-sector stakeholder collaboration and standards creation.

7 Pathways to diversified product sectors

As an agricultural commodity, hemp produces multiple materials with attributes suited for use in the manufacture of a wide range of commercial products. Hemp has a historic connection with marijuana regulations and social injustice. In this sense, hemp comes with significant cultural dynamics that affect its acceptance. As we conceive and develop industrial applications, a concerted effort is needed to inform the public, industry leaders, financiers, and policymakers that industrial hemp is not marijuana and as well as to differentiate it from cannabinoid hemp and medicinal use marijuana. Such distinctions are important as state, tribal, and federal government policies are developed that affect hemp farming, processing, transportation, and manufacturing components of supply chains. At the present, much of the hemp industry is still overshadowed by marijuana regulatory frameworks, much of which is regressive.

As a new commodity crop and with a fledgling supply chain infrastructure to support its utilization, the overall cost of scaling hemp production and its utilization in manufacturing will be relatively high compared to present investments in already established crops and their products and will accrue greater risks. As seen a decade ago, first of a kind biofuel manufacturing facilities saw initial capitalization expenditures far exceed operational expenditures, but the assumption was that as the industry scaled, cost ratios would shift from capital expenditures being reduced as more facilities were built. These expenses greatly differed depending on the technology analyzed (IREA, 2016).

To attract the investment needed to develop a viable hemp-based industry will require targeted policy support and investment, innovative ways to conceive of value-added hemp-based manufactured products and their entry into established markets, validating claims of performance compared to established products, and determination of its commercial feasibility. Such aspects of system feasibility that must be met to establish a successful industry include environmental, technical, management, economic, market, and financial elements. These are no different than what the emergent advanced biofuels industry faced more than a decade ago (The White House, 2010). As with any new agricultural commodity, much is still unknown of its needed manufacturing infrastructure, facilities placement, and supporting supply chains. It will be necessary to develop an understanding of all elements of commercial feasibility so viable supply chains and predictable markets can develop.

At the same time, there are also important social forces that need to be addressed to support the development of hemp-based products industries. First, although hemp has putative potential environmental benefits, there are strong popular beliefs about the growing of hemp, specifically compared to other natural fibers, requires significantly less amounts of chemical, synthetic fertilizer, and supplemental irrigation water inputs. However, there is scant substantiated data because of the recency of hemp being allowed to be grown in the U.S. and across the wide range of agricultural production regions where it is adapted. and so research has not caught up with the opportunity. Second, domestic manufacturing of hemp-based products such as textile fibers and biobased graphene carbon may have strategic national security importance, but comparisons have not been made to products produced from other sourced materials. Third, leveraging hemp-based materials present a wide range of technological challenges as all of the fragmented supply chain components and players are connected. Tackling such challenges always drives policy and investment decisions because they are the basis for continued innovation.

8 An agenda for hemp

As policy barriers are further lowered, we hope, for hemp production and utilization, more countries will start to incorporate hemp into their agricultural sectors. As hemp production gains momentum, we foresee that international trade of hemp will increase as well. At this juncture, industry leaders need to develop supply chains that benefit all stakeholders and communities. We call out the following six-point agenda for practitioners, researchers, and policymakers to consider and to use in developing strategies for sustained growth of hemp biomass to support the manufacture of biobased products: (1) Complementary biobased materials, (2) Hemp industry standards, (3) Hemp opportunities, (4) New cultural meaning for hemp, (5) Hemp value chain, and 6) Protect genetic intellectual property.

8.1 Complementary biobased materials

Compared with other natural fibers, hemp is a fraction of the total production and utilization of cotton and wood. Hemp should be viewed as a complement to instead of a competitive replacement for already established natural fiber product markets. We need to incorporate hemp into existing agricultural and natural landscapes and leverage existing supply chain infrastructure and business networks that include processing, manufacturing, transportation, and distribution components. To scale commercial hemp production, industry leaders need to seek out broad horizontal collaborations across company and institutional boundaries and partner vertically among supply chain partners to identify synergies, economies of scale, and customer-driven product solutions. Flax (*Linum usitatissimum* L.) is another minor natural fiber suited for textile manufacturing that was historically significant in North America and that was displaced by cotton and synthetic fibers. Industry leaders have pointed out that hemp and flax sectors can learn from each other in areas of material processing and seek ways to coordinate their materials being utilized by the apparel sector.

A blend of hemp and flax could be more economically viable complement to the cotton material sector.

8.2 Hemp industry standards

Industry leaders, public research institutions, and government agencies need to work together to establish industry standards for hemp that will facilitate product development, commercialization, and international trade for not only textile fibers but also materials utilized to manufacture other products. Industry standards facilitate research and development, commercialization, and international trade by allowing sellers and buyers to be assured that materials exchanged in transactions meet required specifications. For instance, progress in measurement science directly leads to the creation of new advanced machinery, tools, databases, and pre-competitive technology platforms. Such advances create and support linkages across value chains and encourage stability that allows new entrants into the market who can compete on their own merit. Publicly accessible databases also provide access to new opportunities in computer modeling and simulation that can help reduce innovation costs. This provides opportunities to systematically and holistically close technical gaps instead of doing so in a piecemeal fashion, which helps realize economic impact in a compressed development timeframe. Hemp businesses and the rest of the domestic biobased manufacturing sector have an opportunity to secure a strong international position by developing, owning, and distributing the grading, standardization, and classification technologies that must be developed for hemp fiber and other natural materials.

8.3 Hemp opportunities

Hemp presents an opportunity for radical innovation in the areas of building hyper-local and regional production networks, embrace sustainable development including job creation, and further expand the biobased manufacturing economy. Recent USDA-supported projects provide examples of economic development approaches to scale hemp production by identifying hemp fiber processing and manufacturing bottlenecks in the supply chain. Collaboration among partnering researchers, existing processors, manufacturers, and customers can coalesce disparate practitioners who otherwise are not connected. Inclusion of workforce training in collaboration with community and technical colleges and universities with the goal of providing diversified employment opportunities to learners for developing manufacturing expertise and expanding a skilled workforce. Note that a hyper-local supply network does not mean everything must be made locally.

Supply chains will always be adaptive. Tax policies, tariffs, risks and operational performance of suppliers, and technologies will invariably change. These factors will determine the location of production and associated value-added activities of supply chain members.

8.4 New cultural meaning for hemp

Industry leaders need to integrate culture into product value proposition. Hemp has significant new cultural meanings and offers unique physical and chemical properties to innovatively explore. Its connection to tradition, social movement, and new stories of sustainability are told through products and production practices. Such stories are an integral part of the values of products. For instance, construction materials such as hemp lime are a story not only about a new alternative natural building material produced from the earth that is energy efficient, but also about community development based on local production that can in parallel help meet housing needs of disadvantaged, rural, and Indigenous populations. These functions invariably connect back to tradition, justice, and rebellious innovations. Authentic value proposition draws institutional support and customers, and the public recognizes and responds to such values. New cultural meaning may also require a new vocabulary as a part of story telling away from terminology, especially that associated with the terms hemp, cannabis, and marijuana.

8.5 Hemp value chain

As we have discussed earlier, to build resilient regional and domestic production capabilities, business leaders and policymakers need to identify the process and infrastructure leverage points of the value chain where there are investment opportunities to build out and scale-up capacities and capabilities. Conducting technology roadmaps and supply network mapping exercises allows business leaders, financiers, and policymakers to identify critical value-added processes and to benchmark existing domestic supply network operations and infrastructure against existing global ones to compete. Technology road mapping need to involve all stakeholders to ensure investment decisions and project execution needs have legitimacy and committed support from all supply chain and institutional actors.

8.6 Protect genetic intellectual property

Industry leaders, policymakers, and research institutions must prioritize the development and protection of intellectual property (IP) for superior-performing hemp genetics and seed varieties. Establishing robust IP frameworks will ensure fair compensation for innovation while incentivizing further research and development. Protecting IP also enables the creation of tailored hemp varieties to meet the specific requirements of biobased product manufacturers. Additionally, a transparent IP system can facilitate equitable access to superior genetics by providing licensing agreements for planting seeds that facilitate a linkage between agricultural producers and processors and manufacturers, fostering a cohesive, efficient supply chains based on contracted prices for delivery of materials of guaranteed performance to end users. This should be an extension of publically available genetic materials that are for the public good and which are maintained in the USDA National Plant Germplasm System, just as with other agriculturally important commodities.

9 Conclusion

The potential for industrial hemp to be integrated into the expanding biobased manufacturing landscape offers a unique opportunity to redefine material production in ways that align with the evolving needs of a sustainable economy. By systematically addressing entire supply chains from production to processing and manufacturing, we can unlock the versatility of hemp and its utilization to compete with traditional materials, both natural and synthetic. The systems approach we have outlined provides a pathway to leverage hemp's distinctive properties, not only to meet current market demands but also to drive innovation in product design and improved manufacturing efficiencies.

The success of hemp-based industrial products will depend heavily on developing high-throughput, integrated production systems that focus on creating value-added end-uses for multiple product streams and doing so with no-waste approaches. This requires coordinated efforts across multiple sectors including agriculture, manufacturing, research, and policymaking to ensure that production systems are optimized for optimal efficiency and competitiveness. Moreover, hemp must be disassociated from its controversial cousin, marijuana, to gain full public acceptance in mainstream markets. Regulatory adjustments, public perception, and investor confidence will be key factors in achieving this shift.

As global supply chains continue to experience external disruptions and absorb continual technological innovations, the case for building robust domestic manufacturing capacity for hemp becomes even stronger. Industrial hemp has the potential to fill critical gaps in the consumer demand for more biobased materials and products, offering a renewable alternative to fossil-based products. Moreover, by embedding sustainability and level playing fields within these emerging supply chains, we can ensure that the benefits of hemp-based manufacturing are widely shared and contribute to a resilient and prosperous domestic economy.

This chapter provides an understanding of the complex supply chain components and market dynamics that need to be considered to establish hemp as a new commodity for use in the manufacture of biobased products. The challenge now lies in strategic planning, implementing these strategies, investing in the necessary infrastructure, and fostering collaborations among supply chain participants to propel hemp into the next era of biobased manufacturing as a sustainable, competitive, and forward-thinking biobased manufacturing industrial sector.

10 References

Bloom, J. M. and Blair, S. S. (Eds.) (2009), *The Grove Encyclopedia of Islamic Art and Architecture*. Oxford University Press. www.oxfordreference.com/view/10.1093/acref/9780195309911.001.0001/acref-9780195309911 (accessed: 19 December 2024).

Bourguignon, D. (1993), The history of cannabis and its preparations in saga, science, and sobriquet. *Journal of Psychoactive Drugs*, 25(3), 215–222. www.tandfonline.com/doi/abs/10.1080/02791072.1993.10472238 (accessed: 19 December 2024).

Clarke, R. C. and Merlin, M. D. (2013), Cannabis domestication, breeding history, present-day genetic diversity, and future prospects. *Critical Reviews in Plant Sciences*, 35(5–6), 293–327. www.tandfonline.com/doi/full/10.1080/07352689.2013.847834 (accessed: 19 December 2024).

Cutler, T. (2020), *Moving Ships Through Water*. U.S. Naval Institute. August. https://www.usni.org/magazines/naval-history-magazine/2020/august/moving-ships-through-water#:~:text=Eventually%2C%20manpower%20and%20wind%20were,nuclear%20technology%20to%20drive%20vessels (accessed: 19 December 2024).

Golden, J. S. and Handfield, R. B. (2014), *Why Biobased? Opportunities in the Emerging Bioeconomy*. USDA BioPreferred. www.biopreferred.gov/BPResources/files/WhyBiobas ed.pdf (accessed: 19 December 2024).

Grouiez, P., et al. (2023), The complex relationships between non-food agriculture and the sustainable bioeconomy: the French case. *Ecological Economics*. https://doi.org/10.1016/j.ecolecon.2023.107974.

Herer, J. (1985), *The Emperor Wears No Clothes: Hemp and the Marijuana Conspiracy*. Queen of Clubs Publishing.

International Renewable Energy Agency (2016), Innovation outlook advanced liquid biofuels. https://www.irena.org/-/media/Files/IRENA/Agency/Publication/2016/IRENA_Innovation_Outlook_Advanced_Liquid_Biofuels_2016.pdf (accessed: 19 December 2024).

Johnson, R. (2014), *Hemp as an Agricultural Commodity*. Congressional Research Service, CRS Report of Congress, 7-5700. www.crs.gov.

Kadolph, S. J. (2010), *Textiles*, 11th ed. Pearson, 581 p.

Kaplan, R. S. and McMillan, D. (2020), *Updating the Balanced Scorecard for Triple Bottom Line Strategies*. Harvard Business Review, Working Paper 21-028. https://papers.ssrn.com/sol3/papers.cfm?abstract_id=3682788 (accessed: 19 December 2024).

Kentucky History Society (2024), First hemp crop. https://www.kyhempsters.com/kentucky-hemp-history (accessed: 19 December 2024).

National Institute of Standards and Technology (2024), Advanced manufacturing technology and Industry 4.0 services, Manufacturing Extension Partnership (MEP). https://www.nist.gov/mep/advanced-manufacturing-technology-and-industry-40-services (accessed: 19 December 2024).

Phillips, G. H. (2017), *Indians of the Tulares, Adaptation, Relocation, and Subjugation in Central California 1771-1917*. Great Oak Press.

Popular Mechanics (1938), New billion-dollar crop. *Popular Mechanics*, 69(2).

Small, E. and Marcus, D. (2002), Hemp: a new crop with new uses for North America. In J. Janick and A. Whipkey (Eds.), *Trends in New Crops and New Uses* (pp. 284–326). ASHS Press. https://www.hort.purdue.edu/newcrop/ncnu02/v5-284.html.

Statista (2020), Distribution of global chemical fiber production by region. https://www.statista.com/statistics/271653/distribution-of-global-chemical-fiber-production-by-region (accessed: 19 December 2024).

Textile Exchange (2023), Growing hemp for the future, a global fiber guide, https://textileexchange.org/knowledge-center/reports/growing-hemp-for-the-future-a-global-fiber-guide/.

The White House (2010), Growing America's fuel: an innovation approach to achieving the President's Biofuels target. obamawhitehouse.archives.gov/sites/default/files/rss_viewer/growing_americas_fuels.PDF.

The White House (2023), www.usda.gov/sites/default/files/documents/bioeconomy-2023-accomplishments-fact-sheet.pdf (accessed: 19 December 2024).

United States Congress (1937). *Senate. Committee on Finance. Marijuana Tax Act of 1937*. Government Printing Office.

United States Department of Agriculture (1942), Hemp for Victory [Film]. Retrieved from National Archives: https://catalog.archives.gov/id/24527671 (accessed: 19 December 2024).

United States Department of Agriculture (2023), USDA bioeconomy accomplishments. www.usda.gov/sites/default/files/documents/bioeconomy-2023-accomplishments-fact-sheet.pdf (accessed: 19 December 2024).

United States Government Printing Office (1911), Hearings before the Committee on Ways and Means of the House of Representatives, on the Tariff Act of 1913.

Van der Werf, H.M.G. and L. Turunen (2008), The environmental impacts of the production of hemp and flax textile yarn. *Industrial Crops and Products*, 27, 1–10.

Velasquez-Manoff, M. (2019), Can CBD really do all that? *The New York Times Magazine*. https://www.nytimes.com/interactive/2019/05/14/magazine/cbd-cannabis-cure.html (accessed: 19 December 2024).

Index

www.ingramcontent.com/pod-product-compliance
Ingram Content Group UK Ltd.
Pitfield, Milton Keynes, MK11 3LW, UK
UKHW021825100926
8556IPUK00003B/156

9 781835 450833